VOLUME THREE HUNDRED AND SIXTY SEVEN

INTERNATIONAL REVIEW OF CELL AND MOLECULAR BIOLOGY

One, No one, One Hundred Thousand - The Multifaceted Role of Macrophages in Health and Disease - Part A

INTERNATIONAL REVIEW OF CELL AND MOLECULAR BIOLOGY

VOLUME THREE HUNDRED AND SIXTY SEVEN

INTERNATIONAL REVIEW OF CELL AND MOLECULAR BIOLOGY

One, No one, One Hundred Thousand - The Multifaceted Role of Macrophages in Health and Disease - Part A

Edited by

SAMANTA A. MARIANI
University of Edinburgh,
Edinburgh, United Kingdom

LUCA CASSETTA
University of Edinburgh,
Edinburgh, United Kingdom

LORENZO GALLUZZI
Weill Cornell Medical College,
New York, NY, United States

ELSEVIER

ACADEMIC PRESS
An imprint of Elsevier

Academic Press is an imprint of Elsevier
50 Hampshire Street, 5th Floor, Cambridge, MA 02139, United States
525 B Street, Suite 1650, San Diego, CA 92101, United States
The Boulevard, Langford Lane, Kidlington, Oxford OX5 1GB, United Kingdom
125 London Wall, London, EC2Y 5AS, United Kingdom

First edition 2022

Notices
Knowledge and best practice in this field are constantly changing. As new research and experience broaden our understanding, changes in research methods, professional practices, or medical treatment may become necessary.

Practitioners and researchers must always rely on their own experience and knowledge in evaluating and using any information, methods, compounds, or experiments described herein. In using such information or methods they should be mindful of their own safety and the safety of others, including parties for whom they have a professional responsibility.

To the fullest extent of the law, neither the Publisher nor the authors, contributors, or editors, assume any liability for any injury and/or damage to persons or property as a matter of products liability, negligence or otherwise, or from any use or operation of any methods, products, instructions, or ideas contained in the material herein.

ISBN: 978-0-323-90741-5
ISSN: 1937-6448

For information on all Academic Press publications
visit our website at https://www.elsevier.com/books-and-journals

Publisher: Zoe Kruze
Acquisitions Editor: Leticia M. Lima
Developmental Editor: Cindy Angelita Pe Benito-Gardose
Production Project Manager: Vijayaraj Purushothaman
Cover Designer: Matthew Limbert

Typeset by STRAIVE, India

Contents

Contributors *vii*

1. Pulmonary macrophages and SARS-Cov2 infection **1**

Calum C. Bain, Adriano G. Rossi, and Christopher D. Lucas

1. Introduction 2
2. Macrophages in the healthy lung 3
3. Inflammation as a driver of COVID-19 severity 9
4. Macrophage as mediators of organ damage in COVID-19? 10
5. Targeting the macrophage in severe COVID-19: Current and future therapeutics? 17

Acknowledgments 18
References 19

2. Macrophage control of Crohn's disease **29**

Siu Ling Tai and Arthur Mortha

1. Introduction 29
2. Development of intestinal macrophages 30
3. Localization and heterogeneity of intestinal macrophages 34
4. Functional properties of intestinal macrophages 36
5. Maintenance of homeostatic macrophage function 40
6. Macrophages in Crohn's disease 42
7. Macrophages contribute to in intestinal inflammation in CD 43
8. Macrophage functions and dysfunctions in Crohn's disease 45
9. Targeting macrophages to improve Crohn's disease 50

Acknowledgments 53
References 53

3. Metabolism in tumor-associated macrophages **65**

Jie Li, Gina M. DeNicola, and Brian Ruffell

1. Introduction 66
2. Glucose metabolism 69
3. Amino acid metabolism 76
4. Lipid metabolism in TAM 81
5. Targeting TAM metabolism for therapy 85
6. Conclusion 90

References 90

4. **Macrophages in the kidney in health, injury and repair** **101**
Rachel M.B. Bell and Bryan R. Conway

1. Introduction 102
2. Identification of renal macrophages 103
3. Ontology of renal macrophage and their role in the developing kidney 107
4. Macrophage phenotypic heterogeneity 112
5. Role of macrophages in renal immunosurveillance 119
6. Macrophages in renal injury 120
7. Macrophages in renal repair 128
8. Therapeutic prospects 130
9. Conclusions 134
References 134

5. **Heterogeneity and function of macrophages in the breast during homeostasis and cancer** **149**
Eva Hadadi, Sofie Deschoemaeker, Gerard Vicente Venegas, and Damya Laoui

1. Introduction 150
2. Mammary gland macrophages 151
3. Breast cancer 161
4. Macrophages in breast cancer 162
5. Concluding remarks and future perspectives 170
Competing interests 172
Funding 172
References 172

6. **Endometrial macrophages in health and disease** **183**
Erin Brown, Rocío Martínez-Aguilar, Jacqueline A. Maybin, and Douglas A. Gibson

1. Introduction 184
2. Endometrial physiology 184
3. Endometrial macrophages—Origin, location and abundance 186
4. Identification and characterization of endometrial macrophages 187
5. Regulation of macrophage function during the menstrual cycle 189
6. Endometrial macrophages in women's reproductive health disorders 194
7. Future directions 199
8. Conclusions 200
Acknowledgments 202
Conflict of interest 202
References 202

Contributors

Calum C. Bain
University of Edinburgh Centre for Inflammation Research, Queen's Medical Research Institute; Institute for Regeneration and Repair, University of Edinburgh, Edinburgh BioQuarter, Edinburgh, United Kingdom

Rachel M.B. Bell
Centre for Cardiovascular Science, University of Edinburgh, Edinburgh, United Kingdom

Erin Brown
Centre for Inflammation Research, University of Edinburgh, Edinburgh, United Kingdom

Bryan R. Conway
Centre for Cardiovascular Science, University of Edinburgh, Edinburgh, United Kingdom

Gina M. DeNicola
Department of Cancer Physiology, Lee Moffitt Cancer Center and Research Institute, Tampa, FL, United States

Sofie Deschoemaeker
Myeloid Cell Immunology Lab, VIB Center for Inflammation Research; Lab of Cellular and Molecular Immunology, Vrije Universiteit Brussel, Brussels, Belgium

Douglas A. Gibson
Centre for Inflammation Research; MRC Centre for Reproductive Health, University of Edinburgh, Edinburgh, United Kingdom

Eva Hadadi
Myeloid Cell Immunology Lab, VIB Center for Inflammation Research; Lab of Cellular and Molecular Immunology, Vrije Universiteit Brussel, Brussels, Belgium

Damya Laoui
Myeloid Cell Immunology Lab, VIB Center for Inflammation Research; Lab of Cellular and Molecular Immunology, Vrije Universiteit Brussel, Brussels, Belgium

Jie Li
Department of Immunology, H. Lee Moffitt Cancer Center and Research Institute; Cancer Biology PhD Program, University of South Florida, Tampa, FL, United States

Christopher D. Lucas
University of Edinburgh Centre for Inflammation Research, Queen's Medical Research Institute; Institute for Regeneration and Repair, University of Edinburgh, Edinburgh BioQuarter, Edinburgh, United Kingdom

Rocío Martínez-Aguilar
MRC Centre for Reproductive Health, University of Edinburgh, Edinburgh, United Kingdom

Jacqueline A. Maybin
MRC Centre for Reproductive Health, University of Edinburgh, Edinburgh, United Kingdom

Arthur Mortha
Department of Immunology, University of Toronto, Toronto, ON, Canada

Adriano G. Rossi
University of Edinburgh Centre for Inflammation Research, Queen's Medical Research Institute; Institute for Regeneration and Repair, University of Edinburgh, Edinburgh BioQuarter, Edinburgh, United Kingdom

Brian Ruffell
Department of Immunology; Department of Breast Oncology, H. Lee Moffitt Cancer Center and Research Institute, Tampa, FL, United States

Siu Ling Tai
Department of Immunology, University of Toronto, Toronto, ON, Canada

Gerard Vicente Venegas
Myeloid Cell Immunology Lab, VIB Center for Inflammation Research; Lab of Cellular and Molecular Immunology, Vrije Universiteit Brussel, Brussels, Belgium

CHAPTER ONE

Pulmonary macrophages and SARS-Cov2 infection

Calum C. Bain[a,b,*], Adriano G. Rossi[a,b,*], and Christopher D. Lucas[a,b,*]

[a]University of Edinburgh Centre for Inflammation Research, Queen's Medical Research Institute, Edinburgh BioQuarter, Edinburgh, United Kingdom
[b]Institute for Regeneration and Repair, University of Edinburgh, Edinburgh BioQuarter, Edinburgh, United Kingdom
*Corresponding authors: e-mail address: calum.bain@ed.ac.uk; adriano.rossi@ed.ac.uk; christopher.lucas@ed.ac.uk

Contents

1. Introduction 2
2. Macrophages in the healthy lung 3
   2.1 Alveolar macrophages 4
   2.2 Interstitial macrophages 8
3. Inflammation as a driver of COVID-19 severity 9
4. Macrophage as mediators of organ damage in COVID-19? 10
   4.1 Defining the macrophage compartment in COVID-19 10
   4.2 The role of macrophage subsets in SARS-CoV-2 infection 14
5. Targeting the macrophage in severe COVID-19: Current and future therapeutics? 17
Acknowledgments 18
References 19

Abstract

The emergence of severe acute respiratory syndrome coronavirus 2 (SARS-CoV-2) has led to the largest global pandemic in living memory, with between 4.5 and 15M deaths globally from coronavirus disease 2019 (COVID-19). This has led to an unparalleled global, collaborative effort to understand the pathogenesis of this devastating disease using state-of-the-art technologies. A consistent feature of severe COVID-19 is dysregulation of pulmonary macrophages, cells that under normal physiological conditions play vital roles in maintaining lung homeostasis and immunity. In this article, we will discuss a selection of the pivotal findings examining the role of monocytes and macrophages in SARS-CoV-2 infection and place this in context of recent advances made in understanding the fundamental immunobiology of these cells to try to understand how key homeostatic cells come to be a central pathogenic component of severe COVID-19 and key cells to target for therapeutic gain.

International Review of Cell and Molecular Biology, Volume 367
ISSN 1937-6448
https://doi.org/10.1016/bs.ircmb.2022.01.001

1. Introduction

Severe acute respiratory syndrome coronavirus 2 (SARS-CoV-2) emerged in late 2019 and has to date (late summer 2021) led to over 200M infections and between 4.5 and 15M deaths globally from coronavirus disease 2019 (COVID-19), the clinical illness caused by SARS-CoV-2 infection (The Economist, 2021). Most people have a mild course of disease, but around 15% of cases manifest severe illness (defined by presence of hypoxia) and 5% become critically unwell with acute respiratory distress syndrome (ARDS), usually necessitating intubation and mechanical ventilation as supportive measures for respiratory failure (COVID-19 Treatment Guidelines, 2022). While several organs have the potential to be involved during COVID-19, clinical and pathophysiological abnormalities of the respiratory system, characterized by bilateral pulmonary infiltrates and progressive hypoxia, are a near universal manifestation of severe and critical disease. Principally transmitted by airborne spread, it is thought that SARS-CoV-2 gains access to the lower airways and alveoli via inhalation/aspiration from the upper airways. SARS-CoV-2 can enter host cells by the viral spike protein binding to ACE2 (Angiotensin converting enzyme 2) followed by cleavage by host proteases. In humans, ACE2 is highest expressed within nasal epithelial cells, followed by airway epithelium (particularly secretory and ciliated cells) with only low level expression in alveolar epithelium (<1% of alveolar epithelial cells (Sungnak et al., 2020)), despite the alveolus being the site of an extensive inflammatory reaction and epithelial injury in severe cases. This paradox of low level ACE2 expression within alveolar epithelium is most likely explained by the presence of additional attachment receptors at this site, with recent description of lectins fulfilling this role (Lempp et al., 2021).

Although severe respiratory disease secondary to pathogen infection is a well-recognized phenomenon and shared with many other pathogens (e.g., influenza, respiratory syncytial virus (RSV), bacterial pneumonia, etc.), some features of COVID-19 are sufficiently different to suggest that both shared and unique features were occurring in this disease. Unlike influenza, that mainly causes mortality in extremes of age (both young and old), the main risk factor for development of severe COVID-19 is increasing age, with younger age groups having a much lower (but not absent) risk of hospitalization and mortality. Several other host factors also influence the risk of severe disease, including male sex and the presence of co-morbidities such as obesity, hypertension and diabetes (Budinger et al., 2021). In addition, the duration from onset of symptoms to development of severe disease is particularly long in COVID-19 infection being in the order of 10 days (nearly a week longer than the kinetics in influenza). These observations raised early suspicion that host factors were principally responsible for driving deleterious pulmonary inflammation and exacerbating organ damage, something that is now borne out by numerous studies of the host immune response as well as the reductions in mortality afforded by anti-inflammatory treatment with the glucocorticoid dexamethasone, as well as interleukin-6 receptor blockade (NEJM, 2021; The Lancet, 2021). There is now compelling evidence that dysregulated monocytes and macrophages are a key feature of severe COVID-19 (Velu et al., 2021). How and why these cells come to play such a deleterious role is incompletely understood. Here we will first review recent advances in our understanding of lung macrophage biology before discussing the role of monocytes and macrophages in SARS-CoV-2 infection. Due to the vast number of publications and pre-prints in this area, we focus on the pivotal findings that have informed our understanding of monocytes/macrophages in COVID-19.

2. Macrophages in the healthy lung

Before discussing how the macrophage compartment changes in context of SARS-CoV-2 infection, it is important to understand the composition of this compartment in health. The healthy lung harbors at least two populations of macrophages that occupy distinct anatomical niches. Alveolar (or airway) macrophages (AMs) are present in the bronchoalveolar space and are phenotypically, transcriptionally and functionally distinct from their counterparts in the lung interstitium (interstitial macrophage; IM). Importantly, neither lung macrophage population can be defined using the M1/M2 nomenclature that has become rooted in the macrophage biology field in the last 20 years. In the early 2000s, in an attempt to simplify the nomenclature of macrophage activation states, it was proposed that macrophages could be classified as M1 or M2, similar to the classification for helper T cells (Th1 vs. Th2) (Mills et al., 2000). M1 (or "classically" activated) macrophages were proposed to represent highly bactericidal, pro-inflammatory macrophages induced by IFNγ and/or endotoxin. In contrast, M2 (or "alternatively activated") macrophages were said to arise in response to IL-4 and adopt a "tissue repair" phenotype. As an extension of this, different flavors of M2 macrophages were proposed to differentiate in response to other cytokines, including IL-13, IL-10 or TGFβ (Mantovani et al., 2004). However, because macrophages are highly plastic and these "activation states" were mostly defined using *in vitro* culture systems involving a single stimulus and often defined by a single cell surface marker, it is rather unsurprising that this binary classification system fails to capture the degree of macrophage diversity seen *in vivo*. Indeed, recent transcriptional profiling of macrophages across tissues shows that tissue resident macrophages can have both "M1" and "M2" features in the absence of any injury or insult (Gibbings et al., 2017; Lavin et al., 2014), again highlighting the inadequacy of this system. We agree with Nahrendorf and Swirski that the continued use of the M1/M2 nomenclature stifles discovery rather than promoting it and should be abandoned by the field (Nahrendorf and Swirski, 2016).

2.1 Alveolar macrophages

Much of what is known about the biology of lung macrophages has come from studies using mouse models. There is consensus that AMs in the healthy mouse lung can be defined by their high expression of CD11c, SiglecF and CD169, along with CD64 (high affinity Fcγr1) and Mer-tyrosine kinase (MerTK), a receptor that allows recognition and uptake of apoptotic cells (Gautier et al., 2012a; Mowat et al., 2017). In humans, AMs are defined as HLA-DR$^+$CD11b$^+$CD64$^+$CD169$^+$CD206$^+$ cells (Bharat et al., 2016). Despite the limited overlap of surface marker expression between mouse and human AMs, recent cross-species transcriptomic analyses have shown that there is conservation of large parts of the transcriptome between mouse and man. In particular, genes involved the lipid catabolism (e.g., *PPARG, FABP4, FFAR4, CES1*) are prominent features of this conserved program (Leach et al., 2020; McQuattie-Pimentel et al., 2021). This is consistent with the homeostatic role of AMs in the regulation of pulmonary surfactant, the protein-lipid rich complex produced by alveolar type II epithelial (AT2) cells to lubricate the luminal surface of the lung. AMs are indispensable for this function, as evidenced by accumulation of surfactant and the development of pulmonary alveolar proteinosis (PAP) in humans and mice in whom AM function is defective (Dranoff et al., 1994; Martinez-Moczygemba et al., 2008; Stanley et al., 1994; Suzuki et al., 2014; Willinger et al., 2011). Similarly, AMs are key for the removal of effete cells and particulate matter from the airways (Felton et al., 2018; Poon et al., 2014). AMs are in constant communication with the alveolar epithelium, dialogue that is mediated by a series of receptor ligand interactions (Fig. 1). These include CD200 (Snelgrove et al., 2008), CD172a (Janssen et al., 2008) and IL-10 (Fernandez et al., 2004) expression

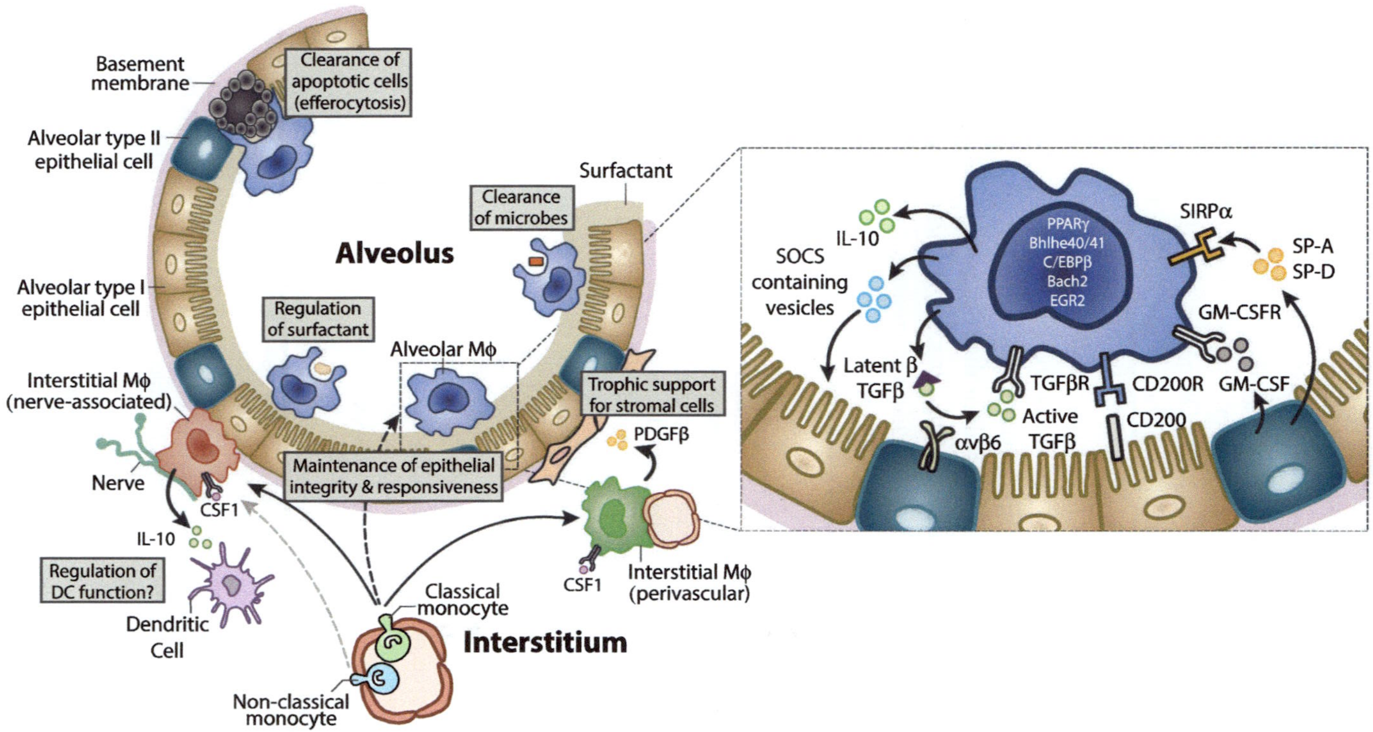

Fig. 1 See figure legend on next page.

by the alveolar epithelium that acts on corresponding receptors on AMs to regulate their responsiveness to exogenous stimulation to prevent unwanted and potentially damaging responses to innocuous material. Regulation is reciprocal with AMs releasing suppressor of cytokine signalling (SOCS)-containing vesicles which dampen the responsiveness of alveolar epithelial cells (Bourdonnay et al., 2015; Draijer et al., 2020). AMs may also help maintain the integrity of the epithelial barrier. For instance, under normal physiological conditions, AMs in both mouse and man produce fibronectin, which can act as a pro-survival and proliferative factor for airway epithelium (Han and Roman, 2006). Another conserved feature is the expression of key scavenger receptors, such as macrophage receptor with collagenous structure (MARCO) which is crucial for the recognition and uptake of a variety of respiratory pathogens, including *Streptococcus pneumoniae, Cryptococcus neoformans* and *Mycobacterium tuberculosis* (Arredouani et al., 2004; Dorrington et al., 2013; Xu et al., 2017). Genetic ablation of key apparatus for pathogen recognition or selective depletion of AMs in mouse models renders the host

Fig. 1 Composition, homeostatic functions and regulation of pulmonary macrophages. Alveolar macrophages (AMs) in the bronchoalveolar space of the lung are crucial for maintaining patency of the alveolar space, where they regulate surfactant levels and phagocytose inhaled microbes and other particulate materials. AMs are intimately associated with the alveolar epithelium and control the renewal and integrity of the barrier, as well as remove dead or senescent cells. AMs release suppressor of cytokines (SOCS)-containing vesicles that act to control the responsiveness of alveolar epithelial cells to exogenous stimulation. In turn, alveolar type II (AT2) epithelial cells produce GM-CSF to control the development and maintenance of AMs. Together with autocrine TGFβ, which is activated by epithelial αvβ6 integrin, GM-CSF induces expression of the transcription factor PPARγ, which in turn induces EGR2 and C/EBPβ, which drive particular aspects of alveolar macrophage differentiation. Bhlhe40 and Bhlhe41 control the identity and the self-renewal capacity of alveolar macrophages. The transcription factor Bach2 is also implicated in AM development. Epithelial-derived surfactant can also alter alveolar macrophage function by, e.g., altering phagocytosis. The interstitial macrophage (IM) compartment is heterogeneous with nerve-associated and blood vessel-associated (perivascular) macrophages, both of which appear to depend on CSF1, although the source of CSF1 is unclear. The homeostatic functions of interstitial macrophages are poorly understood, although they constitutively produce high levels of IL-10 which may alter the behavior of conventional dendritic cells (cDC). They also produce growth factors such as PDGFβ which may support fibroblasts. The transcription factors controlling IM differentiation and function remain largely unexplored. While originally seeded by embryonic progenitors, the IMs are replaced by circulating monocytes progressively with age. The contribution of monocytes to the steady state maintenance of alveolar macrophages remains controversial and to be determined with certainty.

more susceptible to a variety of infectious agents, demonstrating the key role these cells play in protective immunity.

AMs develop from fetal liver-derived monocytes in the first few days of life in parallel with alveolarization of the lung (Bharat et al., 2016; Guilliams et al., 2013; Tan and Krasnow, 2016), in a process highly dependent on granulocyte-macrophage colony stimulating factor (GM-CSF, also known as CSF-2) from AT2 cells (Gschwend et al., 2021; Schneider et al., 2014a). As such, mice or humans in whom the GM-CSF-GM-CSFR axis has been disrupted have considerably fewer mature alveolar macrophages and those present are dysfunctional, resulting in PAP development (Dranoff et al., 1994; Robb et al., 1995; Stanley et al., 1994; Suzuki et al., 2008, 2014). Autocrine transforming growth factor beta (TGFβ) also plays a crucial role in AM development (McCowan et al., 2021; Yu et al., 2017). GM-CSFR and TGFβR signalling induces expression of key transcription factors, including PPARγ (Baker et al., 2010; Gautier et al., 2012b; Malur et al., 2011; Schneider et al., 2014a), PU.1 (Shibata et al., 2001), C/EBPβ (Cain et al., 2013), Bhlhe40/41 (Rauschmeier et al., 2019) and Bach2 (Nakamura et al., 2013) to drive the niche-specific identity and function of AMs. Moreover, we have recently shown the transcription factor early growth response 2 (EGR2) to be indispensable for the identity and immune protective features of AMs (McCowan et al., 2021).

As discussed elsewhere in this review series, the ontogeny of tissue macrophages has received considerable attention in recent years with the realization that some tissue macrophages can derive from yolk sac mesenchyme and persist for long periods of time with little or no replenishment from blood monocytes. This has led to a revision of the long-accepted linear model of the mononuclear phagocyte system in which tissue macrophages were thought to be maintained exclusively through replenishment by blood monocytes, although it is important to note that some early studies reported the long-lived nature of tissue macrophages, including lung alveolar macrophages, and proposed that they may maintain themselves autonomously (Tarling et al., 1987). Indeed, alveolar macrophages can and do proliferate under normal physiological conditions (Hashimoto et al., 2013). However, whether *in situ* self-renewal is sufficient to sustain the alveolar macrophage compartment remains somewhat controversial. Studies using monocytopenic *Ccr2*$^{-/-}$ mice and parabiotic mice support the idea that alveolar macrophages maintain themselves independent of blood monocytes throughout adulthood (Hashimoto et al., 2013), findings supported by tracing of alveolar macrophage origin in human sex mismatched lung transplant studies (Eguíluz-Gracia et al., 2016). However, this is at odds with longitudinal fate mapping using *Flt3*^{Cre}- and *Ms4a3*^{Cre}-based reporter mice, which allow tracing of hematopoietic stem cell (HSC)-derived and granulocyte-monocyte progenitors (GMPs)-derived cells, respectively (Gomez Perdiguero et al., 2015; Liu et al., 2019), as well as fate mapping using tissue protected bone marrow chimeric mice (Bain et al., 2016). Moreover, a more recent study using single cell RNA sequencing (scRNA-seq) to examine AM repopulation dynamics in sex mismatched lung transplant tissue has shown that AMs are largely replaced by recipient monocyte-derived cells (Byrne et al., 2020). Thus, more work is needed to understand the parameters that control alveolar macrophage maintenance even in the absence of inflammation, infection or injury.

2.2 Interstitial macrophages

Interstitial macrophages (IMs) are embedded in the lung parenchyma between the alveoli and capillary beds. Their high phagocytic and bactericidal capacity suggests they may represent an additional layer of innate protection below the basement membrane (Gibbings et al., 2017; Schyns et al., 2019). Furthermore, they may provide trophic support for fibroblasts, endothelia and epithelia through production of mediators such as platelet-derived growth factor beta (PDGFβ) and amphiregulin (Brody et al., 1992; Schyns et al., 2019). While the AM compartment appears to be relatively homogenous in health, the IM compartment is more heterogeneous. Consensus is yet to reached on the best approach to define discrete subsets of these cells, but it is clear that, in mice, they all express high levels of CD11b, CD64 and MerTK, with at least two independent populations defined by their expression of CD206 (mannose receptor). CD206^{-} IMs express high levels of MHCII and associate with nerve fibers. In contrast, CD206^{+} IMs generally lack MHCII expression, but express CD36 and Lyve-1, and are enriched in the perivascular regions of the lung parenchyma (Chakarov et al., 2019; Gibbings et al., 2017). The phenotype and function of human lung IMs is less well-defined owing to difficulty in obtain healthy lung tissue for these analyses. Nevertheless, a limited number of studies have shown that these cells are defined by their HLA-DR^{+}CD68^{+}CD11b^{+}CD206^{+} profile (Bharat et al., 2016; Leach et al., 2020), consistent with recent work by the Willinger group using so-called 'MISTRG' humanized mice (Evren et al., 2021). Moreover, discrete IM subsets may exist in the human lung defined by their levels of HLA-DR and CD36 expression and there appears to be a degree of conservation of gene expression between mouse and human (Leach et al., 2020). Whether these subsets occupy distinct anatomical niches in the lung interstitium is not yet known. Functionally, studies in mice have shown that $CD206^{-}$ IMs are more capable of supporting the differentiation of regulatory T cells, which is consistent with their expression of MHCII and likely aided by their constitutive production of IL-10 (Sabatel et al., 2017; Schyns et al., 2019).

In contrast to AMs, IMs originally develop from yolk sac-derived progenitors (Tan and Krasnow, 2016) but these are displaced, and by adulthood all interstitial macrophages derive from HSC in the adult mouse (Gomez Perdiguero et al., 2015). When combined with the fact that there is only low-level replenishment of interstitial macrophages in $Cx3cr1^{\text{Cre-ERT2}}$. *Rosa26* reporter mice (Schyns et al., 2019), in the context of parabiosis (Tan and Krasnow, 2016) or tissue protected bone marrow chimeric mice (Gibbings et al., 2017; Hawley et al., 2018), this suggests HSC-derived macrophages seed the lung in late embryonic development or early postnatal development and become relatively long-lived. Notably, despite the heterogeneity in surface phenotype, transcriptional profile and anatomical positioning of interstitial macrophage subsets, they appear to have similar replenishment kinetics (Chakarov et al., 2019; Gibbings et al., 2017; Schyns et al., 2019).

Finally, in addition to AMs and IMs, the mammalian lung also appears to have a population of intravascular macrophages, although notably these are not present in rodents (Brain et al., 1999; Dehring and Wismar, 1989; Evren et al., 2021; Longworth, 1997). Transcriptional profiling of these macrophages from humanized mice has shown them to express molecules involved in iron recycling (e.g., *CD163, SLC40A1, SLC48A1*) and the transcription factor *SPIC* (Evren et al., 2021), which controls the differentiation program of red pulp macrophages in the spleen. Thus, intravascular macrophages may allow filtering of the blood in the lung, similar to that occurring in the spleen and liver. Whether they play additional roles in homeostasis, and indeed disease, remains to be investigated.

3. Inflammation as a driver of COVID-19 severity

Early during the pandemic it was appreciated that the development of severe COVID-19 did not relate solely to an increased viral presence within the airways (Merad and Martin, 2020; Rubin et al., 2021). Indeed, a report from our own group, analyzing autopsy collected tissue from decedents that had succumbed to severe disease, explored SARS-CoV-2 organotropism by multiplex PCR, viral sequencing and *in situ* viral S protein immunohistochemistry. This revealed that the duration of illness (from symptom onset to death) was not correlated with the number of organs with viral RNA present, and within lung tissue viral presence was not directly linked to the presence and severity of pulmonary inflammation (Dorward et al., 2021). This evidence of virus-independent immunopathology, rather than direct pathogen-mediated organ inflammation and injury, as a primary driver of severe COVID-19 was swiftly followed by the discovery that immunosuppression with corticosteroids (dexamethasone) limits death in severe cases with established hypoxic respiratory failure (NEJM, 2021). Interestingly, that report demonstrated that patients with a longer duration of symptoms (who were more likely to have had established respiratory failure) and those receiving invasive mechanical ventilation had the greatest benefit, while in patients not on oxygen there was a trend towards worse outcomes with dexamethasone treatment. Overall, these data are consistent with a requirement for early robust inflammation in response to infection, but in later stage disease inflammation directly contributing to disease severity (Velu et al., 2021). Understanding what "type" of inflammation drives disease pathogenesis remains critical to understanding how best to intervene with therapeutics in clinical trials and has been a major focus of COVID-19 research.

4. Macrophage as mediators of organ damage in COVID-19?

4.1 Defining the macrophage compartment in COVID-19

While many groups rapidly delineated the cellular and soluble mediator profile of circulating blood from COVID-19 patients, it is clear that these peripheral abnormalities are not always reflective of the changes occurring within inflamed lung tissue (Dress and Ginhoux, 2021). As described above, whole lung tissue is exceedingly challenging to acquire (especially out with autopsy sampling of fatal cases), whereas sampling of the airways via acquisition of bronchoalveolar lavage fluid (BALF) allows for analysis of luminal cells, with repeated sampling also possible. An early report detailing scRNA-seq of BALF cells from three patients with moderate COVID-19 and six patients with severe disease demonstrated significant abnormalities within the macrophage compartment (Liao et al., 2020). Most notably, tissue resident (*FABP4*$^+$, *MARCO*$^+$) AMs were depleted during COVID-19, with severe disease typified by a marked expansion of elicited macrophages (Liao et al., 2020). These were divided into three clusters based on transcriptional profiles; group 1 expressed typical monocyte markers (e.g., *S100A8/MRP8, FCN1* and *CD14*), group 2 had high level expression of chemokines (e.g., *CCL2, CCL3, CXCL10*) and group 3 expressed immunoregulatory and pro-fibrotic genes (e.g., *SPP1, TREM2*). Analysis of whole lung tissue from autopsy cases of fatal COVID-19 has also demonstrated marked changes in macrophage compartment of non-luminal pulmonary tissue. Specifically, there is an accumulation of monocytes and macrophages frequently expressing MRP8/S100A8 (a subunit of the calcium binding protein calprotectin) consistent with monocytic origin and an IM rather than AM phenotype (Dorward et al., 2021). In addition, a pulmonary arteritis (inflammation of blood vessel walls) was also observed in around half of analyzed cases, with mononuclear cells being the predominant infiltrating cell type. Whether this contributes to the endothelial dysfunction and high rate of thrombosis seen in COVID-19 (Bing et al., 2021; Dolby et al., 2021), or is merely a secondary response to endothelial injury, remains to be clarified. In keeping with a key role for monocytes and macrophages in severe COVID-19, an elegant study analyzing host genetic variants associated with critical illness identified that high genotype-inferred expression of the monocyte–macrophage chemotactic receptor CCR2 is associated with severe COVID-19 (Pairo-Castineira, 2020). Since these early reports, numerous studies have implicated macrophages and their products in the pathogenesis of severe SARS-CoV-2 infection, and some of the more notable studies will now be discussed (Fig. 2).

A longitudinal analysis of immune cells and soluble factors in blood and BALF fluid, investigating different severities of COVID-19, found that disease severity was associated with increased abundance of chemokines involved in myeloid/monocyte recruitment including CCL4, CCL2 and CXCL9 and also soluble complement factor C5a generation (Carvelli et al., 2020). The C5a receptor (C5aR1/CD88) was highly expressed on both blood and BALF neutrophils and monocytes and multiplex IHC of lung tissue from fatal COVID-19 showed expansion of macrophages ($CD68^{+}/CD163^{+}$ cells), many of which also expressed C5aR1. In addition, arteritis was also observed in this study and was associated with infiltration of these $C5aR1^{+}$ macrophages around the arteries and within thrombus. The potential therapeutic benefits of targeting the C5a–C5aR1 axis was also tested in a pre-clinical model of lung injury with the monoclonal antibody avdoralimab that prevents C5a binding to C5aR1. Although experimental lung injury was induced in

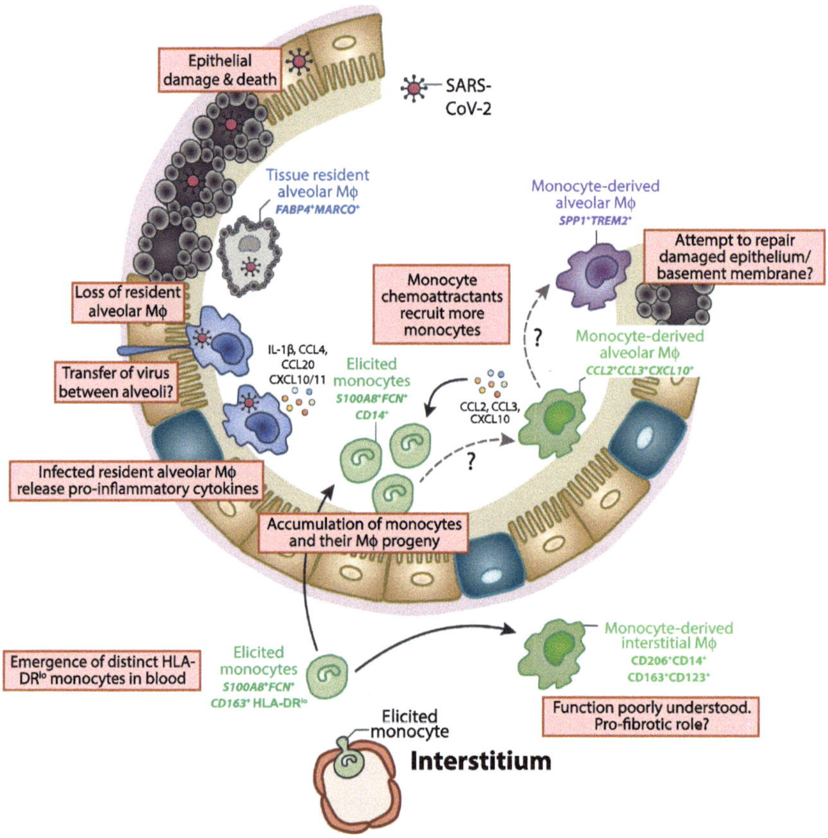

Fig. 2 Composition of pulmonary macrophages during severe COVID-19. In severe COVID-19, tissue resident alveolar macrophages (AMs), defined by expression of *FABP4* and *MARCO*, are diminished in number. Whether this occurs in response to direct infection or due to the hyperinflammatory environment is unclear. Alternatively, because there is epithelial cell damage and death, this could reflect a loss of survival signals for resident AMs. Infected tissue resident AMs express higher levels of *IL1B, CCL4, CCL20, CXCL10* and *CXCL11* than their uninfected counterparts, which leads to monocyte (and T cell; not shown) recruitment. Given that AMs have been shown to be able to transit between alveoli, it has been suggested that AMs may facilitate the spread of SARS-Cov-2 within the lung. Monocytes accumulate in vast numbers in the bronchoalveolar space, as do transcriptionally distinct macrophage subsets. Current evidence supports the idea that these macrophages are derived from recruited monocytes, although the development relationship, if any, between macrophages displaying pro-inflammatory features (*CCL2, CCL3, CXCL10*) and pro-fibrotic features (*SPP1, TREM2*) is unclear. Elevated chemokine production creates a positive feedback loop to recruit more monocytes, thereby maintaining the inflammatory response. The significance of *SPP1*$^+$*TREM2*$^+$ macrophages is unclear, although they could represent an attempted response to repair the damaged barrier. Monocytes and their progeny also accumulate in the lung parenchyma, although their role in COVID-19 pathogenesis and disease outcome is unclear. Importantly, circulating monocytes in severe COVID-19 are phenotypically and transcriptionally distinct from their counterparts in health, suggestion changes in monocyte output at the level of hematopoiesis in the bone marrow.

human C5aR1 knock-in mice (as avdoralimab is human specific) and in response to local delivery of recombinant C5a (and thus limiting the generalizability of the findings), several parameters of lung injury were attenuated with avdoralimab treatment (Carvelli et al., 2020). In an elegantly planned and executed study, Szabo and colleagues performed longitudinal profiling of paired peripheral blood and BALF cells from patients with severe COVID-19 (average 6-7 paired samples per patient) and complemented this with analysis of autopsy lung tissue (Szabo et al., 2021). Spectral flow cytometry based phenotyping alongside transcriptomic and cytokine profiling revealed that aberrant $CD163^+$ and $HLA\text{-}DR^{lo}$ monocytes predominated in COVID-19 peripheral blood, with these populations expressing CCR2. In health, myeloid populations in blood and airways were phenotypically distinct, whereas COVID-19 airway myeloid cells displayed a significant phenotypic overlap with circulating blood myeloid subsets, suggesting myeloid cell infiltration from blood to airways. Consistent with this, CCL2, CCL3 and CCL4 (among other cytokines and chemokines) were elevated in airways, with CCL2 and CCL3 undetectable in blood. Furthermore, airway myeloid cells themselves expressed high levels of CCL2, CCL3, and CCL4 transcripts and airway monocytes and macrophages (along with airway T cells) exhibited hyperinflammatory transcriptional signatures. Analysis of whole lung tissue confirmed a significant increase in numbers of $CD163^+$ monocytes and macrophages, but airway transcriptomics did not demonstrate high expression of genes associated with proliferation (e.g., Ki67). Overall these data supported active myeloid recruitment from blood to lung as a major mechanism that may directly exacerbate inflammation and organ injury in severe COVID-19 (Szabo et al., 2021). These findings have been independently replicated by another group, whereby BALF macrophages had increased transcriptional similarity to circulating monocytes in COVID-19, again suggesting a differential origin of alveolar macrophages (monocyte derived) in SARS-CoV-2 infection (Stephenson et al., 2021). Similarly, targeted proteomics of autopsy lung and splenic tissue from our own group has demonstrated elevated abundance of pulmonary CCL7 (MCP-3), along with enhanced splenic CSF-1 in COVID-19, consistent with monocyte/macrophage production at extrapulmonary sites followed by recruitment into lung tissue (Russell et al., 2022).

One of the major outstanding questions in COVID-19 is how much of the observed alterations in immune responses is a response to widespread epithelial injury, and therefore likely to be shared between other causes of severe lung injury/inflammation (e.g., influenza, sepsis, bacterial pneumonia), and how much is truly specific and unique to SARS-CoV-2 infection. In one of the few studies that has directly analyzed and compared whole lung tissue from different pulmonary insults, Rendeiro et al. performed high-parameter imaging mass cytometry to allow cellular and spatial profiling of pulmonary tissue at single cell resolution (Rendeiro et al., 2021). They compared a small cohort of 23 patients that died from either ARDS after influenza, ARDS after bacterial infection, acute bacterial pneumonia, early or late COVID-19 or non-pulmonary disease (control/normal lung tissue) to understand the cellular changes in the lungs during COVID-19 progression. Imaging mass cytometry was performed using a panel of 36 metal-labeled antibodies to delineate a spatial map of more than 660k cells at 1 μm resolution and was complemented by spatial transcriptomics. This revealed that immune infiltration in the COVID-19 lungs was quantitatively comparable to the other lung infections. However, qualitative differences were observed, with a significant increase in myeloid cells within COVID-19 lungs, with IMs (defined as $CD14^+/CD16^+/CD206^+/CD163^+/CD123^+$ cells) having the greatest expansion in COVID-19. Interestingly, neutrophil numbers were similar in early COVID-19 and the healthy lung with reduced neutrophil numbers in late COVID-19. How this reduction in neutrophils in late disease relates to ongoing inflammation and/or failed inflammation resolution and pulmonary repair requires further investigation. Additional immune-specific phenotyping showed increased levels of the calprotectin subunit S100A9/MRP14 in macrophages, neutrophils and alveolar epithelial cells, and complements the other studies discussed previously which reported increased calprotectin S100A8/MRP8 within monocytes and macrophages (Dorward et al., 2021; Liao et al., 2020). Out with immune lineage cells a reduction in endothelial cells was observed (consistent with the previously observed concurrent endothelial injury (NEJM, 2020)) and increased numbers of fibroblasts and collagen deposition noted in the COVID-19 groups, with obvious implications on recovery and pulmonary fibrosis from severe COVID-19.

4.2 The role of macrophage subsets in SARS-CoV-2 infection

Most of what is known about human pulmonary immunity in response to SARS-CoV-2 relates to severe or fatal disease, whereas most infections lead to mild disease. Thus, it remains incompletely understood if monocyte-macrophage expansion also occurs in mild disease and if these elicited macrophages have direct anti-viral effects and/or if they are responsible for orchestrating epithelial damage that has occurred. However, limited analysis of "mild" COVID-19 shows that tissue resident AMs are lost, albeit to a lower extent than in severe COVID-19, and replaced by monocyte-derived macrophages. The loss of tissue resident macrophages in the setting of inflammation is a well described phenomenon, often referred to as the macrophage "disappearance reaction." Indeed, a common feature of murine models of lung bacterial, viral and fungal infection, as well as sterile injury, is the transient loss of AMs (Ghoneim et al., 2013; McCowan et al., 2021). In the context of SARS-CoV-2 infection, this could be an effect of infection of AM since viral S protein has been detected with macrophages and both S positive epithelial cells and S positive macrophages appear to display elevated levels of cleaved caspase-3 (Dorward et al., 2021; Grant et al., 2021; Rendeiro et al., 2021). Infection of AMs may occur following uptake of infected epithelial cells or through direct infection. Although *in vitro* studies suggest that macrophages do not support productive viral replication (Yang et al., 2020), recent scRNA-seq studies in which viral transcripts can be identified suggest AMs can and do become directly infected (Grant et al., 2021). While the loss of AM seems somewhat counterintuitive given that tissue resident macrophages are thought to be immune effector cells, apoptosis of AMs is required for efficient killing of certain bacterial pathogens (e.g., *Streptococcus pneumoniae, Haemophilus influenzae*) (Preston et al., 2019). Whether loss of tissue resident AMs in the context SARS-CoV-2 infection represents a killing mechanism is unknown, but something that warrants further investigation. Alternatively, loss of resident AMs may occur as a collateral result of hyper-inflammation, for instance in response to elevated IFNγ levels, as has been described in murine models of influenza A infection (Califano et al., 2018). Indeed, the presence of IFNγ-producing T cells is another defining feature of severe COVID-19. However, it is important to note that AMs are not bystanders in the anti-viral response. SARS-CoV-2-infected AMs produce higher levels of pro-inflammatory cytokines *IL1B* and monocyte and T cell chemoattractants, *CCL4* and *CXCL10,* than their uninfected counterparts (Grant et al., 2021). This is consistent with the role of resident AMs in other viral infections, such as influenza and RSV (Goritzka et al., 2015; Schneider et al., 2014b), where they are responsible for the recruitment of anti-viral monocytes through production of type I interferons and monocyte chemoattractants (Goritzka et al., 2015; (Kumagai et al., 2007)). Whether monocytes and their progeny are directly anti-viral in COVID-19 is also unclear. Observations from mouse models may help to unravel this, with multiple mouse models of SARS-Cov-2 infection having been developed (Bao et al., 2020; Hassan et al., 2020; Israelow et al., 2020; Jiang et al., 2020; Sun et al., 2020). Importantly, monocyte/macrophage dynamics replicate observations in humans, with loss of tissue resident AMs and accumulation of monocytes and their progeny in response to elevated monocyte chemoattractants induced by type I interferons (Israelow et al., 2020). That monocytopenic *Ccr2*$^{-/-}$ mice have elevated viral burden when infected with mouse adapted SARS-CoV-2, suggests that elicited monocytes/macrophages may have a direct anti-viral role (Vanderheiden et al., 2021). However, this is complicated by the fact that monocytes and macrophages fail to accumulate in *Ifnar1*$^{-/-}$ mice yet viral clearance is unaffected. Instead, *Ifnar1*-deficiency is associated with significantly reduced inflammation, demonstrating the key role of type I interferons in driving the excessive inflammatory response through monocyte recruitment, consistent with earlier work in SARS-CoV-1 infection (Channappanavar et al., 2016). Continual recruitment of inflammatory monocytes may hinder the regeneration process. In support of this, previous studies had identified that ongoing inflammation, especially mediated by IL-1β, is capable of impairing alveolar epithelial regeneration by inhibiting the maturation of alveolar type I (AT1) cells and arresting them in in a damage-associated transient progenitor state (DATPs) (Choi et al., 2020; Kobayashi et al., 2020). Transcriptional profiling of alveolar epithelial cells during COVID-19 has revealed that alongside extensive epithelial destruction, initiation of regeneration also occurs. However, an incomplete transition of AT2 to AT1 cells is observed with increased numbers of epithelial cells with a transcriptional signature of DATPs (KRT8, CLDN4 and CDKN1A) in COVID-19 lungs (Melms et al., 2021). IL-1β has been noted to be highest expressed in pulmonary monocytes in "early" COVID-19 (up to 30 days from infection onset) (Rendeiro et al., 2021) and peripheral blood monocytes isolated from COVID-19 patients release mature IL-1β in response to viral S protein (alongside nigericin and after differentiation into macrophages by CSF-1) unlike healthy blood monocytes (Theobald et al., 2021). Thus, in early disease inflammatory monocytes and macrophages may directly worsen epithelial injury by liberation of inflammatory mediators such as IL-1β which subsequently impacts upon epithelial regeneration and exacerbates inflammation.

Until now, we have discussed the evidence that monocytes and their derivatives play a key role in lung inflammation and pathology in COVID-19. However, lessons from other pre-clinical models of lung injury, including our own work, supports the notion that monocytes can adopt distinct functional profiles across a disease course. For instance, monocyte-derived AMs are proposed to be central to fibrogenesis following exposure to bleomycin, yet these same cells are indispensable for efficient lung repair in this model (Cui et al., 2020; Joshi et al., 2020; McCowan et al., 2021; Misharin et al., 2017). Indeed, the presence of transcriptionally distinct populations of monocyte-derived macrophages in the COVID-19 BALF suggests that monocytes can assume different effector profiles as they mature into macrophages. It could be that these cells occupy distinct niches, some of which support the differentiation of inflammatory macrophages, with others supporting a tissue repair phenotype. Indeed, as mentioned above, monocyte-derived macrophages with key pro-fibrotic characteristics, including *SPP1, TREM2, TGFB1* and *FGL2* expression have been identified (Liao et al., 2020; Wauters et al., 2021). Alternatively, these subsets of macrophages could reflect cells at different stages of a differentiation trajectory or cells that have arrived at distinct stages of the infection course. Whether *SPP1*$^{+}$ macrophages are involved in tissue repair or if they are indicative of an overactive fibrotic response following severe lung injury is not clear. However, given that pulmonary fibrosis appears to be a lasting effect of COVID-19 in a subset of individuals, the balance of these subsets may have a role in determining outcome from COVID-19.

5. Targeting the macrophage in severe COVID-19: Current and future therapeutics?

How does what we have learnt from studying macrophages in COVID-19 inform rationale choice of therapeutic agents for large scale clinical trials, and do the drugs that have already shown mortality benefits in severe COVID-19 act via effects on macrophages? While the broad spectrum anti-inflammatory glucocorticoid dexamethasone was the first pharmaceutical to show improved mortality in severe cases, the exact mechanism of action remains to be defined. We do know that dexamethasone can have several effects on macrophages including upregulation of the efferocytosis receptor MerTK which can enhance clearance of dead and dying cells (efferocytosis) generated during injury and can impact upon tissue repair (Bosurgi et al., 2017; Felton et al., 2018; Liu et al., 1999; McColl et al., 2009; Zagórska et al., 2014). In addition, dexamethasone can suppress pro-inflammatory cytokine production by macrophages including IL-6 and IL-1 (Heasman et al., 2004) by well investigated but as yet not completely understood mechanisms (Barnes, 2017). Indeed, it has been proposed that glucocorticoid treatment of macrophages results in a unique pro-resolution type phenotype (McColl et al., 2007). Similarly, IL-6 receptor blockade reduces mortality in hospitalized COVID-19 patients with hypoxia and systemic inflammation; whether this acts on macrophages or macrophage-liberated IL-6 as a primary mechanism remains to be determined. But can more targeted therapeutics aiming to disrupt inflammatory macrophages be beneficial? This could take the approach of interfering with recruitment, their longevity within inflamed pulmonary tissue or blocking liberation of pro-inflammatory molecules identified in the above studies (Velu et al., 2021). Interestingly, both blocking (Temesgen et al., 2021) and augmenting (Bosteels et al., 2020) GM-CSF has been reported as having promise while preliminary studies on the tolerability of C5 blockade in human COVID-19 are beginning to appear (Zelek, 2020), but further studies to test therapeutic efficacy are awaited. Supporting the notion that blocking monocyte/macrophage pro-inflammatory molecules could be beneficial, a recent report evaluated the use of the IL-1α/β inhibitor anakinra (a recombinant version of the human interleukin 1 receptor antagonist protein) (Kyriazopoulou et al., 2021). By identifying patients at greatest risk of disease progression (using circulating soluble urokinase-type plasminogen activator receptor levels (suPAR)), and then intervening with anakinra (or placebo) alongside usual standard of care including dexamethasone, this demonstrated reduced mortality and length of hospital stay in the anakinra treated group.

It is likely that the approaches being taken to understand the basic biology of COVID-19 immunopathology will assist with prioritization and repurposing of existing therapeutics for ongoing clinical trials. The combined approaches of using human disease samples and pre-clinical models married to state-of-the-art technologies has led to an unprecedented pace at deciphering pulmonary immunity in SARS-CoV-2 infection. This serves as a paradigm to approach other infection-driven and pulmonary inflammatory diseases, especially when these continue to cause substantial global disability and premature mortality (WHO, n.d).

Acknowledgments

C.D.L. is funded by a Wellcome Trust Clinical Career Development Fellowship (206566/Z/17/Z). C.C.B. is funded by a Sir Henry Dale Fellowship from the Wellcome Trust/Royal Society (206234/Z/17/Z). Servier Medical Art (smart.servier.com) was used for the generation of some of the graphics.

References

Arredouani, M., Yang, Z., Ning, Y., Qin, G., Soininen, R., Tryggvason, K., Kobzik, L., 2004. The scavenger receptor MARCO is required for lung defense against pneumococcal pneumonia and inhaled particles. J. Exp. Med. 200, 267–272.

Bain, C.C., Hawley, C.A., Garner, H., Scott, C.L., Schridde, A., Steers, N.J., Mack, M., Joshi, A., Guilliams, M., Mowat, A.M.I., Geissmann, F., Jenkins, S.J., 2016. Long-lived self-renewing bone marrow-derived macrophages displace embryo-derived cells to inhabit adult serous cavities. Nat. Commun. 7. ncomms11852.

Baker, A.D., Malur, A., Barna, B.P., Kavuru, M.S., Malur, A.G., Thomassen, M.J., 2010. PPARgamma regulates the expression of cholesterol metabolism genes in alveolar macrophages. Biochem. Biophys. Res. Commun. 393, 682–687.

Bao, L., Deng, W., Huang, B., Gao, H., Liu, J., Ren, L., Wei, Q., Yu, P., Xu, Y., Qi, F., Qu, Y., Li, F., Lv, Q., Wang, W., Xue, J., Gong, S., Liu, M., Wang, G., Wang, S., Song, Z., Zhao, L., Liu, P., Zhao, L., Ye, F., Wang, H., Zhou, W., Zhu, N., Zhen, W., Yu, H., Zhang, X., Guo, L., Chen, L., Wang, C., Wang, Y., Wang, X., Xiao, Y., Sun, Q., Liu, H., Zhu, F., Ma, C., Yan, L., Yang, M., Han, J., Xu, W., Tan, W., Peng, X., Jin, Q., Wu, G., Qin, C., 2020. The pathogenicity of SARS-CoV-2 in hACE2 transgenic mice. Nature 583, 830–833.

Barnes, P.J., 2017. Glucocorticosteroids. Handb. Exp. Pharmacol. 237, 93–115.

Bharat, A., Bhorade, S.M., Morales-Nebreda, L., McQuattie-Pimentel, A.C., Soberanes, S., Ridge, K., DeCamp, M.M., Mestan, K.K., Perlman, H., Budinger, G.R.S., Misharin, A.V., 2016. Flow cytometry reveals similarities between lung macrophages in humans and mice. Am. J. Respir. Cell Mol. Biol. 54, 147–149.

Bing, R., Andrews, J.P., Williams, M.C., van Beek, E.J.R., Lucatelli, C., MacNaught, G., Clark, T., Koglin, N., Stephens, A.W., MacAskill, M.G., Tavares, A.A., Dhaliwal, K., Dorward, D.A., Lucas, C.D., Dweck, M.R., Newby, D.E., 2021. In vivo thrombosis imaging in patients recovering from COVID-19 and pulmonary embolism. Am. J. Respir. Crit. Care Med. https://doi.org/10.1164/rccm.202011-4182IM.

Bosteels, C., Maes, B., Van Damme, K., De Leeuw, E., Declercq, J., Delporte, A., Demeyere, B., Vermeersch, S., Vuylsteke, M., Willaert, J., Bollé, L., Vanbiervliet, Y., Decuypere, J., Libeer, F., Vandecasteele, S., Peene, I., Lambrecht, B., 2020. Sargramostim to treat patients with acute hypoxic respiratory failure due to COVID-19 (SARPAC): a structured summary of a study protocol for a randomised controlled trial. Trials 21, 491.

Bosurgi, L., Cao, Y.G., Cabeza-Cabrerizo, M., Tucci, A., Hughes, L.D., Kong, Y., Weinstein, J.S., Licona-Limon, P., Schmid, E.T., Pelorosso, F., Gagliani, N., Craft, J.E., Flavell, R.A., Ghosh, S., Rothlin, C.V., 2017. Macrophage function in tissue repair and remodeling requires IL-4 or IL-13 with apoptotic cells. Science 356, 1072–1076.

Bourdonnay, E., Zasłona, Z., Penke, L.R.K., Speth, J.M., Schneider, D.J., Przybranowski, S., Swanson, J.A., Mancuso, P., Freeman, C.M., Curtis, J.L., Peters-Golden, M., 2015. Transcellular delivery of vesicular SOCS proteins from macrophages to epithelial cells blunts inflammatory signaling. J. Exp. Med. 212, 729–742.

Brain, J.D., Molina, R.M., DeCamp, M.M., Warner, A.E., 1999. Pulmonary intravascular macrophages: their contribution to the mononuclear phagocyte system in 13 species. Am. J. Physiol. 276, L146–L154.

Brody, A.R., Bonner, J.C., Overby, L.H., Badgett, A., Kalter, V., Kumar, R.K., Bennett, R.A., 1992. Interstitial pulmonary macrophages produce platelet-derived growth factor that stimulates rat lung fibroblast proliferation in vitro. J. Leukoc. Biol. 51, 640–648.

Budinger, G.R.S., Misharin, A.V., Ridge, K.M., Singer, B.D., Wunderink, R.G., 2021. Distinctive features of severe SARS-CoV-2 pneumonia. J. Clin. Invest. 131. https://doi.org/10.1172/JCI149412.

Byrne, A.J., Powell, J.E., O'Sullivan, B.J., Ogger, P.P., Hoffland, A., Cook, J., Bonner, K.L., Hewitt, R.J., Wolf, S., Ghai, P., Walker, S.A., Lukowski, S.W., Molyneaux, P.L., Saglani, S., Chambers, D.C., Maher, T.M., Lloyd, C.M., 2020. Dynamics of human monocytes and airway macrophages during healthy aging and after transplant. J. Exp. Med. 217, e20191236.

Cain, D.W., O'Koren, E.G., Kan, M.J., Womble, M., Sempowski, G.D., Hopper, K., Gunn, M.D., Kelsoe, G., 2013. Identification of a tissue-specific, C/EBPβ-dependent pathway of differentiation for murine peritoneal macrophages. J. Immunol. 191, 4665–4675.

Califano, D., Furuya, Y., Metzger, D.W., 2018. Effects of influenza on alveolar macrophage viability are dependent on mouse genetic strain. J. Immunol. 201, 134–144.

Carvelli, J., Demaria, O., Vély, F., Batista, L., Chouaki Benmansour, N., Fares, J., Carpentier, S., Thibult, M.-L., Morel, A., Remark, R., André, P., Represa, A., Piperoglou, C., Cordier, P.Y., Le Dault, E., Guervilly, C., Simeone, P., Gainnier, M., Morel, Y., Ebbo, M., Schleinitz, N., Vivier, E., 2020. Association of COVID-19 inflammation with activation of the C5a–C5aR1 axis. Nature 588, 146–150.

Chakarov, S., Lim, H.Y., Tan, L., Lim, S.Y., See, P., Lum, J., Zhang, X.-M., Foo, S., Nakamizo, S., Duan, K., Kong, W.T., Gentek, R., Balachander, A., Carbajo, D., Bleriot, C., Malleret, B., Tam, J.K.C., Baig, S., Shabeer, M., Toh, S.-A.E.S., Schlitzer, A., Larbi, A., Marichal, T., Malissen, B., Chen, J., Poidinger, M., Kabashima, K., Bajenoff, M., Ng, L.G., Angeli, V., Ginhoux, F., 2019. Two distinct interstitial macrophage populations coexist across tissues in specific subtissular niches. Science 363. https://doi.org/10.1126/science.aau0964.

Channappanavar, R., Fehr, A.R., Vijay, R., Mack, M., Zhao, J., Meyerholz, D.K., Perlman, S., 2016. Dysregulated type I interferon and inflammatory monocyte-macrophage responses cause lethal pneumonia in SARS-CoV-infected mice. Cell Host Microbe 19, 181–193.

Choi, J., Park, J.-E., Tsagkogeorga, G., Yanagita, M., Koo, B.-K., Han, N., Lee, J.-H., 2020. Inflammatory signals induce AT2 cell-derived damage-associated transient progenitors that mediate alveolar regeneration. Cell Stem Cell. 27, 366–382.e7.

COVID-19 Treatment Guidelines, 2022. Clinical Spectrum of SARS-CoV-2 Infection. National Institutes of Health.

Cui, H., Jiang, D., Banerjee, S., Xie, N., Kulkarni, T., Liu, R.-M., Duncan, S.R., Liu, G., 2020. Monocyte-derived alveolar macrophage apolipoprotein E participates in pulmonary fibrosis resolution. JCI Insight 5, e134539.

Dehring, D.J., Wismar, B.L., 1989. Intravascular macrophages in pulmonary capillaries of humans. Am. Rev. Respir. Dis. 139, 1027–1029.

Dolby, H.W., Potey, P., Wilder-Smith, A.B., Clohisey, S., Millar, J.E., Baillie, J.K., Dorward, D.A., Lucas, C.D., Russell, C.D., 2021. Histological evidence of pulmonary microthrombosis and vasculitis in life-threatening respiratory virus diseases. Open Forum Infect. Dis. 8. https://doi.org/10.1093/ofid/ofaa640.

Dorrington, M.G., Roche, A.M., Chauvin, S.E., Tu, Z., Mossman, K.L., Weiser, J.N., Bowdish, D.M.E., 2013. MARCO is required for TLR2- and NOD2-mediated responses to Streptococcus pneumoniae and clearance of pneumococcal colonization in the murine nasopharynx. J. Immunol. 190, 250–258.

Dorward, D.A., Russell, C.D., Um, I.H., Elshani, M., Armstrong, S.D., Penrice-Randal, R., Millar, T., Lerpiniere, C.E.B., Tagliavini, G., Hartley, C.S., Randle, N.P., Gachanja, N.N., Potey, P.M.D., Dong, X., Anderson, A.M., Campbell, V.L., Duguid, A.J., Al Qsous, W., BouHaidar, R., Baillie, J.K., Dhaliwal, K.,

Wallace, W.A., Bellamy, C.O.C., Prost, S., Smith, C., Hiscox, J.A., Harrison, D.J., Lucas, C.D., 2021. Tissue-specific immunopathology in fatal COVID-19. Am. J. Respir. Crit. Care Med. 203, 192–201.

Draijer, C., Speth, J.M., Penke, L.R.K., Zaslona, Z., Bazzill, J.D., Lugogo, N., Huang, Y.J., Moon, J.J., Peters-Golden, M., 2020. Resident alveolar macrophage-derived vesicular SOCS3 dampens allergic airway inflammation. FASEB J. 34, 4718–4731.

Dranoff, G., Crawford, A.D., Sadelain, M., Ream, B., Rashid, A., Bronson, R.T., Dickersin, G.R., Bachurski, C.J., Mark, E.L., Whitsett, J.A., 1994. Involvement of granulocyte-macrophage colony-stimulating factor in pulmonary homeostasis. Science 264, 713–716.

Dress, R.J., Ginhoux, F., 2021. Monocytes and macrophages in severe COVID-19—friend, foe or both? Immunol. Cell Biol. 99, 561–564.

Eguíluz-Gracia, I., Schultz, H.H.L., Sikkeland, L.I.B., Danilova, E., Holm, A.M., Pronk, C.J.H., Agace, W.W., Iversen, M., Andersen, C., Jahnsen, F.L., Baekkevold, E.S., 2016. Long-term persistence of human donor alveolar macrophages in lung transplant recipients. Thorax 71, 1006–1011.

Evren, E., Ringqvist, E., Tripathi, K.P., Sleiers, N., Rives, I.C., Alisjahbana, A., Gao, Y., Sarhan, D., Halle, T., Sorini, C., Lepzien, R., Marquardt, N., Michaëlsson, J., Smed-Sörensen, A., Botling, J., Karlsson, M.C.I., Villablanca, E.J., Willinger, T., 2021. Distinct developmental pathways from blood monocytes generate human lung macrophage diversity. Immunity 54, 259–275.e7.

Felton, J.M., Lucas, C.D., Dorward, D.A., Duffin, R., Kipari, T., Vermeren, S., Robb, C.T., MacLeod, K.G., Serrels, B., Schwarze, J., Haslett, C., Dransfield, I., Rossi, A.G., 2018. Mer-mediated eosinophil efferocytosis regulates resolution of allergic airway inflammation. J. Allergy Clin. Immunol. 142, 1884–1893.e6.

Fernandez, S., Jose, P., Avdiushko, M.G., Kaplan, A.M., Cohen, D.A., 2004. Inhibition of IL-10 receptor function in alveolar macrophages by Toll-like receptor agonists. J. Immunol. 172, 2613–2620.

Gautier, E.L., Shay, T., Miller, J., Greter, M., Jakubzick, C., Ivanov, S., Helft, J., Chow, A., Elpek, K.G., Gordonov, S., Mazloom, A.R., Ma'ayan, A., Chua, W.-J., Hansen, T.H., Turley, S.J., Merad, M., Randolph, G.J., The Immunological Genome Consortium, 2012a. Gene-expression profiles and transcriptional regulatory pathways that underlie the identity and diversity of mouse tissue macrophages. Nat. Immunol. 13, 1118–1128.

Gautier, E.L., Chow, A., Spanbroek, R., Marcelin, G., Greter, M., Jakubzick, C., Bogunovic, M., Leboeuf, M., van Rooijen, N., Habenicht, A.J., Merad, M., Randolph, G.J., 2012b. Systemic analysis of PPARγ in mouse macrophage populations reveals marked diversity in expression with critical roles in resolution of inflammation and airway immunity. J. Immunol. 189, 2614–2624.

Ghoneim, H.E., Thomas, P.G., McCullers, J.A., 2013. Depletion of alveolar macrophages during influenza infection facilitates bacterial superinfections. J. Immunol. 191, 1250–1259.

Gibbings, S.L., Thomas, S.M., Atif, S.M., McCubbrey, A.L., Desch, A.N., Danhorn, T., Leach, S.M., Bratton, D.L., Henson, P.M., Janssen, W.J., Jakubzick, C.V., 2017. Three unique interstitial macrophages in the murine lung at steady state. Am. J. Respir. Cell Mol. Biol. 57, 66–76.

Gomez Perdiguero, E., Klapproth, K., Schulz, C., Busch, K., Azzoni, E., Crozet, L., Garner, H., Trouillet, C., de Bruijn, M.F., Geissmann, F., Rodewald, H.-R., 2015. Tissue-resident macrophages originate from yolk-sac-derived erythro-myeloid progenitors. Nature 518, 547–551.

Goritzka, M., Makris, S., Kausar, F., Durant, L.R., Pereira, C., Kumagai, Y., Culley, F.J., Mack, M., Akira, S., Johansson, C., 2015. Alveolar macrophage-derived type I interferons orchestrate innate immunity to RSV through recruitment of antiviral monocytes. J. Exp. Med. 212, 699–714.

Grant, R.A., Morales-Nebreda, L., Markov, N.S., Swaminathan, S., Querrey, M., Guzman, E.R., Abbott, D.A., Donnelly, H.K., Donayre, A., Goldberg, I.A., Klug, Z.M., Borkowski, N., Lu, Z., Kihshen, H., Politanska, Y., Sichizya, L., Kang, M., Shilatifard, A., Qi, C., Lomasney, J.W., Argento, A.C., Kruser, J.M., Malsin, E.S., Pickens, C.O., Smith, S.B., Walter, J.M., Pawlowski, A.E., Schneider, D., Nannapaneni, P., Abdala-Valencia, H., Bharat, A., Gottardi, C.J., Budinger, G.R.S., Misharin, A.V., Singer, B.D., Wunderink, R.G., NU SCRIPT Study Investigators, 2021. Circuits between infected macrophages and T cells in SARS-CoV-2 pneumonia. Nature 590, 635–641.

Gschwend, J., Sherman, S.P.M., Ridder, F., Feng, X., Liang, H.-E., Locksley, R.M., Becher, B., Schneider, C., 2021. Alveolar macrophages rely on GM-CSF from alveolar epithelial type 2 cells before and after birth. J. Exp. Med. 218, e20210745.

Guilliams, M., De Kleer, I., Henri, S., Post, S., Vanhoutte, L., De Prijck, S., Deswarte, K., Malissen, B., Hammad, H., Lambrecht, B.N., 2013. Alveolar macrophages develop from fetal monocytes that differentiate into long-lived cells in the first week of life via GM-CSF. J. Exp. Med. 210, 1977–1992.

Han, S.W., Roman, J., 2006. Fibronectin induces cell proliferation and inhibits apoptosis in human bronchial epithelial cells: pro-oncogenic effects mediated by PI3-kinase and NF-kappa B. Oncogene 25, 4341–4349.

Hashimoto, D., Chow, A., Noizat, C., Teo, P., Beasley, M.B., Leboeuf, M., Becker, C.D., See, P., Price, J., Lucas, D., Greter, M., Mortha, A., Boyer, S.W., Forsberg, E.C., Tanaka, M., van Rooijen, N., García-Sastre, A., Stanley, E.R., Ginhoux, F., Frenette, P.S., Merad, M., 2013. Tissue-resident macrophages self-maintain locally throughout adult life with minimal contribution from circulating monocytes. Immunity 38, 792–804.

Hassan, A.O., Case, J.B., Winkler, E.S., Thackray, L.B., Kafai, N.M., Bailey, A.L., McCune, B.T., Fox, J.M., Chen, R.E., Alsoussi, W.B., Turner, J.S., Schmitz, A.J., Lei, T., Shrihari, S., Keeler, S.P., Fremont, D.H., Greco, S., McCray, P.B., Perlman, S., Holtzman, M.J., Ellebedy, A.H., Diamond, M.S., 2020. A SARS-CoV-2 infection model in mice demonstrates protection by neutralizing antibodies. Cell 182, 744–753.e4.

Hawley, C.A., Rojo, R., Raper, A., Sauter, K.A., Lisowski, Z.M., Grabert, K., Bain, C.C., Davis, G.M., Louwe, P.A., Ostrowski, M.C., Hume, D.A., Pridans, C., Jenkins, S.J., 2018. Csf1r-mApple transgene expression and ligand binding in vivo reveal dynamics of CSF1R expression within the mononuclear phagocyte system. J. Immunol. 200, 2209–2223.

Heasman, S.J., Giles, K.M., Rossi, A.G., Allen, J.E., Haslett, C., Dransfield, I., 2004. Interferon gamma suppresses glucocorticoid augmentation of macrophage clearance of apoptotic cells. Eur. J. Immunol. 34, 1752–1761.

Israelow, B., Song, E., Mao, T., Lu, P., Meir, A., Liu, F., Alfajaro, M.M., Wei, J., Dong, H., Homer, R.J., Ring, A., Wilen, C.B., Iwasaki, A., 2020. Mouse model of SARS-CoV-2 reveals inflammatory role of type I interferon signaling. J. Exp. Med. 217, e20201241.

Janssen, W.J., McPhillips, K.A., Dickinson, M.G., Linderman, D.J., Morimoto, K., Xiao, Y.Q., Oldham, K.M., Vandivier, R.W., Henson, P.M., Gardai, S.J., 2008. Surfactant proteins A and D suppress alveolar macrophage phagocytosis via interaction with SIRP alpha. Am. J. Respir. Crit. Care Med. 178, 158–167.

Jiang, R.-D., Liu, M.-Q., Chen, Y., Shan, C., Zhou, Y.-W., Shen, X.-R., Li, Q., Zhang, L., Zhu, Y., Si, H.-R., Wang, Q., Min, J., Wang, X., Zhang, W., Li, B., Zhang, H.-J., Baric, R.S., Zhou, P., Yang, X.-L., Shi, Z.-L., 2020. Pathogenesis of SARS-CoV-2 in transgenic mice expressing human angiotensin-converting enzyme 2. Cell 182, 50–58.e8.

Joshi, N., Watanabe, S., Verma, R., Jablonski, R.P., Chen, C.-I., Cheresh, P., Markov, N.S., Reyfman, P.A., McQuattie-Pimentel, A.C., Sichizya, L., Lu, Z., Piseaux-Aillon, R., Kirchenbuechler, D., Flozak, A.S., Gottardi, C.J., Cuda, C.M., Perlman, H., Jain, M., Kamp, D.W., Budinger, G.R.S., Misharin, A.V., 2020. A spatially restricted fibrotic niche in pulmonary fibrosis is sustained by M-CSF/M-CSFR signalling in monocyte-derived alveolar macrophages. Eur. Respir. J. 55, 1900646.

Kobayashi, Y., Tata, A., Konkimalla, A., Katsura, H., Lee, R.F., Ou, J., Banovich, N.E., Kropski, J.A., Tata, P.R., 2020. Persistence of a regeneration-associated, transitional alveolar epithelial cell state in pulmonary fibrosis. Nat. Cell Biol. 22, 934–946.

Kumagai, Y., Takeuchi, O., Kato, H., Kumar, H., Matsui, K., Morii, E., Aozasa, K., Kawai, T., Akira, S., 2007. Alveolar macrophages are the primary interferon-alpha producer in pulmonary infection with RNA viruses. Immunity 27, 240–252.

Kyriazopoulou, E., Poulakou, G., Milionis, H., Metallidis, S., Adamis, G., Tsiakos, K., Fragkou, A., Rapti, A., Damoulari, C., Fantoni, M., Kalomenidis, I., Chrysos, G., Angheben, A., Kainis, I., Alexiou, Z., Castelli, F., Serino, F.S., Tsilika, M., Bakakos, P., Nicastri, E., Tzavara, V., Kostis, E., Dagna, L., Koufargyris, P., Dimakou, K., Savvanis, S., Tzatzagou, G., Chini, M., Cavalli, G., Bassetti, M., Katrini, K., Kotsis, V., Tsoukalas, G., Selmi, C., Bliziotis, I., Samarkos, M., Doumas, M., Ktena, S., Masgala, A., Papanikolaou, I., Kosmidou, M., Myrodia, D.-M., Argyraki, A., Cardellino, C.S., Koliakou, K., Katsigianni, E.-I., Rapti, V., Giannitsioti, E., Cingolani, A., Micha, S., Akinosoglou, K., Liatsis-Douvitsas, O., Symbardi, S., Gatselis, N., Mouktaroudi, M., Ippolito, G., Florou, E., Kotsaki, A., Netea, M.G., Eugen-Olsen, J., Kyprianou, M., Panagopoulos, P., Dalekos, G.N., Giamarellos-Bourboulis, E.J., 2021. Early treatment of COVID-19 with anakinra guided by soluble urokinase plasminogen receptor plasma levels: a double-blind, randomized controlled phase 3 trial. Nat. Med., 1–9.

Lavin, Y., Winter, D., Blecher-Gonen, R., David, E., Keren-Shaul, H., Merad, M., Jung, S., Amit, I., 2014. Tissue-resident macrophage enhancer landscapes are shaped by the local microenvironment. Cell 159, 1312–1326.

Leach, S.M., Gibbings, S.L., Tewari, A.D., Atif, S.M., Vestal, B., Danhorn, T., Janssen, W.J., Wager, T.D., Jakubzick, C.V., 2020. Human and mouse transcriptome profiling identifies cross-species homology in pulmonary and lymph node mononuclear phagocytes. Cell Rep. 33, 108337.

Lempp, F.A., Soriaga, L., Montiel-Ruiz, M., Benigni, F., Noack, J., Park, Y.-J., Bianchi, S., Walls, A.C., Bowen, J.E., Zhou, J., Kaiser, H., Joshi, A., Agostini, M., Meury, M., Dellota, E., Jaconi, S., Cameroni, E., Martinez-Picado, J., Vergara-Alert, J., Izquierdo-Useros, N., Virgin, H.W., Lanzavecchia, A., Veesler, D., Purcell, L., Telenti, A., Corti, D., 2021. Lectins enhance SARS-CoV-2 infection and influence neutralizing antibodies. Nature. https://doi.org/10.1038/s41586-021-03925-1.

Liao, M., Liu, Y., Yuan, J., Wen, Y., Xu, G., Zhao, J., Cheng, L., Li, J., Wang, X., Wang, F., Liu, L., Amit, I., Zhang, S., Zhang, Z., 2020. Single-cell landscape of bronchoalveolar immune cells in patients with COVID-19. Nat. Med. 26, 842–844.

Liu, Y., Cousin, J.M., Hughes, J., Damme, J.V., Seckl, J.R., Haslett, C., Dransfield, I., Savill, J., Rossi, A.G., 1999. Glucocorticoids promote nonphlogistic phagocytosis of apoptotic leukocytes. J. Immunol. 162, 3639–3646.

Liu, Z., Gu, Y., Chakarov, S., Bleriot, C., Kwok, I., Chen, X., Shin, A., Huang, W., Dress, R.J., Dutertre, C.-A., Schlitzer, A., Chen, J., Ng, L.G., Wang, H., Liu, Z., Su, B., Ginhoux, F., 2019. Fate mapping via Ms4a3-expression history traces monocyte-derived cells. Cell 178, 1509–1525.e19.

Longworth, K.E., 1997. The comparative biology of pulmonary intravascular macrophages. Front. Biosci. 2, d232–d241.

Malur, A., Baker, A.D., McCoy, A.J., Wells, G., Barna, B.P., Kavuru, M.S., Malur, A.G., Thomassen, M.J., 2011. Restoration of PPARγ reverses lipid accumulation in alveolar macrophages of GM-CSF knockout mice. Am. J. Physiol. Lung Cell Mol. Physiol. 300, L73–L80.
Mantovani, A., Sica, A., Sozzani, S., Allavena, P., Vecchi, A., Locati, M., 2004. The chemokine system in diverse forms of macrophage activation and polarization. Trends Immunol. 25, 677–686.
Martinez-Moczygemba, M., Doan, M.L., Elidemir, O., Fan, L.L., Cheung, S.W., Lei, J.T., Moore, J.P., Tavana, G., Lewis, L.R., Zhu, Y., Muzny, D.M., Gibbs, R.A., Huston, D.P., 2008. Pulmonary alveolar proteinosis caused by deletion of the GM-CSFRalpha gene in the X chromosome pseudoautosomal region 1. J. Exp. Med. 205, 2711–2716.
McColl, A., Michlewska, S., Dransfield, I., Rossi, A.G., 2007. Effects of glucocorticoids on apoptosis and clearance of apoptotic cells. ScientificWorldJournal 7, 1165–1181.
McColl, A., Bournazos, S., Franz, S., Perretti, M., Morgan, B.P., Haslett, C., Dransfield, I., 2009. Glucocorticoids induce protein S-dependent phagocytosis of apoptotic neutrophils by human macrophages. J. Immunol. 183, 2167–2175.
McCowan, J., Fercoq, F., Kirkwood, P.M., T'Jonck, W., Hegarty, L.M., Mawer, C.M., Cunningham, R., Mirchandani, A.S., Hoy, A., Humphries, D.C., Jones, G.R., Hansen, C.G., Hirani, N., Jenkins, S.J., Henri, S., Malissen, B., Walmsley, S.R., Dockrell, D.H., Saunders, P.T.K., Carlin, L.M., Bain, C.C., 2021. The transcription factor EGR2 is indispensable for tissue-specific imprinting of alveolar macrophages in health and tissue repair. Sci. Immunol. 6 (65), eabj2132. https://doi.org/10.1126/sciimmunol.abj2132. Epub 2021 Nov 19. PMID: 34797692.
McQuattie-Pimentel, A.C., Ren, Z., Joshi, N., Watanabe, S., Stoeger, T., Chi, M., Lu, Z., Sichizya, L., Aillon, R.P., Chen, C.-I., Soberanes, S., Chen, Z., Reyfman, P.A., Walter, J.M., Anekalla, K.R., Davis, J.M., Helmin, K.A., Runyan, C.E., Abdala-Valencia, H., Nam, K., Meliton, A.Y., Winter, D.R., Morimoto, R.I., Mutlu, G.M., Bharat, A., Perlman, H., Gottardi, C.J., Ridge, K.M., Chandel, N.S., Sznajder, J.I., Balch, W.E., Singer, B.D., Misharin, A.V., Budinger, G.R.S., 2021. The lung microenvironment shapes a dysfunctional response of alveolar macrophages in aging. J. Clin. Invest. 131. https://doi.org/10.1172/JCI140299.
Melms, J.C., Biermann, J., Huang, H., Wang, Y., Nair, A., Tagore, S., Katsyv, I., Rendeiro, A.F., Amin, A.D., Schapiro, D., Frangieh, C.J., Luoma, A.M., Filliol, A., Fang, Y., Ravichandran, H., Clausi, M.G., Alba, G.A., Rogava, M., Chen, S.W., Ho, P., Montoro, D.T., Kornberg, A.E., Han, A.S., Bakhoum, M.F., Anandasabapathy, N., Suárez-Fariñas, M., Bakhoum, S.F., Bram, Y., Borczuk, A., Guo, X.V., Lefkowitch, J.H., Marboe, C., Lagana, S.M., Del Portillo, A., Zorn, E., Markowitz, G.S., Schwabe, R.F., Schwartz, R.E., Elemento, O., Saqi, A., Hibshoosh, H., Que, J., Izar, B., 2021. A molecular single-cell lung atlas of lethal COVID-19. Nature, 1–9.
Merad, M., Martin, J.C., 2020. Pathological inflammation in patients with COVID-19: a key role for monocytes and macrophages. Nat. Rev. Immunol. 20, 355–362.
Mills, C.D., Kincaid, K., Alt, J.M., Heilman, M.J., Hill, A.M., 2000. M-1/M-2 macrophages and the Th1/Th2 paradigm. J. Immunol. 164, 6166–6173.
Misharin, A.V., Morales-Nebreda, L., Reyfman, P.A., Cuda, C.M., Walter, J.M., McQuattie-Pimentel, A.C., Chen, C.-I., Anekalla, K.R., Joshi, N., Williams, K.J.N., Abdala-Valencia, H., Yacoub, T.J., Chi, M., Chiu, S., Gonzalez-Gonzalez, F.J., Gates, K., Lam, A.P., Nicholson, T.T., Homan, P.J., Soberanes, S., Dominguez, S., Morgan, V.K., Saber, R., Shaffer, A., Hinchcliff, M., Marshall, S.A., Bharat, A., Berdnikovs, S., Bhorade, S.M., Bartom, E.T., Morimoto, R.I., Balch, W.E., Sznajder, J.I., Chandel, N.S., Mutlu, G.M., Jain, M., Gottardi, C.J., Singer, B.D.,

Ridge, K.M., Bagheri, N., Shilatifard, A., Budinger, G.R.S., Perlman, H., 2017. Monocyte-derived alveolar macrophages drive lung fibrosis and persist in the lung over the life span. J. Exp. Med. 214, 2387–2404.

Mowat, A.M., Scott, C.L., Bain, C.C., 2017. Barrier-tissue macrophages: functional adaptation to environmental challenges. Nat. Med. 23, 1258–1270.

Nahrendorf, M., Swirski, F.K., 2016. Abandoning M1/M2 for a network model of macrophage function. Circ. Res. 119, 414–417.

Nakamura, A., Ebina-Shibuya, R., Itoh-Nakadai, A., Muto, A., Shima, H., Saigusa, D., Aoki, J., Ebina, M., Nukiwa, T., Igarashi, K., 2013. Transcription repressor Bach2 is required for pulmonary surfactant homeostasis and alveolar macrophage function. J. Exp. Med. 210, 2191–2204.

NEJM, 2020. Pulmonary vascular endothelialitis, thrombosis, and angiogenesis in Covid-19. NEJM. Available at: https://www.nejm.org/doi/full/10.1056/nejmoa2015432.

NEJM, 2021. Dexamethasone in hospitalized patients with Covid-19. NEJM. Available at: https://www.nejm.org/doi/full/10.1056/NEJMoa2021436.

Pairo-Castineira, E., 2020. Genetic mechanisms of critical illness in COVID-19. Nature. Available at: https://www.nature.com/articles/s41586-020-03065-y.

Poon, I.K.H., Lucas, C.D., Rossi, A.G., Ravichandran, K.S., 2014. Apoptotic cell clearance: basic biology and therapeutic potential. Nat. Rev. Immunol. 14, 166–180.

Preston, J.A., Bewley, M.A., Marriott, H.M., McGarry Houghton, A., Mohasin, M., Jubrail, J., Morris, L., Stephenson, Y.L., Cross, S., Greaves, D.R., Craig, R.W., van Rooijen, N., Bingle, C.D., Read, R.C., Mitchell, T.J., Whyte, M.K.B., Shapiro, S.D., Dockrell, D.H., 2019. Alveolar macrophage apoptosis-associated bacterial killing helps prevent murine pneumonia. Am. J. Respir Crit. Care Med. 200 (1), 84–97. https://doi.org/10.1164/rccm.201804-0646OC. PMID: 30649895.

Rauschmeier, R., Gustafsson, C., Reinhardt, A., Gonzalez, N.A., Tortola, L., Cansever, D., Subramanian, S., Taneja, R., Rossner, M.J., Sieweke, M.H., Greter, M., Månsson, R., Busslinger, M., Kreslavsky, T., 2019. Bhlhe40 and Bhlhe41 transcription factors regulate alveolar macrophage self-renewal and identity. EMBO J. 38, e101233.

Rendeiro, A.F., Ravichandran, H., Bram, Y., Chandar, V., Kim, J., Meydan, C., Park, J., Foox, J., Hether, T., Warren, S., Kim, Y., Reeves, J., Salvatore, S., Mason, C.E., Swanson, E.C., Borczuk, A.C., Elemento, O., Schwartz, R.E., 2021. The spatial landscape of lung pathology during COVID-19 progression. Nature 593, 564–569.

Robb, L., Drinkwater, C.C., Metcalf, D., Li, R., Köntgen, F., Nicola, N.A., Begley, C.G., 1995. Hematopoietic and lung abnormalities in mice with a null mutation of the common beta subunit of the receptors for granulocyte-macrophage colony-stimulating factor and interleukins 3 and 5. Proc. Natl. Acad. Sci. U. S. A. 92, 9565–9569.

Rubin, E.J., Longo, D.L., Baden, L.R., 2021. Interleukin-6 receptor inhibition in Covid-19—cooling the inflammatory soup. N. Engl. J. Med. 384, 1564–1565.

Russell, C.D., Valanciute, A., Gachanja, N.N., Stephen, J., Penrice-Randal, R., Armstrong, S.D., Clohisey, S., Wang, B., Al Qsous, W., Wallace, W.A., Oniscu, G.C., Stevens, J., Harrison, D.J., Dhaliwal, K., Hiscox, J.A., Baillie, J.K., Akram, A.R., Dorward, D.A., Lucas, C.D., 2022. Tissue proteomic analysis identifies mechanisms and stages of immunopathology in fatal COVID-19. Am. J. Respir. Cell Mol. Biol. 66 (2), 196–205. https://doi.org/10.1165/rcmb.2021-0358OC. PMID: 34710339.

Sabatel, C., Radermecker, C., Fievez, L., Paulissen, G., Chakarov, S., Fernandes, C., Olivier, S., Toussaint, M., Pirottin, D., Xiao, X., Quatresooz, P., Sirard, J.-C., Cataldo, D., Gillet, L., Bouabe, H., Desmet, C.J., Ginhoux, F., Marichal, T., Bureau, F., 2017. Exposure to bacterial CpG DNA protects from airway allergic inflammation by expanding regulatory lung interstitial macrophages. Immunity 46, 457–473.

Schneider, C., Nobs, S.P., Kurrer, M., Rehrauer, H., Thiele, C., Kopf, M., 2014a. Induction of the nuclear receptor PPAR-γ by the cytokine GM-CSF is critical for the differentiation of fetal monocytes into alveolar macrophages. Nat. Immunol. 15, 1026–1037.

Schneider, C., Nobs, S.P., Heer, A.K., Kurrer, M., Klinke, G., van Rooijen, N., Vogel, J., Kopf, M., 2014b. Alveolar macrophages are essential for protection from respiratory failure and associated morbidity following influenza virus infection. PLoS Pathog. 10, e1004053.

Schyns, J., Bai, Q., Ruscitti, C., Radermecker, C., De Schepper, S., Chakarov, S., Farnir, F., Pirottin, D., Ginhoux, F., Boeckxstaens, G., Bureau, F., Marichal, T., 2019. Non-classical tissue monocytes and two functionally distinct populations of interstitial macrophages populate the mouse lung. Nat. Commun. 10, 3964.

Shibata, Y., Berclaz, P.Y., Chroneos, Z.C., Yoshida, M., Whitsett, J.A., Trapnell, B.C., 2001. GM-CSF regulates alveolar macrophage differentiation and innate immunity in the lung through PU.1. Immunity 15, 557–567.

Snelgrove, R.J., Goulding, J., Didierlaurent, A.M., Lyonga, D., Vekaria, S., Edwards, L., Gwyer, E., Sedgwick, J.D., Barclay, A.N., Hussell, T., 2008. A critical function for CD200 in lung immune homeostasis and the severity of influenza infection. Nat. Immunol. 9, 1074–1083.

Stanley, E., Lieschke, G.J., Grail, D., Metcalf, D., Hodgson, G., Gall, J.A., Maher, D.W., Cebon, J., Sinickas, V., Dunn, A.R., 1994. Granulocyte/macrophage colony-stimulating factor-deficient mice show no major perturbation of hematopoiesis but develop a characteristic pulmonary pathology. Proc. Natl. Acad. Sci. U. S. A. 91, 5592–5596.

Stephenson, E., Reynolds, G., Botting, R.A., Calero-Nieto, F.J., Morgan, M.D., Tuong, Z.K., Bach, K., Sungnak, W., Worlock, K.B., Yoshida, M., Kumasaka, N., Kania, K., Engelbert, J., Olabi, B., Spegarova, J.S., Wilson, N.K., Mende, N., Jardine, L., Gardner, L.C.S., Goh, I., Horsfall, D., McGrath, J., Webb, S., Mather, M.W., Lindeboom, R.G.H., Dann, E., Huang, N., Polanski, K., Prigmore, E., Gothe, F., Scott, J., Payne, R.P., Baker, K.F., Hanrath, A.T., Schim van der Loeff, I.C.D., Barr, A.S., Sanchez-Gonzalez, A., Bergamaschi, L., Mescia, F., Barnes, J.L., Kilich, E., de Wilton, A., Saigal, A., Saleh, A., Janes, S.M., Smith, C.M., Gopee, N., Wilson, C., Coupland, P., Coxhead, J.M., Kiselev, V.Y., van Dongen, S., Bacardit, J., King, H.W., Rostron, A.J., Simpson, A.J., Hambleton, S., Laurenti, E., Lyons, P.A., Meyer, K.B., Nikolić, M.Z., Duncan, C.J.A., Smith, K.G.C., Teichmann, S.A., Clatworthy, M.R., Marioni, J.C., Göttgens, B., Haniffa, M., 2021. Single-cell multi-omics analysis of the immune response in COVID-19. Nat. Med., 1–13.

Sun, S.-H., Chen, Q., Gu, H.-J., Yang, G., Wang, Y.-X., Huang, X.-Y., Liu, S.-S., Zhang, N.-N., Li, X.-F., Xiong, R., Guo, Y., Deng, Y.-Q., Huang, W.-J., Liu, Q., Liu, Q.-M., Shen, Y.-L., Zhou, Y., Yang, X., Zhao, T.-Y., Fan, C.-F., Zhou, Y.-S., Qin, C.-F., Wang, Y.-C., 2020. A mouse model of SARS-CoV-2 infection and pathogenesis. Cell Host Microbe 28, 124–133.e4.

Sungnak, W., Huang, N., Bécavin, C., Berg, M., Queen, R., Litvinukova, M., Talavera-López, C., Maatz, H., Reichart, D., Sampaziotis, F., Worlock, K.B., Yoshida, M., Barnes, J.L., 2020. SARS-CoV-2 entry factors are highly expressed in nasal epithelial cells together with innate immune genes. Nat. Med. 26, 681–687.

Suzuki, T., Sakagami, T., Rubin, B.K., Nogee, L.M., Wood, R.E., Zimmerman, S.L., Smolarek, T., Dishop, M.K., Wert, S.E., Whitsett, J.A., Grabowski, G., Carey, B.C., Stevens, C., van der Loo, J.C.M., Trapnell, B.C., 2008. Familial pulmonary alveolar proteinosis caused by mutations in CSF2RA. J. Exp. Med. 205, 2703–2710.

Suzuki, T., Arumugam, P., Sakagami, T., Lachmann, N., Chalk, C., Sallese, A., Abe, S., Trapnell, C., Carey, B., Moritz, T., Malik, P., Lutzko, C., Wood, R.E., Trapnell, B.C., 2014. Pulmonary macrophage transplantation therapy. Nature 514, 450–454.

Szabo, P.A., Dogra, P., Gray, J.I., Wells, S.B., Connors, T.J., Weisberg, S.P., Krupska, I., Matsumoto, R., Poon, M.M.L., Idzikowski, E., Morris, S.E., Pasin, C., Yates, A.J., Ku, A., Chait, M., Davis-Porada, J., Guo, X.V., Zhou, J., Steinle, M., Mackay, S., Saqi, A., Baldwin, M.R., Sims, P.A., Farber, D.L., 2021. Longitudinal profiling of respiratory and systemic immune responses reveals myeloid cell-driven lung inflammation in severe COVID-19. Immunity 54, 797–814.e6.

Tan, S.Y.S., Krasnow, M.A., 2016. Developmental origin of lung macrophage diversity. Development 143, 1318–1327.

Tarling, J.D., Lin, H.S., Hsu, S., 1987. Self-renewal of pulmonary alveolar macrophages: evidence from radiation chimera studies. J. Leukoc. Biol. 42, 443–446.

Temesgen, Z., Burger, C.D., Baker, J., Polk, C., Libertin, C., Kelley, C., Marconi, V.C., Orenstein, R., Durrant, C., Chappell, D., Ahmed, O., Chappell, G., Badley, A.D., 2021. Lenzilumab efficacy and safety in newly hospitalized Covid-19 subjects: results from the live-air phase 3 randomized double-blind placebo-controlled trial. medRxiv. https://doi.org/10.1101/2021.05.01.21256470.

Theobald, S.J., Simonis, A., Georgomanolis, T., Kreer, C., Zehner, M., Eisfeld, H.S., Albert, M.-C., Chhen, J., Motameny, S., Erger, F., Fischer, J., Malin, J.J., Gräb, J., Winter, S., Pouikli, A., David, F., Böll, B., Koehler, P., Vanshylla, K., Gruell, H., Suárez, I., Hallek, M., Fätkenheuer, G., Jung, N., Cornely, O.A., Lehmann, C., Tessarz, P., Altmüller, J., Nürnberg, P., Kashkar, H., Klein, F., Koch, M., Rybniker, J., 2021. Long-lived macrophage reprogramming drives spike protein-mediated inflammasome activation in COVID-19. EMBO Mol. Med. 13, e14150.

The Economist, 2021. The pandemic's true death toll. The Economist. Available at: https://www.economist.com/graphic-detail/coronavirus-excess-deaths-estimates.

The Lancet, 2021. Tocilizumab in patients admitted to hospital with COVID-19 (RECOVERY): a randomised, controlled, open-label, platform trial. The Lancet 397, 1637–1645. Available at: https://www.thelancet.com/journals/lancet/article/PIIS0140-6736(21)00676-0/fulltext.

Vanderheiden, A., Thomas, J., Soung, A.L., Davis-Gardner, M.E., Floyd, K., Jin, F., Cowan, D.A., Pellegrini, K., Creanga, A., Pegu, A., Derrien-Colemyn, A., Shi, P.-Y., Grakoui, A., Klein, R.S., Bosinger, S.E., Kohlmeier, J.E., Menachery, V.D., Suthar, M.S., 2021. CCR2-dependent monocyte-derived cells restrict SARS-CoV-2 infection. bioRxiv. https://doi.org/10.1101/2021.05.03.442538. in press.

Velu, P.P., Lucas, C.D., Conway Morris, A., 2021. Post-mortem dissection of COVID-19: a pathogenic role for macrophages? Intensive Care Med. https://doi.org/10.1007/s00134-021-06509-9.

Wauters, E., Van Mol, P., Garg, A.D., Jansen, S., Van Herck, Y., Vanderbeke, L., Bassez, A., Boeckx, B., Malengier-Devlies, B., Timmerman, A., Van Brussel, T., Van Buyten, T., Schepers, R., Heylen, E., Dauwe, D., Dooms, C., Gunst, J., Hermans, G., Meersseman, P., Testelmans, D., Yserbyt, J., Tejpar, S., De Wever, W., Matthys, P., Neyts, J., Wauters, J., Qian, J., Lambrechts, D., 2021. Discriminating mild from critical COVID-19 by innate and adaptive immune single-cell profiling of bronchoalveolar lavages. Cell Res. 31, 272–290.

WHO, n.d. Chronic Respiratory Diseases. Available at: https://www.who.int/health-topics/chronic-respiratory-diseases. (Accessed 2022).

Willinger, T., Rongvaux, A., Takizawa, H., Yancopoulos, G.D., Valenzuela, D.M., Murphy, A.J., Auerbach, W., Eynon, E.E., Stevens, S., Manz, M.G., Flavell, R.A., 2011. Human IL-3/GM-CSF knock-in mice support human alveolar macrophage development and human immune responses in the lung. Proc. Natl. Acad. Sci. U. S. A. 108, 2390–2395.

Xu, J., Flaczyk, A., Neal, L.M., Fa, Z., Eastman, A.J., Malachowski, A.N., Cheng, D., Moore, B.B., Curtis, J.L., Osterholzer, J.J., Olszewski, M.A., 2017. Scavenger receptor MARCO orchestrates early defenses and contributes to fungal containment during cryptococcal infection. J. Immunol. 198, 3548–3557.

Yang, D., Chu, H., Hou, Y., Chai, Y., Shuai, H., Lee, A.C.-Y., Zhang, X., Wang, Y., Hu, B., Huang, X., Yuen, T.T.-T., Cai, J.-P., Zhou, J., Yuan, S., Zhang, A.J., Chan, J.F.-W., Yuen, K.-Y., 2020. Attenuated interferon and proinflammatory response in SARS-CoV-2-infected human dendritic cells is associated with viral antagonism of STAT1 phosphorylation. J. Infect. Dis. 222, 734–745.

Yu, X., Buttgereit, A., Lelios, I., Utz, S.G., Cansever, D., Becher, B., Greter, M., 2017. The cytokine TGF-β promotes the development and homeostasis of alveolar macrophages. Immunity 47. https://doi.org/10.1016/j.immuni.2017.10.007.

Zagórska, A., Través, P.G., Lew, E.D., Dransfield, I., Lemke, G., 2014. Diversification of TAM receptor tyrosine kinase function. Nat. Immunol. 15, 920–928.

Zelek, W.M., et al., 2020. Complement inhibition with the C5 blocker LFG316 in severe COVID-19. Am. J. Respir. Crit. Care Med. 202 (9), 1304–1308. PMID: 32897730.

CHAPTER TWO

Macrophage control of Crohn's disease

Siu Ling Tai and Arthur Mortha*
Department of Immunology, University of Toronto, Toronto, ON, Canada
*Corresponding author: e-mail address: arthur.mortha@utoronto.ca

Contents

1. Introduction 29
2. Development of intestinal macrophages 30
3. Localization and heterogeneity of intestinal macrophages 34
4. Functional properties of intestinal macrophages 36
5. Maintenance of homeostatic macrophage function 40
6. Macrophages in Crohn's disease 42
7. Macrophages contribute to in intestinal inflammation in CD 43
8. Macrophage functions and dysfunctions in Crohn's disease 45
9. Targeting macrophages to improve Crohn's disease 50
Acknowledgments 53
References 53

Abstract

The intestinal tract is the body's largest mucosal surface and permanently exposed to microbial and environmental signals. Maintaining a healthy intestine requires the presence of sentinel grounds keeper cells, capable of controlling immunity and tissue homeostasis through specialized functions. Intestinal macrophages are such cells and important players in steady-state functions and during acute and chronic inflammation. Crohn's disease, a chronic inflammatory condition of the intestinal tract is proposed to be the consequence of an altered immune system through microbial and environmental stimulation. This hypothesis suggests an involvement of macrophages in the regulation of this pathology. Within this chapter, we will discuss intestinal macrophage development and highlight data suggesting their implication in chronic intestinal pathologies like Crohn's disease.

1. Introduction

Found in every organ, macrophages are known for their ability to engulf debris and pathogens to sustain tissue and immune homeostasis. With the largest population of macrophages found in the gut, these cells

International Review of Cell and Molecular Biology, Volume 367
ISSN 1937-6448
https://doi.org/10.1016/bs.ircmb.2022.01.003

localize within specific microanatomic niches across the intestinal tract and fulfill numerous functions as gatekeeper and groundskeeper of the organ (Lavin and Merad, 2013). The intestinal tract contains the body's largest mucosal surface, rich in luminal content composed of nutrients, metabolites, and the most abundant and diverse collection of microbes of the body, commonly known as the microbiota. Separated from the lumen by a single layer of epithelial cells along with a barrier of antimicrobial peptides, macrophages sit in a privileged position to respond to local intrusions and cues (Bain and Schridde, 2018). Gut macrophages have been shown to support host-defense, barrier integrity, tissue cell differentiation, oral tolerance, mucosal antibody responses, homeostasis of enteric neurons, and tissue repair (Chiaranunt et al., 2021; Wynn and Vannella, 2016). Their deficiency or functional perturbation is associated with an increased susceptibility to chronic inflammation, infection, and tissue damage in the gut (Han et al., 2021). Inflammatory bowel disease (IBD), a chronic inflammatory condition of the gastrointestinal tract, is broadly separated into two forms: ulcerative colitis (UC) and Crohn's disease (CD). Each manifestation is characterized by distinct anatomic and inflammatory features. The current paradigm suggests that CD pathogenesis is multifactorial. However, the causal factors, with the exception of very-early onset IBD (an acute form of IBD commonly associated and driven by a genetic variation in pediatric patients), remain poorly understood and may vastly differ between patients (Torres et al., 2017). For CD, it is proposed that unique combinations of environmental factors, including lifestyle, geographic location, genetic mutations, immune activation, and the composition of the intestinal microbiota, act in concert to trigger CD pathogenesis (Sartor, 2006). Steady-state macrophage functions and development are regulated by the aforementioned factors, and thus, understanding the regulation of gut macrophages is important to improve and open new avenues for therapeutic modulation of immune and tissue homeostasis to ameliorate IBD (Na et al., 2019; Viola and Boeckxstaens, 2020). Within this chapter, we will summarize the development and ontogeny of intestinal macrophages and discuss their role and function in the context of CD.

2. Development of intestinal macrophages

Our current understanding of intestinal macrophage origin and function has been primarily driven by mouse studies, while the ontogeny of the human counterparts are still largely unknown. In mice, macrophages arise from myeloid precursor cells during three distinct waves of hematopoiesis. Erythromyeloid yolk sac precursors initiate the first wave of macrophage development, primarily filling niches in the developing central nervous system. As hematopoiesis shifts from the yolk sac to the fetal liver, erythromyeloid precursors and fetal liver monocytes release the second wave of progenitors, which seed all organs including the intestinal tract with macrophages. Once hematopoietic stem cells settle in the bone marrow, classical myelopoiesis provides the third wave of macrophages, arising from circulating monocytes (Hoeffel and Ginhoux, 2018). In barrier organs like the skin or lung, fetal liver-derived macrophages self-renew without major contribution from adult bone marrow-derived monocytes, at least in the steady state. In contrast, the pool of fetal liver-derived intestinal macrophages experiences steady and continuous replenishment by monocytes, leading to an almost complete replacement by adult precursor cells (Bain et al., 2014). Monocyte differentiation into macrophages is gradual and follows a "waterfall"-like transition from $CD11b^{+}$ $CCR2^{+}$ $CX3CR1^{lo}$ $Ly6C^{hi}$ $MHCII^{-}$ monocyte phenotype to a $CD11b^{+}$ $CCR2^{-}$ $CX3CR1^{hi}$ $Ly6C^{-}$ $MHCII^{+}$ macrophage phenotype (Bain et al., 2014; Honda et al., 2020; Schridde et al., 2017; Tamoutounour et al., 2012) (Fig. 1). The egress of $Ly6C^{hi}$ monocytes from the bone marrow has been demonstrated to be critically dependent on CCR2 (Bernardo et al., 2018; Serbina and Pamer, 2006; Zigmond et al., 2012). A similar "monocyte waterfall" has been observed in humans, indicating a transitional development of $CD14^{hi}$ $CCR2^{+}CD11c^{hi}$ monocytes to $CD14^{lo}CCR2^{-}CD11c^{lo}$ macrophages (Bain et al., 2013; Bernardo et al., 2018; Bujko et al., 2018) (Fig. 1). However, unlike in mouse, human gut macrophages are characterized by loss of CX3CR1 expression (Chiaranunt et al., 2021). Studies in human duodenal transplantation have revealed a variation in replenishment of macrophage subsets in the gut, suggesting heterogeneity within macrophage niches or macrophage origin. Seminal work by Bujko et al. described four distinct human MP populations (Mf1–4) in the small intestine (Bujko et al., 2018) (Table 1a). Comparable to experiments in mice, human resident macrophage subsets expressing high levels of CD14 and CCR2 (Mf1 and Mf2) within the transplant were rapidly replaced by host-derived infiltrating monocytes, while long-lived macrophages in the small intestine (Mf3 and Mf4) displayed a slower turnover with full replacement within 1 year (Bujko et al., 2018; De Schepper et al., 2019; Liu et al., 2019) (Table 1a). Collectively, these findings suggest that monocytes are precursors of MPs in both the mouse and human small and large intestine during steady state

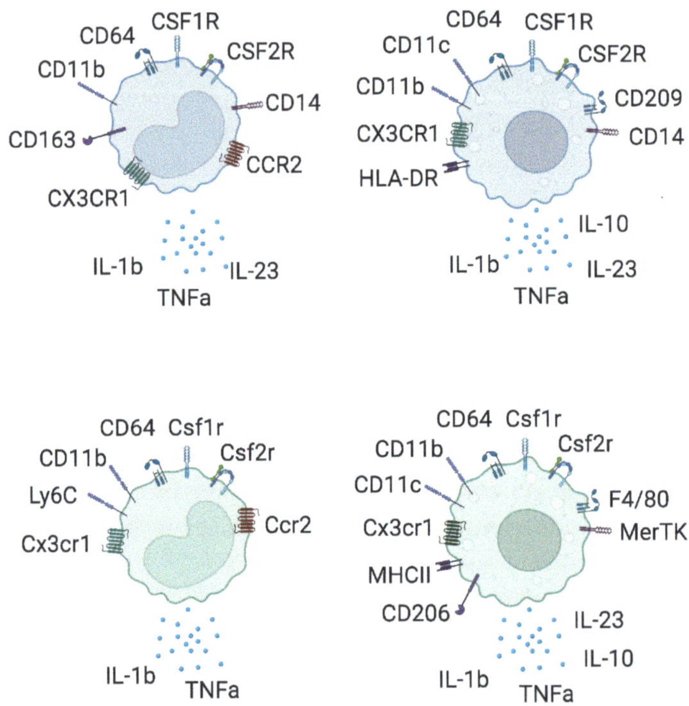

Fig. 1 Core markers for the identification of mouse and human gut monocytes and macrophages. A comparison of commonly used surface marker and released cytokines helpful to identify human (light blue) and mouse (light green) gut monocytes (left) and macrophages (right).

and display similar diversity and maintenance of this immune population across species. These recent advances are supplemented by findings in mice that reveal further heterogeneity by describing two distinct co-existing subsets: the long-lived, self-renewing $CD4^+$ $Tim4^+$ macrophages and the $Tim4^-$ macrophages, with the latter constantly replaced by infiltrating monocytes (De Schepper et al., 2019; Shaw et al., 2018) (Table 1b). This replacement of macrophages is dependent on the presence of the microbiota, implicating a role for the local environment in driving this process (Bain et al., 2014; Honda et al., 2020; Kang et al., 2020). The combination of elegant fate-mapping studies, parabiosis, and bone marrow transplant experiments demonstrated that $Tim4^-$ macrophages require CCR2-dependent monocyte infiltration for their maintenance, while $Tim4^+$ macrophages self-maintain within tissues for a prolonged period of time (Bain et al., 2014; De Schepper et al., 2019; Yona et al., 2013). While these reports

Table 1 Selection of human and murine macrophage subset markers.

(a)

	$CD14^+$ monocytes	Mf1	Mf2	MF3	Mf4
CCR2	++	+	+	–	–
CD14	+	++	+	+	++
CD163	++	–	–	+	+
CD11B	+	+	+	+	+
CSF2R	+	+	+	+	+
CD209	–	+	++	++	++
CD64	++	++	++	+	+
CSF1R	++	+	+	+	+
CX3CR1	++	++	++	+	+
HLA-DR	–	–/+	++	++	++
CD11C	–	+	+	+	+
IL-23	++	++	+	+	–
IL10	–	–	–	+	+
IL1B	+	++	+	+	+
TNFA	+	+	+	–	–

(b)

	$Ly6C^{hi}$ mono	DN MP	SP MP	DP MP
Ccr2	++	+	–	–
Cd163	–	–/+	+	–/+
Csf1r	+	++	++	++
Csf2rb2	+	+	++	++
Cx3cr1	+	++	++	++
Fcgr1	–	+	+	+
H2-Ab1	–	++	+	+
Il10rb	–	+	+	+
Itgam	++	++	++	++
Itgax	–	–/+	–/+	–/+

Continued

Table 1 Selection of human and murine macrophage subset markers.—cont'd
(b)

	Ly6C^{hi} mono	DN MP	SP MP	DP MP
Lyve1	–	–	ND	–/+
Lyz1	++	+	–	–
Mertk	–	+	+	+
Tgfbr1	–	+	+	++
Vcam1	–	+	+	++
Il10	–	+	+	++
Il1b	–	++	+	+
Tnf	–	++	+	+

Tables show a selection of markers that allow the classification of macrophages into subsets in (a) human and (b) mice. Markers listed in the table are based on data published in Bujko et al. and Shaw et al. (Bujko et al., 2018; Shaw et al., 2018). Abbreviations of cells in the table are based on nomenclature used in Bujko et al. and Shaw et al. For (a) Mf = macrophages. For (b) MP = macrophages. DN, SP, DP refers to double negative, single positive and double positive, respectively. Shaw et al. referred to the expression of CD4 and Tim-4 on macrophages in their study.

are an excellent starting point to understand macrophage heterogeneity, further spatial and single cell transcriptome analysis paired with a thorough description of human macrophage ontogeny and functional adaptation during inflammation is necessary to bridge the gap in our understanding of macrophage heterogeneity and development in the healthy and diseased human intestine.

3. Localization and heterogeneity of intestinal macrophages

There is broad heterogeneity of macrophages between organs and even within tissue, adding another layer of complexity to the ecosystem. Within the bowel, macrophages are localized in different microanatomic niches. A large population of tissue resident macrophages is found within the mucosa of the villi where they possibly interact with the epithelium, lacteal, blood vasculature, or neuronal protrusions (Chiaranunt et al., 2021). Macrophages also populate gut-associated lymphoid structures, such as mesenteric lymph nodes, Peyer's Patches, isolated lymphoid follicles, and cryptopatches, while both the submucosa and muscularis contain macrophages in close proximity to neurons and vasculature (Koscso et al., 2020; Muller et al., 2014; Shou et al., 2021). Within the muscularis compartments, macrophages are distributed in every layer, displaying unique morphological characteristics compared to macrophages in the other intestinal compartments (Muller et al., 2014). Seminal studies by DeSchepper et al. have begun to unfold the heterogeneity of macrophages across these microanatomic locations to uncover possible unique functional contributions to tissue cell homeostasis (De Schepper et al., 2019). The association of intestinal pathologies with distinct locations across the intestine warrant a careful consideration of anatomic and even microanatomic localization of macrophages at steady state and during disease. This is supported by reports indicating that relative abundance of macrophages increases toward the descending part of the mouse gut (Denning et al., 2011). Despite sharing important functions across these regions, macrophages within specific segments of the intestine show functional specialization (Denning et al., 2011). For example, ileal macrophages extend protrusions across the intestinal epithelium to sample luminal microbes, while rectal macrophages absorb microbial factors through balloon-like protrusions across the epithelium (Chikina et al., 2020; Niess et al., 2005). Such interactions have not yet been made in other parts of the intestine (Persson et al., 2013). Furthermore, macrophage localization requires chemoattraction and is suggested to be critical for host-defense and immune tolerance (Hadis et al., 2011; Medina-Contreras et al., 2011). A landmark study investigating the adaptation of newly generated, monocyte-derived gut macrophages to the ileal and colonic lamina propria at steady state indicated significant transcriptional differences between newly generated macrophages from each location in the mouse gut (Gross-Vered et al., 2020). These findings demonstrate that the local microenvironment of macrophages substantially impacts their phenotype and function and should be taken into consideration for future studies. These suggestions should also be applied in human studies, as altered distribution of macrophages even within the ileal lamina propria have been documented in CD patients. Enrichment of lamina propria macrophages toward the distal regions of the ileum indicates a change in the distribution of macrophages in ileal CD (Mahida et al., 1989). As the ileum is often affected by severe inflammation during this disease, a closer examination of macrophages localization within this segment could delineate the functional and developmental differences compared to macrophages in the non-inflamed mucosa.

4. Functional properties of intestinal macrophages

Phagocytosis, a hallmark of macrophage function, supports tissue homeostasis and host defense by clearance of pathogens, elimination of cellular debris, and tissue repair (Cummings et al., 2016; Muller et al., 2012; Nagashima et al., 1996). Gut macrophages are avidly phagocytic and support the continuous renewal of the epithelial barrier through efferocytosis, the clearance of dead cells (A-Gonzalez et al., 2017; Bain et al., 2013; Bujko et al., 2018; Cummings et al., 2016; Pull et al., 2005; Smythies et al., 2005). This function also imprints distinct transcriptional profiles in macrophages across tissues (A-Gonzalez et al., 2017; Casanova-Acebes et al., 2018). Engulfment or sampling of apoptotic intestinal epithelial cells and neutrophils has been demonstrated to alter gut macrophage gene expression programming, stimulating upregulation of genes associated with phagocytosis, lipid processing, and amino acid catabolism, and downregulation of proinflammatory genes and transforming growth factor β (TGFβ) secretion (Casanova-Acebes et al., 2018; Cummings et al., 2016; Fadok et al., 1998). In the context of inflammation, this imprinting supports immune tolerance, resolution of inflammation, tissue repair, and return to homeostasis. Alterations in these processes may thus promote chronic inflammation of the gut.

The gene expression profile of gut macrophages at homeostasis in both mice and humans suggest hyporesponsiveness as default state (Bujko et al., 2018; Rugtveit et al., 1997; Smith et al., 2001; Smythies et al., 2005). This is indicated, for example, by reduced expression of the costimulatory molecules CD80 and CD86 (Rogler et al., 1998; Smythies et al., 2005). In mice, colonic macrophages are refractory to stimulation with Toll-like receptor (TLR) ligands (Bain et al., 2013; Hirotani et al., 2005). Disrupted interleukin (IL)-10 production in mice results in increased response to TLR stimulation and CD80/86 expression, suggesting a role for IL-10 signaling in programming this phenotype (Hirotani et al., 2005; Kayama et al., 2012). Human gut macrophages also downregulate expression of *Myeloid differentiation primary response 88* (MyD88) and *TIR domain containing adapter-inducing IFN B adapter proteins* (TRIF), molecules that drive nuclear factor kappa-light-chain-enhancer of activated B cells (NF-κB)-mediated signaling, which may contribute to their unique phenotype (Smythies et al., 2010). Although the molecular mechanisms that drive this signature profile are yet to be fully elucidated, this highlights a potential functional adaptation of macrophages to the antigen-rich gastrointestinal environment to prevent unnecessary inflammation and maintain tissue homeostasis.

Despite their hyporesponsive nature to microbial encroachment, gut macrophages are sentinels, endowed with receptors that enable them to detect microbes, including TLRs, Nod-like receptors (NLRs), and C-type lectin receptors (CLRs). A vast network of lamina propria macrophages resides beneath the single layer of intestinal epithelium, acting as groundskeepers to uphold barrier homeostasis and defend against microbial invasion. In mice, CX3CR1$^+$ lamina propria macrophages have been shown to sample luminal antigens, such as microbes, through extending transepithelial dendrites into the lumen (Chieppa et al., 2006; Medina-Contreras et al., 2011; Niess et al., 2005; Rescigno et al., 2001). Of note, this process is only seen in the ileum of specific mouse strains (Persson et al., 2013). Their capacity to recognize fungal elements further enables immune surveillance and control of the murine gut fungal community (Leonardi et al., 2018). Upon encountering microbial molecular patterns, gut macrophages secrete soluble mediators to support intestinal homeostasis in both human and mice (Grainger et al., 2013; Longman et al., 2014). These can range from anti-inflammatory IL-10, pro-resolving Prostaglandin E_2 (PGE2), to pro-inflammatory molecules such as IL-1β, tumor necrosis factor-like cytokine 1A (TL1A), tumor necrosis factor (TNF)-α or IL-23 (Zigmond and Jung, 2013). CX3CR1$^+$ macrophages are potent producers of IL-1β and IL-23, drivers of gut-resident group 3 innate lymphocyte (ILC3)-derived IL-22, which has been shown to be integral in intestinal barrier integrity and host protection through induction of anti-microbial peptides and chemokines (Manta et al., 2013; Mortha et al., 2014; Sanos et al., 2009). IL-1β and PGE2 additionally promote IL-22 production from murine ILC3s, and thus may be another avenue by which macrophages contribute to gut health (Duffin et al., 2016; Mortha et al., 2014). These factors, while often described as acting in trans, may influence the functions of macrophages in an autocrine manner. An interesting example is PGE2, which has been shown to support macrophage anti-inflammatory responses independent of IL-10 (Na et al., 2021).

Macrophages express high levels of MHCII, enabling them to act as antigen-presenting cells and potentially influence T cell polarization. While migration of these cells to gut-draining lymph nodes remains controversial, their ability to uptake microbes and present the processed antigen is apparent (Niess et al., 2005; Schulz et al., 2009). For example, using *Salmonella* infections, mouse CX3CR1^{hi} phagocytes (including subsets of dendritic cells) were shown to constitutively sample antigens in the lamina propria (Muller et al., 2012). However, it remains to be determined whether intestinal macrophages are a prominent subset in uptake and presentation of antigen, which could also involve antigen-transfer from macrophages or epithelial cells directly to migratory dendritic cells (Chieppa et al., 2006; McDole et al., 2012; Niess et al., 2005). In the murine intestinal mucosa, macrophages have been implicated in promoting T helper type 1 (Th1) and Th17 cell functions *via* production of IL-12 and IL-1β, respectively (Panea et al., 2015; Schreiber et al., 2013; Shaw et al., 2012). While a Th17 response drives host protection against extracellular pathogens, this response is also implicated in mediating chronic inflammation (Aujla et al., 2007; Weaver et al., 2013). Although macrophages can support Th17 function, some evidence suggests that IL-10 derived from macrophages is necessary to control myeloid cell recruitment, regulate Th17 response, and restrain excessive inflammation in mouse models of colitis (Krause et al., 2015). Th17-dependent secretory IgA production in the gut is yet another aspect of host defense that has been demonstrated to require macrophage-derived factors through promoting B cell class-switch recombination in a T cell-dependent fashion (Hirota et al., 2013; Kim et al., 2018; Koscso et al., 2020).

Due to the constant exposure to dietary antigens, commensals and other metabolites, establishing immune tolerance in the gut is crucial to maintain homeostasis. Regulatory T cells (T_{reg}) are major players in tolerance and suppression of inflammatory responses (Hadis et al., 2011; Murai et al., 2009). In mice, macrophages support the expansion of T_{reg} through IL-10, TGF-β, and retinoic acid (RA) production, implicating gut macrophages in preservation of immune tolerance (Denning et al., 2007; Mortha et al., 2014). Upon maturation, mouse colonic macrophages upregulate IL-10 expression (Schridde et al., 2017). Macrophage-derived IL-10 has been shown to be required for wound healing following colitis and is produced during the acute phase of injury (Morhardt et al., 2019; Quiros et al., 2017). Constitutive secretion of IL-10 by gut macrophages in healthy tissues is microbiota-dependent and has been shown to facilitate epithelial integrity and the induction of T_{reg} (Bain et al., 2013; Hadis et al., 2011; Hirota et al., 2013; Kim et al., 2018; Morhardt et al., 2019). However, genetic deletion of IL-10 in murine macrophages alone does not ablate T_{reg} frequencies, but alters macrophage function, suggesting other sources of IL-10 at play (Kim et al., 2018; Zigmond and Jung, 2013). Microbiota-mediated IL-1β production by macrophages also stimulates

ILC3 secretion of colony stimulating factor 2 (Csf2) in mice (Longman et al., 2014; Mortha et al., 2014). Driving a local circuit of crosstalk, Csf2, in turn, potentiates macrophage functions and induces the release of RA and IL-10, both supporting T_{reg} homeostasis (Mortha et al., 2014). RA also upregulates IL-10 production, phagocytosis, and bactericidal activity in macrophages, promoting homeostatic functions in the steady state (Wang et al., 2007). IL-10 and macrophage-intrinsic WNT signaling do not only regulate homeostasis but are vital to promote wound healing following dextran sodium sulfate (DSS) or biopsy-induced injury in mice (Cosin-Roger et al., 2016; Pull et al., 2005; Seno et al., 2009). Steady-state Csf2-release promotes homeostasis and repair, but was conversely reported to counterbalance excessive wound healing by suppressing tissue repair in mouse macrophages (Castro-Dopico et al., 2020). These findings may explain why Csf2-deficient mice have increased susceptibility to enteric infections and intestinal inflammation (Hirata et al., 2010; Sainathan et al., 2008; Song et al., 2015).

Although our current understanding of gut macrophage functions beyond the lamina propria remains incomplete, studies in mice have begun to unveil the underappreciated functions of gut macrophages. For example, muscularis macrophages located in close proximity to the enteric nervous system have been shown to adapt their functional program to support immunity through immune cell and neuronal-derived stimulation. This crosstalk supports intestinal motility and secretion of stem cell niche factors (De Schepper et al., 2019; Gabanyi et al., 2016; Muller et al., 2014). In the context of infection, muscularis macrophages suppress neuronal inflammasome activation and limited cell death, mediating protection of host cells (Ahrends et al., 2021; Matheis et al., 2020). This process has recently been shown to require reprogramming of muscularis macrophages in an IL-4- and IL-13-dependent manner (Ahrends et al., 2021). Intestinal macrophages form a network around the enteric vasculature, supporting homeostasis of endothelial cells and preventing bacterial translocation into the blood stream in a microbiota-dependent fashion (Honda et al., 2020; Nagashima et al., 1996). While the majority of research has been in mice due to scarcity of human samples, analogous macrophage heterogeneity between mucosal and muscularis subsets has been found in human tissues (Bujko et al., 2018).

Collectively, these reports suggest that gut macrophages are key players in orchestrating intestinal health. They integrate local microbial and environmental signals to exert host protection as well as tissue homeostasis.

Continued efforts to understand the intricacies of gut macrophage development, heterogeneity, and function in context of steady-state and inflammation could thus reveal new therapeutic targets for the treatment of chronic gastrointestinal disorders.

5. Maintenance of homeostatic macrophage function

The unique features of gut macrophages, including their replacement and functional phenotypes, distinguish them from this cell type in other organs. Classical Ly6C^{hi} monocytes seeds different organs yet display distinctive profiles upon maturation across murine tissues (Bain et al., 2014; Schridde et al., 2017; Tamoutounour et al., 2012). Even within the lamina propria, macrophages vary in frequency along the different sections of the gastrointestinal tract (Denning et al., 2011; Nagashima et al., 1996). These differences are suspected to be driven by the surrounding environment, suggesting a role for tissue-specific factors in imprinting macrophage development or accumulation.

Development and maintenance of intestinal macrophages relies on soluble growth factors such as colony stimulating factor 1 (CSF1). Mice deficient in *Csf1* or its cognate receptor develop significantly lower macrophage frequencies across all tissues except the brain, epidermis, and peripheral nerves (Dai et al., 2002). Treatment with blocking anti-CSF1R antibodies strongly reduces gut macrophage numbers (MacDonald et al., 2010; Mortha et al., 2014). However, this depletion is not complete, suggesting alternative growth factors for their local maintenance (Mikkelsen and Thuneberg, 1999). Although only partially required for macrophage development, *Csf2*-deficient mice display lower gut macrophage numbers, implying a role for this cytokine in the maintenance of at least a fraction of the intestinal macrophage pool (Mortha et al., 2014). TGF-β plays a vital role in the terminal differentiation of macrophages. This growth factor drives CX3CR1 expression and regulates macrophage turnover in the murine gut (Schridde et al., 2017). Loss of function of TGF-β receptor (TGFβR) on macrophages renders mice more susceptible to DSS-colitis with delayed epithelial regeneration and reduced IL-10 secretion (Papadakis et al., 2015; Rani et al., 2011). *In vitro* work supports the suppressive action of TGF-β on proinflammatory cytokine secretion in LPS-stimulated human gut macrophages (Maheshwari et al., 2011; Smythies et al., 2010). However, this finding does not extend to mice (Schridde et al., 2017). In contrast, IL-10-mediated conditioning of gut macrophages in both species potentiates the hyporesponsive and anti-inflammatory properties of gut macrophages (Kayama et al., 2012; Shouval et al., 2014; Zigmond and Jung, 2013). Conclusively, IL-10 deficiency in mice results in spontaneous colitis, involving the MyD88 pathway in macrophages (Hoshi et al., 2012; Kuhn et al., 1993). Systemic and macrophage-specific ablation of STAT3, a molecule downstream in the IL-10R signaling pathway, recapitulates these findings, implicating STAT3 as a downstream transcriptional regulator of intestinal tolerance (Takeda et al., 1999). Furthermore, colonic macrophages from mice deficient in IL-10 or STAT3 display increased expression of inflammatory mediators, such as IL-6 and TNF-α, in response to LPS stimulation (Ueda et al., 2010). Human macrophages lacking IL-10R derived *in vitro* from blood monocytes display a similar phenotype and single nucleotide polymorphisms in the human *Il10R* gene are associated with early onset IBD (Kotlarz et al., 2012; Shouval et al., 2014). Collectively, these studies highlight the role of TGF-β and IL-10 in regulating inflammatory activity of gut macrophages and emphasize the central role of these cells in sustaining gut homeostasis. Interestingly, IL-10 impacts the production of PGE2, which is produced by gut macrophages. The corresponding prostaglandin E2 receptor (PTGER4) is expressed on a subset of gut macrophages and has been shown to be indispensable for tissue repair in mice, suggesting a role for PGE2 in modulating macrophage function in an autocrine fashion (Miyoshi et al., 2017; Na et al., 2021). In line with this, PGE2 signaling also induces anti-inflammatory functions in mouse colonic macrophages (Chinen et al., 2011). In contrast, PGE2 has been shown to be able to promote type I interferon signaling in macrophages, perturbing T_{reg} function, and increasing susceptibility to intestinal inflammation (Chinen et al., 2011). The distinct effects of PGE2 on macrophages may depend on context, reflecting the pleiotropic features of this mediator.

The vicinity of macrophages to the lumen, their capacity for antigen sampling, and their interactions with the microbiota or their metabolites influences the local macrophage landscape. Germ-free mice display reduced frequencies of mature macrophages compared to conventional SPF-control mice. In addition, diminishing the microbial load in the gut through continuous antibiotics treatment perturbs the replenishment of the local colonic macrophages by infiltrating monocytes (Bain et al., 2014). The colonization of the intestinal tract by commensal bacteria promotes hyporesponsiveness and IL-10 production in colonic macrophages, either through bacterial metabolites or direct interactions of bacteria with the gut epithelium (Danne et al., 2017; Ueda et al., 2010). Several lines of evidence suggest that short-chain fatty acids produced by bacteria modulate macrophage function. Butyrate, in particular, has been shown to induce activation of STAT6 signaling and downregulation of proinflammatory proteins like IL-6, IL-12, and nitric oxide synthase (Chang et al., 2014). In a recent study by Jain and colleagues, macrophages uptake of the yeast *Debaryomyces hansenii*, commonly found in dairy products, resulted in impaired epithelial regeneration associated with the upregulation of CCL5. Intriguingly, *D. hansenii* is enriched in inflamed tissues of CD patients, implicating several potential targets for therapeutic intervention and a possible implication for non-bacterial microbes in promoting CD (Jain et al., 2021). These are in line with the serological analysis of IBD sera, demonstrating an enrichment of antibodies to fungal antigens (Main et al., 1988). Similar to this study, an intriguing report demonstrated the association of the fungal species *Malassezia restricta* with CD, involving a CARD9-dependent pathway to promote intestinal inflammation (Limon et al., 2019). Colonic macrophages in close proximity to the intestinal epithelium maintain epithelial integrity by lowering the absorption of fungal products through fungal recognition by balloon-like protrusions (Chikina et al., 2020). Taken together, these data suggest that the local milieu and localization contributes to shaping and maintaining gut macrophage activation. The availability of survival factors, growth factors, and local interactions within their microanatomic niches may be critical in driving the diversity in macrophage phenotypes to support tissue homeostasis and host defense (Chiaranunt et al., 2021).

6. Macrophages in Crohn's disease

IBD is a complex, devastating condition characterized by chronic inflammation along the gastrointestinal tract that is accompanied by severe symptoms. Without an available cure, the current reports estimate that IBD affects 0.75% of the world's population. IBD manifests in two major forms: UC and CD (Kaplan and Windsor, 2021). Although the exact etiology of both diseases is still unknown, the current consensus is that IBD susceptibility is multifactorial and results from intersecting genetic, environmental, and lifestyle conditions (Franke et al., 2010a; Plevy et al., 2013; Silverberg et al., 2005). Here, we will discuss evidence and functional phenotypes in the literature that support a role of macrophages in CD.

While a multitude of advances suggest the role of genetic, environment, diets, lifestyle, and the microbiome as risk factors for CD, the initial trigger of the disease remains elusive. Theories of CD etiology propose a multi-hit model in which infectious causes, loss of barrier integrity, or loss of intestinal immune tolerance cooperate to promote the development of CD. A common theme in this model is the initiation of an aberrant immune response, leading towards chronic inflammation, mucosal damage, or impaired healing. It remains a possibility that each of the above responses individually promotes CD or results from interactions between genes, diet, or the microbiome (Bain et al., 2014; Kang et al., 2020; Zigmond et al., 2012). Genome-wide association studies (GWAS) of CD patients have identified multiple risk loci with aberrant promoter activity. Several of these loci encode or regulate genes associated with innate immune homeostasis and macrophage function (Baillie et al., 2018; Franke et al., 2010b). Comparing the gene expression of monocytes and monocyte-derived macrophages to microbial stimulation revealed >200 differentially regulated genes with direct or possible implication in CD (Baillie et al., 2017). Macrophages are central sensory entities of the intestinal mucosa; they patrol the tissue and integrate the results of genetic defects, dietary alterations or microbial perturbations into the architecture of the local immune landscape. Therefore, a better understanding of macrophage biology is critical to uncovering the many unknowns in the development of CD.

7. Macrophages contribute to in intestinal inflammation in CD

Experimental evidence from murine studies suggest that as classical monocytes enter the intestinal mucosa, they functionally adapt to the local environment and differentiate into tissue resident macrophages. However, this progressive development is perturbed in the inflamed gut. DSS-induced intestinal injury of the mouse intestine disrupts monocyte differentiation, altering the local macrophage composition and increases the proportion of TNF-α producers (Bain et al., 2013). Following injury, macrophages further contribute to mucosal inflammation by amplifying the recruitment of inflammatory monocytes and neutrophils, exacerbating the inflammatory response (Asano et al., 2015). Consistent with this, CD and UC patients express increased amounts of chemokines that contribute to perpetuating inflammatory responses through the recruitment of monocytes and neutrophils (Banks et al., 2003). In a surgical mouse model of small intestinal inflammation, monocyte differentiation favors accumulation of Ly6C^{+} CX3CR1^{int} cells (Desalegn and Pabst, 2019). In mice, CX3CR1^{int}

macrophages display a proinflammatory profile, secreting higher levels of inflammatory cytokines such as IL-1β, IL-6, IL-23, and inducible nitric oxide synthase (iNOS) compared to CX3CR1^{hi} macrophages (Bain et al., 2013; Zigmond et al., 2012). Similar findings have been observed in patients and are supported by reports that demonstrate an accumulation and aggregation of macrophages in the inflamed CD mucosa, especially in subepithelial regions (Rugtveit et al., 1994; Yao et al., 1996; Yao et al., 2000). Intriguingly, these clusters of macrophages appear to distinguish CD from UC. Macrophages in the inflamed mucosa expressed high levels of CD163, CD68 and Calprotectin, the latter an important fecal biomarker used to assess CD-associated inflammation, implicating phenotypical adaptation or altered ontogeny of resident macrophages during CD (Franze et al., 2013). Furthermore, elevated numbers of CD163^{hi} CD68^{+} CD14^{+} macrophages have been observed in the inflamed CD mucosa which released high levels of IL-23 when exposed to gut commensal microbes *in vitro*, suggesting that macrophage recognition of the intestinal microbiota may play a role in CD (Kamada et al., 2008). These macrophages also promoted the release of IL-17 and TNFα when cocultured with naïve CD4^{+} T cells, implicating crosstalk with the adaptive immune system in promoting chronic inflammation (Ogino et al., 2013). Lastly, the expression of triggering receptor expressed on myeloid cells 1 (TREM-1) on macrophages was shown to identify a population that contributes to intestinal inflammation through the release of inflammatory mediators (e.g., IL-1β or TNFα) (Verstockt et al., 2019). These results suggest that infiltrating monocytes and accumulating inflammatory macrophages exacerbate intestinal inflammation in human and mice. However, the absence of macrophages has also been observed to increase susceptibility to colitis in experimental murine models. Conditional deletion of macrophages and dendritic cells in DSS-treated mice results in increased histopathology, weight loss, rectal bleeding, and shortened colon length, while mice lacking circulating monocytes display a defective resolution of inflammation (Farro et al., 2017; Qualls et al., 2006). In line with this, research proposes that CD (or a form thereof) may be the result of a primary macrophage immunodeficiency (Casanova and Abel, 2009). In contrast, depletion of macrophages in IL-10 knock out mice ameliorates colitis symptoms, suggesting a pathogenic role for macrophages in other forms of CD (Watanabe et al., 2003). These findings suggest that the development of CD in some patients may involve elevated frequencies of activated macrophages, while CD in other patients may involve reduced macrophage numbers or functions. These findings underscore the importance of understanding the phenotypical and functional heterogeneity of macrophages and the multifactorial etiology of CD.

8. Macrophage functions and dysfunctions in Crohn's disease

Multiple limitations hinder the ability to effectively study macrophage heterogeneity in the human gut. These obstacles include the size of the organ, limited tissue accessibility, and confounding factors that impact the immune system, such as demographics, lifestyles, medications, or the microbiota. Sampling biopsies during the monitoring of the intestinal tract and the often disease-associated requirement for tissue resections in IBD patients offer an opportunity for scientists to overcome this hurdle. In addition, large scale GWAS and high-dimensional single cell analyses have substantially enhanced our understanding of human gut macrophage heterogeneity. Advances in single cell transcriptional profiling have allowed mapping of the macrophage landscape using limited samples at high resolution, revealing differential gene signatures in human health and disease (Mulder et al., 2021). Integration of these data with high dimensional proteomics and metabolomics characterization may lead to definitive identification of the pathogenic subsets and their ontogeny. While this work is valuable in advancing our knowledge of human macrophages, it is important to consider potential biases in patient heterogeneity and the statistical power of the data.

Consistent with mouse models of colitis, multiple studies identified accumulation of a pro-inflammatory $CCR2^{+}CX3CR1^{+}$ intestinal macrophages subset in the inflamed colonic mucosa of CD patients (Bernardo et al., 2018; Kamada et al., 2008; Rivollier et al., 2012; Zigmond et al., 2012). High-dimensional single-cell analyses in patients revealed striking similarities, reporting an increase in $CD14^{+}CD64^{+}CD163^{-}$ macrophages in inflamed lesions from CD patients (Martin et al., 2019). This increase is accompanied by an altered profile of plasma cells, inflammatory macrophages and dendritic cells, activated T cells, and stromal cells, alluding to the complex immune interactions in gastrointestinal diseases (Martin et al., 2019). This detailed analysis has just begun to uncover the role of previously unappreciated interactions of inflammatory cytokines and non-hematopoietic intestinal tissue cells in the context of a genetic risk factors, environment, and microbiota (Nayar et al., 2021).

Functionally, macrophages present with a broad differential transcriptional profile and functional response to pathogens in CD patients

(Autschbach et al., 2002; Smith et al., 2009). Characterization of the CD intestinal mucosa identified an enrichment in $CD14^{+}$ $CD163^{-}$ macrophages that secrete elevated levels of interferon (IFN)-γ, IL-1β, TNF-α, IL-23, and reactive oxygen species when compared to healthy or UC patients at steady state (Autschbach et al., 2002; Chapuy et al., 2019; Kamada et al., 2008; McAlindon et al., 1999; Rugtveit et al., 1997). Co-cultures of $CD163^{-}$ macrophages from CD patients with autologous $CD4^{+}$ T cells promoted Th17 polarization in an IL-1β dependent fashion, further reflecting the pro-inflammatory phenotype of these macrophages in CD (Chang et al., 2014; Ogino et al., 2013). Microbial stimulation of macrophages from CD patients with the commensal bacteria, *Escherichia coli* and *Enterococcus faecalis*, led to an increase in the release of pro-inflammatory cytokines, highlighting a distinct phenotype that differs from hyporesponsive macrophages that are found in healthy intestinal tissues (Kamada et al., 2008). The colonization of germ-free mice with the fecal microbiome of CD patients recapitulated an elevation in Th17 polarization and indicates the microbiota as a driving force of the altered macrophage profile (Britton et al., 2019). Serological analysis of active CD patients identified 13 elevated analytes involved in macrophage migration and colorectal disease location, suggesting that intestinal markers of inflammation involving macrophages may be traceable in the blood (Boucher et al., 2021). It remains to be shown if these markers can be used to predict disease severity and whether they are detectable prior to the manifestation of CD. Alternatively, such serology may allow monitoring of ongoing intestinal inflammation during disease progression.

In addition to the shift in composition and effector functions, mucosal macrophages from IBD patients display an upregulation of the costimulatory molecules, CD80 and CD86 compared to non-inflamed and tissues from healthy subjects, indicative of their elevated activation and ability to interact with T cells (Rugtveit et al., 1997). While TLR and NLR activation of innate cells is critical for microbial recognition and host defense, aberrant activation of these pathways results in exaggerated response and triggering of chronic mucosal inflammation. *NOD2/CARD15* was the first gene identified through GWAS with a strong association to CD (Jostins et al., 2012; Liu et al., 2015; Ogura et al., 2001). Mutations in NOD2, a cytosolic sensor required for recognition of muramyl dipeptide derived from the bacterial cell wall, has been implicated in intestinal dysfunction in both mouse and human (Biswas et al., 2010; Hugot et al., 2001; Ogura et al., 2001). In mice, *Nod2*-deficiency results in defective anti-microbial function in Paneth cells and increased susceptibility to opportunistic and pathogenic intestinal infections, reflecting its function in antimicrobial host defense (Biswas et al., 2010; Kobayashi et al., 2005). Similarly, *in vitro* stimulation of primary human cells and cell lines carrying *NOD2* mutations display defective NOD2-mediated signaling and NF-κB activation (Bonen et al., 2003; Girardin et al., 2003; Inohara et al., 2003; Li et al., 2004; Mukherjee et al., 2019). Recent work from Nayar and colleagues demonstrated that deficiency in *NOD2* in CD patients leads to dysregulated homeostasis and a pathogenic signature in macrophages, identifying a functional link between NOD2 polymorphisms and macrophage dysfunction in CD (Nayar et al., 2021). The proposed mechanism of pathogenesis is centered around interactions between inflammatory macrophages and intestinal stromal cells *via* a gp130-dependent signaling pathway, resulting in exacerbated disease.

As discussed above, pathogen clearance and dead cell removal are critical functions of macrophages. Autophagy, an intracellular phagocytic process, plays a role in the homeostasis of mitochondria, cytosolic pathogen clearance, secretion of anti-microbial peptides, and the inflammatory response (Germic et al., 2019; Maculins et al., 2021). A missense variant in *ATG16L1*, a gene involved in the formation of a functional autophagosome, is associated with an increased risk of CD (Hampe et al., 2007). Mice expressing the *ATG16L1* risk allele (T316A) display elevated proinflammatory cytokine secretion and defective clearance of the enteric pathogen *Yersinia enterocolitica* (Murthy et al., 2014). In line with this report, mice lacking *Atg16l1* in myeloid cells (including macrophages) have increased disease severity during DSS-induced colitis (Zhang et al., 2017). Functional analyses of both murine and human macrophages revealed that ATG16L1-deficiency results in defective mitophagy, lysosomal trafficking, and anti-microbial activity accompanied by an enhanced oxidative stress response (Maculins et al., 2021; Zhang et al., 2017). Together, these data suggest that loss of ATG16L1 induces perturbation in cellular homeostasis and antimicrobial activity, and promotes a shift in the activation of intestinal macrophages that may potentiate CD pathogenesis.

Several GWAS identified single nucleotide polymorphisms (SNPs) in genes encoding for IL12/IL-23R and downstream signaling molecules, such as the kinase JAK2 and the transcription factors, STAT3 and STAT4, suggesting a role for these signaling pathways in CD (Jostins et al., 2012; Liu et al., 2015). Selective depletion of *Stat3* in macrophages, for example, has been found to promote spontaneous gut inflammation in mice (Takeda et al., 1999). IL-12 and IL-23 are heterodimeric proteins composed of either IL-12p35,/IL-12p40 or IL-23p19/IL12p35, and are key effectors in driving the differentiation and function of T cells and ILCs (Tait Wojno et al., 2019). These cytokines have been shown to control the plasticity of ILC3 and the stability of Th17 cells in the intestine of patients (Bernink et al., 2013, 2015; Wilson et al., 2007). Aberrant expression of these cytokines by macrophages may thus drive the proinflammatory immune landscape observed in CD patients (Bernink et al., 2013, 2015; Mazzurana et al., 2021). While myeloid cytokines, including IL12 and IL-23, associate with CD, metabolites released by macrophages may conversely contribute to the balance and homeostatic immune signature. RA, a vitamin A-derived compound, has been demonstrated to be produced by gut macrophages and promotes the generation of mouse and human T_{reg} that suppress Th17 activation (Hoechst et al., 2011; Sanders et al., 2014). Interestingly, RA plays an additional role in sustaining human and murine ILC3 stability as well as polarizing macrophages toward an anti-inflammatory state in the gut, collectively implicating a role for RA in preventing intestinal inflammation (Erkelens et al., 2020; Goverse et al., 2016). Of note, CD patients display an upregulation of RA production in macrophages in a CSF2-dependent manner, suggesting an increased demand for pathogen recognition and host defense (Sanders et al., 2014). In turn, RA upregulates Dectin-1, a pattern recognition receptor expressed on macrophages that plays an important role in the recognition and antimicrobial host defense against pathogens (Erkelens et al., 2020). Although the mechanisms of regulation between these pathways remain unknown, this data highlights the importance of macrophages in maintaining intestinal immune homeostasis.

The historic interest in CSF2 as a modulator of intestinal myeloid cell function experienced a revival, which inspired the exploration into genetic evidence for this molecule and its signaling pathways in regulating CD (Korzenik et al., 2005). Mutations in *CSF2RB*, encoding the common β-chain of the receptor (the signal transducing adaptor mediating CSF2 signaling), confers risk for severe CD (Chuang et al., 2016; Levine et al., 2016). In line with this, whole-exome sequencing of a large cohort of IBD patients identified 18 additional missense polymorphisms in the CSF2 signaling pathway (Denson et al., 2019). Loss of CSF2-driven signaling, regardless of the mutation, reduced antimicrobial and homeostatic monocyte and neutrophil functions (Chuang et al., 2016; Denson et al., 2019; Levine et al., 2016). Similarly, genetic ablation of CSF2 in mice is associated with decreased protection against *Citrobacter rodentium*, non-steroidal anti-inflammatory drug (NSAID)-induced ileitis, and bacterial translocation across the intestinal epithelium (Han et al., 2010; Hirata et al., 2010). As discussed above, CSF2 promotes anti-inflammatory properties and host defense in monocytes, which confers protection in experimental mouse models of colitis (Dabritz et al., 2015). CSF2 also regulates wound healing and tissue repair to prevent excessive fibrosis and tissue scaring in mice (Bernasconi et al., 2014; Castro-Dopico et al., 2020). Given that CSF2 is an important growth, differentiation, and survival factor for macrophages, this data supports macrophages as sentinels in intestinal health. Aside from genetic evidence for CSF2-driven signaling in CD, experimental evidence accumulates that perturbation of CSF2 activity through neutralizing anti-CSF2 autoantibodies correlates with CD-associated complications (Bonneau et al., 2015). Recent investigations analyzing serological biomarkers suggest that anti-CSF2 autoantibodies precede the onset of CD years before diagnosis, suggesting an imbalanced intestinal immune homeostasis prior to disease onset through alterations of macrophage functions (Arthur Mortha et al., 2021).

Macrophages from *Csf2*-deficient mice were reported to produce less RA and IL-10 altering the generation of T_{reg} regulated by a microbiota-ILC3 axis that is critical for the induction of oral tolerance and tissue homeostasis (Mortha et al., 2014). CSF2 acts on IL-10 production in mouse macrophages and the associated impairment in intestinal homeostasis are in line with findings of mutations in IL-10 and IL10R in humans as strong risk factor for CD and early-onset IBD (Begue et al., 2011; Franke et al., 2010a; Glocker et al., 2009). In mouse models, IL-10 and IL-10R-deficient macrophages display a pro-inflammatory profile, producing elevated levels of IL-12 and IL-23 upon stimulation (Kamada et al., 2008; Zigmond et al., 2014). Even at steady state, IL-10-deficient macrophages secrete more IL-6 and TNF-α in comparison to their hyporesponsive wild-type controls (Hirotani et al., 2005). Although macrophages are a source of IL-10, specific deletion of this mediator in mouse macrophages does not result in spontaneous colitis. Rather, macrophages unable to sense IL-10 renders mice susceptible to severe intestinal inflammation (Shouval et al., 2014; Zigmond et al., 2014). Interestingly, absence of IL-10 produced by T_{reg} promoted spontaneous colitis in mice (Rubtsov et al., 2008). IL-10R deficiency in macrophages also resulted in an impaired production of PGE2, the ligand for PTGER4 in the colon (Zigmond et al., 2014). Genetic variants associated with changes in PTGER4 expression were identified as susceptibility factors in CD and studies using macrophage-specific *Ptger4*-deficient mice demonstrated a protective role against DSS-induced colitis due to defective regeneration of the intestinal epithelium (Kabashima et al., 2002). Consistent with these data, IL-10R-deficient patients display increased TNF-α, IL-1, and IL-6 levels, altered differentiation, and functional profiles in their macrophages (Glocker et al., 2009; Shouval et al., 2014). Taken together, this suggests that macrophage-derived IL-10 is dispensable for prevention of CD, but may impact other sources of IL-10 (including T_{reg}) that are critical in eliciting macrophage homeostatic functions to maintain intestinal health.

The elevated expression of TNF-α is a prominent hallmark of IBD and many other autoinflammatory conditions, inspiring the therapeutic targeting of this highly inflammatory cytokine. TNF-α blockade using monoclonal antibodies revolutionized the management of CD as evident by its years of successful application in the clinic (Colombel et al., 2010; Lichtenstein et al., 2018). The rationale for blocking TNF-α was supported by the experimental overexpression of TNF-α in mice ($TNF^{\Delta ARE/+}$), which induced severe and spontaneous ileitis (Bamias et al., 2013). However, studies in mouse models of colitis have also implicated a protective role for TNF-α in the context of gut inflammation. Genetic deletion of *Tnfa* or treatment with antagonistic anti-TNF-α antibodies exacerbate DSS-induced colitis in mice (Kojouharoff et al., 1997; Naito et al., 2003). Intriguingly, the ameliorative effect of anti-TNF-α treatment in IBD patients and mouse models has been shown to require the production of IL-10 in macrophages (Koelink et al., 2020). Consistent with this, mice deficient in both TNF-α and IL-10 develop spontaneous colitis, reduced barrier integrity, and increased secretion of IL-6, IL-17, IL-2 and IL-12/IL-23p40 (Koelink et al., 2020). Given the potency of TNF-α blockade, these findings highlight the importance of identifying anti-TNF-α responders vs non-responders in moving toward a better understanding of patient heterogeneity to advance personalized approaches to manage CD. Collectively, the dynamic regulation of macrophage function as a prominent source of TNF-α and IL-10 is central to intestinal health. Recent advances are illuminating the complex interplay between gut macrophages and inflammation, allowing us to pinpoint the basis of CD pathogenesis and offer prospects in targeted therapies.

9. Targeting macrophages to improve Crohn's disease

IBD poses a significant strain on our healthcare system with the burden compounded by chronicity, incurability, and increasing incidence (Kaplan and Windsor, 2021). Moreover, the currently available regimens are immunosuppressive. Although the available immunosuppressive strategies have proven to be effective in a larger group of patients, these therapies may predispose these patients to increased susceptibility to opportunistic pathogens and diseases that require separate therapeutic regimens (Lichtenstein et al., 2018). Furthermore, a proportion of patients fail to respond to these treatments and lack of therapeutic options pose a drastic challenge for the management of their disease (Hazlewood et al., 2015). Identifying predictive biomarkers and a robust intervention for IBD could not only prevent the escalation of prevalence, but also improve quality of life and revolutionize the method of care for patients. Given the potential role of macrophages in intestinal health, targeting and promoting their function may serve as a valuable therapeutic avenue for IBD management and intervention.

Beyond novel therapeutics, elucidating the role of macrophages in disease and host-environment interactions will shed light on the predictive markers for response to currently available therapeutics. For example, anti-TNF-α is the most commonly used biologic in IBD. Despite robust responses, there is a subset of patients that do not benefit from this therapy (Colombel et al., 2010; Papamichael et al., 2019). Given the powerful immunomodulatory nature of biologics, risk of adverse events, and extremely high costs, it is pertinent to discriminate between responders and non-responders to selectively provide this type of medication to patients (Lichtenstein et al., 2018). Single cell analyses of human CD lesions identified a unique cellular signature of inflammatory monocytes and macrophages associated with resistance to anti-TNF-α therapy, emphasizing the need to understand macrophage dynamics in health and disease (Martin et al., 2019). Responders to anti-TNF-α therapy can also be identified by low TREM-1 expression in their blood and mucosal tissue and decreased expression of oncostatin M (OSM) and TNF receptor in mucosal biopsies in these patients (Verstockt et al., 2019). Characterization of monocytes either expressing high or low levels of TREM-1 revealed a decrease in autophagy and FcγR signaling in TREM^{high} monocytes, indicative of an anti-inflammatory polarization. Conversely, TREM-1^{high} monocytes diminished their inflammatory profile when exposed to a chemical TREM-1 inhibitor, implicating that TREM-1 inhibition may be possible therapeutic option for anti-TNF non-responders (Prins et al., 2021). While these findings are exciting, a more detailed characterization of monocytes and gut macrophages is required before assessing clinical response to TREM-1 inhibition in CD patients (Prins et al., 2021; Verstockt et al., 2019). Another recent report provides promising experimental data that suggests selective targeting of gp130, the transducer of IL-6, IL-11, OSM and LIF signaling across a macrophage-stroma cell axis, can potentially ameliorate CD (Nayar et al., 2021). With data implicating the gut microbiome in governing the release of cytokines, further investigations into a better understanding of host-microbiota interactions in CD will be valuable. Clinical and experimental data support an implication of the gut microbiota in IBD (Clooney et al., 2021). However, it remains to be shown whether dysbiosis is a cause or consequence of IBD development. Mouse experiments demonstrate that the microbiota of CD patients, driving an inflammatory response, can be reverted through transplantation of defined fecal microbial communities (Britton et al., 2019, 2020). These findings partly recapitulate findings in smaller randomized control trials, but will require larger studies for confirmation (Sokol et al., 2020). An investigation into the adaptations of macrophages following fecal microbial transplantation would further our understanding into the microbial regulation of macrophages during CD. These studies may require an analysis of commensal microbes beyond their phylogenetic classification and would benefit from an analysis of their functional specialization on the transcriptomic level. The presence of anti-microbial antibodies in the serum of CD patients precedes the onset of disease and implicates immune-microbiota crosstalk in disease development (Choung et al., 2016). Despite the fact that no single causative pathogen has been definitively linked to CD, increased presence of *E. coli* has been detected in CD-associated granulomas (Cartun et al., 1993; Liu et al., 1995; Ryan et al., 2004). Intestinal granulomas are a hallmark of CD and were reported to contain *E. coli* antigens and DNA (Liu et al., 1995; Ryan et al., 2004). Interestingly, adherent-invasive *E. coli* (AIEC) have been reported to induce granuloma formation *in vitro* and may be responsible for granuloma formation in CD (Meconi et al., 2007). While the interaction of microbes and host may not be causal of disease, possible interactions of commensals and macrophages within CD granulomas could influence the severity of disease, with a microbial-derived molecule serving as an inflammation-driving agent (Buisson et al., 2016).

As is likely true for many other autoinflammatory diseases, the diagnosis of CD may be preceded by a phase of subclinical inflammation that remains unnoticed by patient and physicians. Similar to the occurrence of anti-CSF2 autoantibodies and anti-microbial antibodies years before the diagnosis of disease, other alterations affecting macrophage activation and function (e.g., microbial dysbiosis) may affect tissue and immune homeostasis in patients prior to the development of CD. A detailed understanding of the underlying mechanisms that lead to a perturbed intestinal macrophage pool may inspire innovative strategies to prevent or delay the onset of disease. The multitude of mechanisms regulating macrophage function in CD implies a need for subclassification of CD patients based on disease mechanism to understand the underlying cellular and molecular factors that are dysregulated. Such patient profiles should be of high priority to CD-focused research and aim at providing an optimal, tailored therapy to each individual suffering from CD.

Acknowledgments

S.L.T. is a recipient of the Dr. Edward Ketchum Graduate Student Scholarship and the Canada Graduate Scholarships—Master's (CGS M) award. A.M. is supported by a CIHR-Project Grant (388337) and a NSERC-Discovery Grant (RGPIN-2019-04521). A.M. is the Tier 2 Canadian Research Chair in Mucosal Immunology and supported by the Tier 2 CRC-CIHR program. We would like to thank all members of #theonlylabever for their feedback and discussions on this manuscript.

References

A-Gonzalez, N., Quintana, J.A., Garcia-Silva, S., Mazariegos, M., Gonzalez de la Aleja, A., Nicolas-Avila, J.A., et al., 2017. Phagocytosis imprints heterogeneity in tissue-resident macrophages. J. Exp. Med. 214 (5), 1281–1296.

Ahrends, T., Aydin, B., Matheis, F., Classon, C.H., Marchildon, F., Furtado, G.C., et al., 2021. Enteric pathogens induce tissue tolerance and prevent neuronal loss from subsequent infections. Cell 184 (23), 5715–5727.e12.

Arthur Mortha, R.R., Del Valle, D.M., Chuang, L.-S., Chai, Z., Alves, I., Azevedo, C., Gaifem, J., Martin, J., Tuballes, K., Barcessat, V., Tai, S.L., Huang, H.-H., Laface, I., Jerez, Y.A., Boschetti, G., Villaverde, N., Wang, M.D., Korie, U.M., Murray, J., Choung, R.-S., Sato, T., Laird, R.M., Plevy, S., Rahman, A., Torres, J., Porter, C., Riddle, M.S., Kenigsberg, E., Pinho, S.S., Cho, J.H., Merad, M., Colombel, J.-F., Gnjatic, S., 2021. Anti–GM-CSF autoantibodies promote a "pre-diseased" state in Crohn's disease. MedRvix.

Asano, K., Takahashi, N., Ushiki, M., Monya, M., Aihara, F., Kuboki, E., et al., 2015. Intestinal CD169(+) macrophages initiate mucosal inflammation by secreting CCL8 that recruits inflammatory monocytes. Nat. Commun. 6, 7802.

Aujla, S.J., Dubin, P.J., Kolls, J.K., 2007. Th17 cells and mucosal host defense. Semin. Immunol. 19 (6), 377–382.

Autschbach, F., Giese, T., Gassler, N., Sido, B., Heuschen, G., Heuschen, U., et al., 2002. Cytokine/chemokine messenger-RNA expression profiles in ulcerative colitis and Crohn's disease. Virchows Arch. 441 (5), 500–513.

Baillie, J.K., Arner, E., Daub, C., De Hoon, M., Itoh, M., Kawaji, H., et al., 2017. Analysis of the human monocyte-derived macrophage transcriptome and response to lipopolysaccharide provides new insights into genetic aetiology of inflammatory bowel disease. PLoS Genet. 13 (3), e1006641.

Baillie, J.K., Bretherick, A., Haley, C.S., Clohisey, S., Gray, A., Neyton, L.P.A., et al., 2018. Shared activity patterns arising at genetic susceptibility loci reveal underlying genomic and cellular architecture of human disease. PLoS Comput. Biol. 14 (3), e1005934.

Bain, C.C., Schridde, A., 2018. Origin, differentiation, and function of intestinal macrophages. Front. Immunol. 9, 2733.
Bain, C.C., Scott, C.L., Uronen-Hansson, H., Gudjonsson, S., Jansson, O., Grip, O., et al., 2013. Resident and pro-inflammatory macrophages in the colon represent alternative context-dependent fates of the same Ly6Chi monocyte precursors. Mucosal Immunol. 6 (3), 498–510.
Bain, C.C., Bravo-Blas, A., Scott, C.L., Perdiguero, E.G., Geissmann, F., Henri, S., et al., 2014. Constant replenishment from circulating monocytes maintains the macrophage pool in the intestine of adult mice. Nat. Immunol. 15 (10), 929–937.
Bamias, G., Dahman, M.I., Arseneau, K.O., Guanzon, M., Gruska, D., Pizarro, T.T., et al., 2013. Intestinal-specific TNFalpha overexpression induces Crohn's-like ileitis in mice. PLoS One 8 (8), e72594.
Banks, C., Bateman, A., Payne, R., Johnson, P., Sheron, N., 2003. Chemokine expression in IBD. Mucosal chemokine expression is unselectively increased in both ulcerative colitis and Crohn's disease. J. Pathol. 199 (1), 28–35.
Begue, B., Verdier, J., Rieux-Laucat, F., Goulet, O., Morali, A., Canioni, D., et al., 2011. Defective IL10 signaling defining a subgroup of patients with inflammatory bowel disease. Am. J. Gastroenterol. 106 (8), 1544–1555.
Bernardo, D., Marin, A.C., Fernandez-Tome, S., Montalban-Arques, A., Carrasco, A., Tristan, E., et al., 2018. Human intestinal pro-inflammatory CD11c(high)CCR2 (+)CX3CR1(+) macrophages, but not their tolerogenic CD11c(−)CCR2(−)CX3CR1 (−) counterparts, are expanded in inflammatory bowel disease. Mucosal Immunol. 11 (4), 1114–1126.
Bernasconi, E., D'Angelo, F., Michetti, P., Velin, D., 2014. Critical role of the GM-CSF signaling pathway in macrophage pro-repair activities. Pathobiology 81 (4), 183–189.
Bernink, J.H., Peters, C.P., Munneke, M., te Velde, A.A., Meijer, S.L., Weijer, K., et al., 2013. Human type 1 innate lymphoid cells accumulate in inflamed mucosal tissues. Nat. Immunol. 14 (3), 221–229.
Bernink, J.H., Krabbendam, L., Germar, K., de Jong, E., Gronke, K., Kofoed-Nielsen, M., et al., 2015. Interleukin-12 and -23 control plasticity of CD127(+) group 1 and group 3 innate lymphoid cells in the intestinal lamina propria. Immunity 43 (1), 146–160.
Biswas, A., Liu, Y.J., Hao, L., Mizoguchi, A., Salzman, N.H., Bevins, C.L., et al., 2010. Induction and rescue of Nod2-dependent Th1-driven granulomatous inflammation of the ileum. Proc. Natl. Acad. Sci. U. S. A. 107 (33), 14739–14744.
Bonen, D.K., Ogura, Y., Nicolae, D.L., Inohara, N., Saab, L., Tanabe, T., et al., 2003. Crohn's disease-associated NOD2 variants share a signaling defect in response to lipopolysaccharide and peptidoglycan. Gastroenterology 124 (1), 140–146.
Bonneau, J., Dumestre-Perard, C., Rinaudo-Gaujous, M., Genin, C., Sparrow, M., Roblin, X., et al., 2015. Systematic review: new serological markers (anti-glycan, anti-GP2, anti-GM-CSF ab) in the prediction of IBD patient outcomes. Autoimmun. Rev. 14 (3), 231–245.
Boucher, G., Paradis, A., Chabot-Roy, G., Coderre, L., Hillhouse, E.E., Bitton, A., et al., 2021. Serum analyte profiles associated with Crohn's disease and disease location. Inflamm. Bowel Dis.
Britton, G.J., Contijoch, E.J., Mogno, I., Vennaro, O.H., Llewellyn, S.R., Ng, R., et al., 2019. Microbiotas from humans with inflammatory bowel disease Alter the balance of gut Th17 and RORgammat(+) regulatory T cells and exacerbate colitis in mice. Immunity 50 (1), 212–224 (e4).
Britton, G.J., Contijoch, E.J., Spindler, M.P., Aggarwala, V., Dogan, B., Bongers, G., et al., 2020. Defined microbiota transplant restores Th17/RORgammat(+) regulatory T cell balance in mice colonized with inflammatory bowel disease microbiotas. Proc. Natl. Acad. Sci. U. S. A. 117 (35), 21536–21545.

Buisson, A., Bringer, M.A., Barnich, N., Vazeille, E., 2016. Macrophages versus Escherichia coli: a decisive fight in Crohn's disease. Inflamm. Bowel Dis. 22 (12), 2943–2955.

Bujko, A., Atlasy, N., Landsverk, O.J.B., Richter, L., Yaqub, S., Horneland, R., et al., 2018. Transcriptional and functional profiling defines human small intestinal macrophage subsets. J. Exp. Med. 215 (2), 441–458.

Cartun, R.W., Van Kruiningen, H.J., Pedersen, C.A., Berman, M.M., 1993. An immunocytochemical search for infectious agents in Crohn's disease. Mod. Pathol. 6 (2), 212–219.

Casanova, J.L., Abel, L., 2009. Revisiting Crohn's disease as a primary immunodeficiency of macrophages. J. Exp. Med. 206 (9), 1839–1843.

Casanova-Acebes, M., Nicolas-Avila, J.A., Li, J.L., Garcia-Silva, S., Balachander, A., Rubio-Ponce, A., et al., 2018. Neutrophils instruct homeostatic and pathological states in naive tissues. J. Exp. Med. 215 (11), 2778–2795.

Castro-Dopico, T., Fleming, A., Dennison, T.W., Ferdinand, J.R., Harcourt, K., Stewart, B.J., et al., 2020. GM-CSF calibrates macrophage defense and wound healing programs during intestinal infection and inflammation. Cell Rep. 32 (1), 107857.

Chang, P.V., Hao, L., Offermanns, S., Medzhitov, R., 2014. The microbial metabolite butyrate regulates intestinal macrophage function via histone deacetylase inhibition. Proc. Natl. Acad. Sci. U. S. A. 111 (6), 2247–2252.

Chapuy, L., Bsat, M., Sarkizova, S., Rubio, M., Therrien, A., Wassef, E., et al., 2019. Two distinct colonic CD14(+) subsets characterized by single-cell RNA profiling in Crohn's disease. Mucosal Immunol. 12 (3), 703–719.

Chiaranunt, P., Tai, S.L., Ngai, L., Mortha, A., 2021. Beyond immunity: underappreciated functions of intestinal macrophages. Front. Immunol. 12, 749708.

Chieppa, M., Rescigno, M., Huang, A.Y., Germain, R.N., 2006. Dynamic imaging of dendritic cell extension into the small bowel lumen in response to epithelial cell TLR engagement. J. Exp. Med. 203 (13), 2841–2852.

Chikina, A.S., Nadalin, F., Maurin, M., San-Roman, M., Thomas-Bonafos, T., Li, X.V., et al., 2020. Macrophages maintain epithelium integrity by limiting fungal product absorption. Cell 183 (2), 411–428 (e16).

Chinen, T., Komai, K., Muto, G., Morita, R., Inoue, N., Yoshida, H., et al., 2011. Prostaglandin E2 and SOCS1 have a role in intestinal immune tolerance. Nat. Commun. 2, 190.

Choung, R.S., Princen, F., Stockfisch, T.P., Torres, J., Maue, A.C., Porter, C.K., et al., 2016. Serologic microbial associated markers can predict Crohn's disease behaviour years before disease diagnosis. Aliment. Pharmacol. Ther. 43 (12), 1300–1310.

Chuang, L.S., Villaverde, N., Hui, K.Y., Mortha, A., Rahman, A., Levine, A.P., et al., 2016. A frameshift in CSF2RB predominant among Ashkenazi jews increases risk for Crohn's disease and reduces monocyte signaling via GM-CSF. Gastroenterology 151 (4), 710–723 (e2).

Clooney, A.G., Eckenberger, J., Laserna-Mendieta, E., Sexton, K.A., Bernstein, M.T., Vagianos, K., et al., 2021. Ranking microbiome variance in inflammatory bowel disease: a large longitudinal intercontinental study. Gut 70 (3), 499–510.

Colombel, J.F., Sandborn, W.J., Reinisch, W., Mantzaris, G.J., Kornbluth, A., Rachmilewitz, D., et al., 2010. Infliximab, azathioprine, or combination therapy for Crohn's disease. N. Engl. J. Med. 362 (15), 1383–1395.

Cosin-Roger, J., Ortiz-Masia, D., Calatayud, S., Hernandez, C., Esplugues, J.V., Barrachina, M.D., 2016. The activation of Wnt signaling by a STAT6-dependent macrophage phenotype promotes mucosal repair in murine IBD. Mucosal Immunol. 9 (4), 986–998.

Cummings, R.J., Barbet, G., Bongers, G., Hartmann, B.M., Gettler, K., Muniz, L., et al., 2016. Different tissue phagocytes sample apoptotic cells to direct distinct homeostasis programs. Nature 539 (7630), 565–569.

Dabritz, J., Weinhage, T., Varga, G., Wirth, T., Walscheid, K., Brockhausen, A., et al., 2015. Reprogramming of monocytes by GM-CSF contributes to regulatory immune functions during intestinal inflammation. J. Immunol. 194 (5), 2424–2438.
Dai, X.M., Ryan, G.R., Hapel, A.J., Dominguez, M.G., Russell, R.G., Kapp, S., et al., 2002. Targeted disruption of the mouse colony-stimulating factor 1 receptor gene results in osteopetrosis, mononuclear phagocyte deficiency, increased primitive progenitor cell frequencies, and reproductive defects. Blood 99 (1), 111–120.
Danne, C., Ryzhakov, G., Martinez-Lopez, M., Ilott, N.E., Franchini, F., Cuskin, F., et al., 2017. A large polysaccharide produced by helicobacter hepaticus induces an anti-inflammatory gene signature in macrophages. Cell Host Microbe 22 (6), 733–745 (e5).
De Schepper, S., Verheijden, S., Aguilera-Lizarraga, J., Viola, M.F., Boesmans, W., Stakenborg, N., et al., 2019. Self-maintaining gut macrophages are essential for intestinal homeostasis. Cell 176 (3), 676.
Denning, T.L., Wang, Y.C., Patel, S.R., Williams, I.R., Pulendran, B., 2007. Lamina propria macrophages and dendritic cells differentially induce regulatory and interleukin 17-producing T cell responses. Nat. Immunol. 8 (10), 1086–1094.
Denning, T.L., Norris, B.A., Medina-Contreras, O., Manicassamy, S., Geem, D., Madan, R., et al., 2011. Functional specializations of intestinal dendritic cell and macrophage subsets that control Th17 and regulatory T cell responses are dependent on the T cell/APC ratio, source of mouse strain, and regional localization. J. Immunol. 187 (2), 733–747.
Denson, L.A., Jurickova, I., Karns, R., Shaw, K.A., Cutler, D.J., Okou, D., et al., 2019. Genetic and transcriptomic variation linked to neutrophil granulocyte-macrophage Colony-stimulating factor signaling in pediatric Crohn's disease. Inflamm. Bowel Dis. 25 (3), 547–560.
Desalegn, G., Pabst, O., 2019. Inflammation triggers immediate rather than progressive changes in monocyte differentiation in the small intestine. Nat. Commun. 10 (1), 3229.
Duffin, R., O'Connor, R.A., Crittenden, S., Forster, T., Yu, C., Zheng, X., et al., 2016. Prostaglandin E(2) constrains systemic inflammation through an innate lymphoid cell-IL-22 axis. Science 351 (6279), 1333–1338.
Erkelens, M.N., Goverse, G., Konijn, T., Molenaar, R., Beijer, M.R., Van den Bossche, J., et al., 2020. Intestinal macrophages balance inflammatory expression profiles via vitamin a and Dectin-1-mediated signaling. Front. Immunol. 11, 551.
Fadok, V.A., Bratton, D.L., Konowal, A., Freed, P.W., Westcott, J.Y., Henson, P.M., 1998. Macrophages that have ingested apoptotic cells in vitro inhibit proinflammatory cytokine production through autocrine/paracrine mechanisms involving TGF-beta, PGE2, and PAF. J. Clin. Invest. 101 (4), 890–898.
Farro, G., Stakenborg, M., Gomez-Pinilla, P.J., Labeeuw, E., Goverse, G., Di Giovangiulio, M., et al., 2017. CCR2-dependent monocyte-derived macrophages resolve inflammation and restore gut motility in postoperative ileus. Gut 66 (12), 2098–2109.
Franke, A., Balschun, T., Sina, C., Ellinghaus, D., Hasler, R., Mayr, G., et al., 2010a. Genome-wide association study for ulcerative colitis identifies risk loci at 7q22 and 22q13 (IL17REL). Nat. Genet. 42 (4), 292–294.
Franke, A., McGovern, D.P., Barrett, J.C., Wang, K., Radford-Smith, G.L., Ahmad, T., et al., 2010b. Genome-wide meta-analysis increases to 71 the number of confirmed Crohn's disease susceptibility loci. Nat. Genet. 42 (12), 1118–1125.
Franze, E., Caruso, R., Stolfi, C., Sarra, M., Cupi, M.L., Caprioli, F., et al., 2013. Lesional accumulation of CD163-expressing cells in the gut of patients with inflammatory bowel disease. PLoS One 8 (7), e69839.

Gabanyi, I., Muller, P.A., Feighery, L., Oliveira, T.Y., Costa-Pinto, F.A., Mucida, D., 2016. Neuro-immune interactions drive tissue programming in intestinal macrophages. Cell 164 (3), 378–391.

Germic, N., Frangez, Z., Yousefi, S., Simon, H.U., 2019. Regulation of the innate immune system by autophagy: monocytes, macrophages, dendritic cells and antigen presentation. Cell Death Differ. 26 (4), 715–727.

Girardin, S.E., Boneca, I.G., Viala, J., Chamaillard, M., Labigne, A., Thomas, G., et al., 2003. Nod2 is a general sensor of peptidoglycan through muramyl dipeptide (MDP) detection. J. Biol. Chem. 278 (11), 8869–8872.

Glocker, E.O., Kotlarz, D., Boztug, K., Gertz, E.M., Schaffer, A.A., Noyan, F., et al., 2009. Inflammatory bowel disease and mutations affecting the interleukin-10 receptor. N. Engl. J. Med. 361 (21), 2033–2045.

Goverse, G., Labao-Almeida, C., Ferreira, M., Molenaar, R., Wahlen, S., Konijn, T., et al., 2016. Vitamin a controls the presence of RORgamma + innate lymphoid cells and lymphoid tissue in the small intestine. J. Immunol. 196 (12), 5148–5155.

Grainger, J.R., Wohlfert, E.A., Fuss, I.J., Bouladoux, N., Askenase, M.H., Legrand, F., et al., 2013. Inflammatory monocytes regulate pathologic responses to commensals during acute gastrointestinal infection. Nat. Med. 19 (6), 713–721.

Gross-Vered, M., Trzebanski, S., Shemer, A., Bernshtein, B., Curato, C., Stelzer, G., et al., 2020. Defining murine monocyte differentiation into colonic and ileal macrophages. Elife 9.

Hadis, U., Wahl, B., Schulz, O., Hardtke-Wolenski, M., Schippers, A., Wagner, N., et al., 2011. Intestinal tolerance requires gut homing and expansion of FoxP3 + regulatory T cells in the lamina propria. Immunity 34 (2), 237–246.

Hampe, J., Franke, A., Rosenstiel, P., Till, A., Teuber, M., Huse, K., et al., 2007. A genome-wide association scan of nonsynonymous SNPs identifies a susceptibility variant for Crohn disease in ATG16L1. Nat. Genet. 39 (2), 207–211.

Han, X., Gilbert, S., Groschwitz, K., Hogan, S., Jurickova, I., Trapnell, B., et al., 2010. Loss of GM-CSF signalling in non-haematopoietic cells increases NSAID ileal injury. Gut 59 (8), 1066–1078.

Han, X., Ding, S., Jiang, H., Liu, G., 2021. Roles of macrophages in the development and treatment of gut inflammation. Front. Cell Dev. Biol. 9, 625423.

Hazlewood, G.S., Rezaie, A., Borman, M., Panaccione, R., Ghosh, S., Seow, C.H., et al., 2015. Comparative effectiveness of immunosuppressants and biologics for inducing and maintaining remission in Crohn's disease: a network meta-analysis. Gastroenterology 148 (2), 344–354 (e5; quiz e14-5).

Hirata, Y., Egea, L., Dann, S.M., Eckmann, L., Kagnoff, M.F., 2010. GM-CSF-facilitated dendritic cell recruitment and survival govern the intestinal mucosal response to a mouse enteric bacterial pathogen. Cell Host Microbe 7 (2), 151–163.

Hirota, K., Turner, J.E., Villa, M., Duarte, J.H., Demengeot, J., Steinmetz, O.M., et al., 2013. Plasticity of Th17 cells in Peyer's patches is responsible for the induction of T cell-dependent IgA responses. Nat. Immunol. 14 (4), 372–379.

Hirotani, T., Lee, P.Y., Kuwata, H., Yamamoto, M., Matsumoto, M., Kawase, I., et al., 2005. The nuclear IkappaB protein IkappaBNS selectively inhibits lipopolysaccharide-induced IL-6 production in macrophages of the colonic lamina propria. J. Immunol. 174 (6), 3650–3657.

Hoechst, B., Gamrekelashvili, J., Manns, M.P., Greten, T.F., Korangy, F., 2011. Plasticity of human Th17 cells and iTregs is orchestrated by different subsets of myeloid cells. Blood 117 (24), 6532–6541.

Hoeffel, G., Ginhoux, F., 2018. Fetal monocytes and the origins of tissue-resident macrophages. Cell. Immunol. 330, 5–15.

Honda, M., Surewaard, B.G.J., Watanabe, M., Hedrick, C.C., Lee, W.Y., Brown, K., et al., 2020. Perivascular localization of macrophages in the intestinal mucosa is regulated by Nr4a1 and the microbiome. Nat. Commun. 11 (1), 1329.

Hoshi, N., Schenten, D., Nish, S.A., Walther, Z., Gagliani, N., Flavell, R.A., et al., 2012. MyD88 signalling in colonic mononuclear phagocytes drives colitis in IL-10-deficient mice. Nat. Commun. 3, 1120.

Hugot, J.P., Chamaillard, M., Zouali, H., Lesage, S., Cezard, J.P., Belaiche, J., et al., 2001. Association of NOD2 leucine-rich repeat variants with susceptibility to Crohn's disease. Nature 411 (6837), 599–603.

Inohara, N., Ogura, Y., Fontalba, A., Gutierrez, O., Pons, F., Crespo, J., et al., 2003. Host recognition of bacterial muramyl dipeptide mediated through NOD2. Implications for Crohn's disease. J. Biol. Chem. 278 (8), 5509–5512.

Jain, U., Ver Heul, A.M., Xiong, S., Gregory, M.H., Demers, E.G., Kern, J.T., et al., 2021. Debaryomyces is enriched in Crohn's disease intestinal tissue and impairs healing in mice. Science 371 (6534), 1154–1159.

Jostins, L., Ripke, S., Weersma, R.K., Duerr, R.H., McGovern, D.P., Hui, K.Y., et al., 2012. Host-microbe interactions have shaped the genetic architecture of inflammatory bowel disease. Nature 491 (7422), 119–124.

Kabashima, K., Saji, T., Murata, T., Nagamachi, M., Matsuoka, T., Segi, E., et al., 2002. The prostaglandin receptor EP4 suppresses colitis, mucosal damage and CD4 cell activation in the gut. J. Clin. Invest. 109 (7), 883–893.

Kamada, N., Hisamatsu, T., Okamoto, S., Chinen, H., Kobayashi, T., Sato, T., et al., 2008. Unique CD14 intestinal macrophages contribute to the pathogenesis of Crohn disease via IL-23/IFN-gamma axis. J. Clin. Invest. 118 (6), 2269–2280.

Kang, B., Alvarado, L.J., Kim, T., Lehmann, M.L., Cho, H., He, J., et al., 2020. Commensal microbiota drive the functional diversification of colon macrophages. Mucosal Immunol. 13 (2), 216–229.

Kaplan, G.G., Windsor, J.W., 2021. The four epidemiological stages in the global evolution of inflammatory bowel disease. Nat. Rev. Gastroenterol. Hepatol. 18 (1), 56–66.

Kayama, H., Ueda, Y., Sawa, Y., Jeon, S.G., Ma, J.S., Okumura, R., et al., 2012. Intestinal CX3C chemokine receptor 1(high) (CX3CR1(high)) myeloid cells prevent T-cell-dependent colitis. Proc. Natl. Acad. Sci. U. S. A. 109 (13), 5010–5015.

Kim, Y.I., Song, J.H., Ko, H.J., Kweon, M.N., Kang, C.Y., Reinecker, H.C., et al., 2018. CX3CR1(+) macrophages and CD8(+) T cells control intestinal IgA production. J. Immunol. 201 (4), 1287–1294.

Kobayashi, K.S., Chamaillard, M., Ogura, Y., Henegariu, O., Inohara, N., Nunez, G., et al., 2005. Nod2-dependent regulation of innate and adaptive immunity in the intestinal tract. Science 307 (5710), 731–734.

Koelink, P.J., Bloemendaal, F.M., Li, B., Westera, L., Vogels, E.W.M., van Roest, M., et al., 2020. Anti-TNF therapy in IBD exerts its therapeutic effect through macrophage IL-10 signalling. Gut 69 (6), 1053–1063.

Kojouharoff, G., Hans, W., Obermeier, F., Mannel, D.N., Andus, T., Scholmerich, J., et al., 1997. Neutralization of tumour necrosis factor (TNF) but not of IL-1 reduces inflammation in chronic dextran sulphate sodium-induced colitis in mice. Clin. Exp. Immunol. 107 (2), 353–358.

Korzenik, J.R., Dieckgraefe, B.K., Valentine, J.F., Hausman, D.F., Gilbert, M.J., 2005. Sargramostim in Crohn's disease study G. Sargramostim for active Crohn's disease. N. Engl. J. Med. 352 (21), 2193–2201.

Koscso, B., Kurapati, S., Rodrigues, R.R., Nedjic, J., Gowda, K., Shin, C., et al., 2020. Gut-resident CX3CR1(hi) macrophages induce tertiary lymphoid structures and IgA response in situ. Sci. Immunol. 5 (46).

Kotlarz, D., Beier, R., Murugan, D., Diestelhorst, J., Jensen, O., Boztug, K., et al., 2012. Loss of interleukin-10 signaling and infantile inflammatory bowel disease: implications for diagnosis and therapy. Gastroenterology 143 (2), 347–355.

Krause, P., Morris, V., Greenbaum, J.A., Park, Y., Bjoerheden, U., Mikulski, Z., et al., 2015. IL-10-producing intestinal macrophages prevent excessive antibacterial innate immunity by limiting IL-23 synthesis. Nat. Commun. 6, 7055.

Kuhn, R., Lohler, J., Rennick, D., Rajewsky, K., Muller, W., 1993. Interleukin-10-deficient mice develop chronic enterocolitis. Cell 75 (2), 263–274.

Lavin, Y., Merad, M., 2013. Macrophages: gatekeepers of tissue integrity. Cancer Immunol. Res. 1 (4), 201–209.

Leonardi, I., Li, X., Semon, A., Li, D., Doron, I., Putzel, G., et al., 2018. CX3CR1(+) mononuclear phagocytes control immunity to intestinal fungi. Science 359 (6372), 232–236.

Levine, A.P., Pontikos, N., Schiff, E.R., Jostins, L., Speed, D., Consortium, N.I.B.D.G., et al., 2016. Genetic complexity of Crohn's disease in two large Ashkenazi Jewish families. Gastroenterology 151 (4), 698–709.

Li, J., Moran, T., Swanson, E., Julian, C., Harris, J., Bonen, D.K., et al., 2004. Regulation of IL-8 and IL-1beta expression in Crohn's disease associated NOD2/CARD15 mutations. Hum. Mol. Genet. 13 (16), 1715–1725.

Lichtenstein, G.R., Feagan, B.G., Cohen, R.D., Salzberg, B.A., Safdi, M., Popp Jr., J.W., et al., 2018. Infliximab for Crohn's disease: more than 13 years of real-world experience. Inflamm. Bowel Dis. 24 (3), 490–501.

Limon, J.J., Tang, J., Li, D., Wolf, A.J., Michelsen, K.S., Funari, V., et al., 2019. Malassezia is associated with Crohn's disease and exacerbates colitis in mouse models. Cell Host Microbe 25 (3), 377–388 (e6).

Liu, Y., van Kruiningen, H.J., West, A.B., Cartun, R.W., Cortot, A., Colombel, J.F., 1995. Immunocytochemical evidence of listeria, Escherichia coli, and streptococcus antigens in Crohn's disease. Gastroenterology 108 (5), 1396–1404.

Liu, J.Z., van Sommeren, S., Huang, H., Ng, S.C., Alberts, R., Takahashi, A., et al., 2015. Association analyses identify 38 susceptibility loci for inflammatory bowel disease and highlight shared genetic risk across populations. Nat. Genet. 47 (9), 979–986.

Liu, Z., Gu, Y., Chakarov, S., Bleriot, C., Kwok, I., Chen, X., et al., 2019. Fate mapping via Ms4a3-expression history traces monocyte-derived cells. Cell 178 (6), 1509–1525 (e19).

Longman, R.S., Diehl, G.E., Victorio, D.A., Huh, J.R., Galan, C., Miraldi, E.R., et al., 2014. CX(3)CR1(+) mononuclear phagocytes support colitis-associated innate lymphoid cell production of IL-22. J. Exp. Med. 211 (8), 1571–1583.

MacDonald, K.P., Palmer, J.S., Cronau, S., Seppanen, E., Olver, S., Raffelt, N.C., et al., 2010. An antibody against the colony-stimulating factor 1 receptor depletes the resident subset of monocytes and tissue- and tumor-associated macrophages but does not inhibit inflammation. Blood 116 (19), 3955–3963.

Maculins, T., Verschueren, E., Hinkle, T., Choi, M., Chang, P., Chalouni, C., et al., 2021. Multiplexed proteomics of autophagy-deficient murine macrophages reveals enhanced antimicrobial immunity via the oxidative stress response. Elife 10.

Maheshwari, A., Kelly, D.R., Nicola, T., Ambalavanan, N., Jain, S.K., Murphy-Ullrich, J., et al., 2011. TGF-beta2 suppresses macrophage cytokine production and mucosal inflammatory responses in the developing intestine. Gastroenterology 140 (1), 242–253.

Mahida, Y.R., Patel, S., Gionchetti, P., Vaux, D., Jewell, D.P., 1989. Macrophage subpopulations in lamina propria of normal and inflamed colon and terminal ileum. Gut 30 (6), 826–834.

Main, J., McKenzie, H., Yeaman, G.R., Kerr, M.A., Robson, D., Pennington, C.R., et al., 1988. Antibody to Saccharomyces cerevisiae (bakers' yeast) in Crohn's disease. BMJ 297 (6656), 1105–1106.

Manta, C., Heupel, E., Radulovic, K., Rossini, V., Garbi, N., Riedel, C.U., et al., 2013. CX(3)CR1(+) macrophages support IL-22 production by innate lymphoid cells during infection with Citrobacter rodentium. Mucosal Immunol. 6 (1), 177–188.

Martin, J.C., Chang, C., Boschetti, G., Ungaro, R., Giri, M., Grout, J.A., et al., 2019. Single-cell analysis of Crohn's disease lesions identifies a pathogenic cellular module associated with resistance to anti-TNF therapy. Cell 178 (6), 1493–1508 (e20).

Matheis, F., Muller, P.A., Graves, C.L., Gabanyi, I., Kerner, Z.J., Costa-Borges, D., et al., 2020. Adrenergic signaling in muscularis macrophages limits infection-induced neuronal loss. Cell 180 (1), 64–78 (e16).

Mazzurana, L., Bonfiglio, F., Forkel, M., D'Amato, M., Halfvarson, J., Mjosberg, J., 2021. Crohn's disease is associated with activation of circulating innate lymphoid cells. Inflamm. Bowel Dis. 27 (7), 1128–1138.

McAlindon, M.E., Galvin, A., McKaig, B., Gray, T., Sewell, H.F., Mahida, Y.R., 1999. Investigation of the expression of IL-1beta converting enzyme and apoptosis in normal and inflammatory bowel disease (IBD) mucosal macrophages. Clin. Exp. Immunol. 116 (2), 251–257.

McDole, J.R., Wheeler, L.W., McDonald, K.G., Wang, B., Konjufca, V., Knoop, K.A., et al., 2012. Goblet cells deliver luminal antigen to CD103+ dendritic cells in the small intestine. Nature 483 (7389), 345–349.

Meconi, S., Vercellone, A., Levillain, F., Payre, B., Al Saati, T., Capilla, F., et al., 2007. Adherent-invasive Escherichia coli isolated from Crohn's disease patients induce granulomas in vitro. Cell. Microbiol. 9 (5), 1252–1261.

Medina-Contreras, O., Geem, D., Laur, O., Williams, I.R., Lira, S.A., Nusrat, A., et al., 2011. CX3CR1 regulates intestinal macrophage homeostasis, bacterial translocation, and colitogenic Th17 responses in mice. J. Clin. Invest. 121 (12), 4787–4795.

Mikkelsen, H.B., Thuneberg, L., 1999. Op/op mice defective in production of functional colony-stimulating factor-1 lack macrophages in muscularis externa of the small intestine. Cell Tissue Res. 295 (3), 485–493.

Miyoshi, H., VanDussen, K.L., Malvin, N.P., Ryu, S.H., Wang, Y., Sonnek, N.M., et al., 2017. Prostaglandin E2 promotes intestinal repair through an adaptive cellular response of the epithelium. EMBO J. 36 (1), 5–24.

Morhardt, T.L., Hayashi, A., Ochi, T., Quiros, M., Kitamoto, S., Nagao-Kitamoto, H., et al., 2019. IL-10 produced by macrophages regulates epithelial integrity in the small intestine. Sci. Rep. 9 (1), 1223.

Mortha, A., Chudnovskiy, A., Hashimoto, D., Bogunovic, M., Spencer, S.P., Belkaid, Y., et al., 2014. Microbiota-dependent crosstalk between macrophages and ILC3 promotes intestinal homeostasis. Science 343 (6178), 1249288.

Mukherjee, T., Hovingh, E.S., Foerster, E.G., Abdel-Nour, M., Philpott, D.J., Girardin, S.E., 2019. NOD1 and NOD2 in inflammation, immunity and disease. Arch. Biochem. Biophys. 670, 69–81.

Mulder, K., Patel, A.A., Kong, W.T., Piot, C., Halitzki, E., Dunsmore, G., et al., 2021. Cross-tissue single-cell landscape of human monocytes and macrophages in health and disease. Immunity 54 (8), 1883–1900 (e5).

Muller, A.J., Kaiser, P., Dittmar, K.E., Weber, T.C., Haueter, S., Endt, K., et al., 2012. Salmonella gut invasion involves TTSS-2-dependent epithelial traversal, basolateral exit, and uptake by epithelium-sampling lamina propria phagocytes. Cell Host Microbe 11 (1), 19–32.

Muller, P.A., Koscso, B., Rajani, G.M., Stevanovic, K., Berres, M.L., Hashimoto, D., et al., 2014. Crosstalk between muscularis macrophages and enteric neurons regulates gastrointestinal motility. Cell 158 (5), 1210.

Murai, M., Turovskaya, O., Kim, G., Madan, R., Karp, C.L., Cheroutre, H., et al., 2009. Interleukin 10 acts on regulatory T cells to maintain expression of the transcription factor Foxp3 and suppressive function in mice with colitis. Nat. Immunol. 10 (11), 1178–1184.

Murthy, A., Li, Y., Peng, I., Reichelt, M., Katakam, A.K., Noubade, R., et al., 2014. A Crohn's disease variant in Atg16l1 enhances its degradation by caspase 3. Nature 506 (7489), 456–462.

Na, Y.R., Stakenborg, M., Seok, S.H., Matteoli, G., 2019. Macrophages in intestinal inflammation and resolution: a potential therapeutic target in IBD. Nat. Rev. Gastroenterol. Hepatol. 16 (9), 531–543.

Na, Y.R., Jung, D., Stakenborg, M., Jang, H., Gu, G.J., Jeong, M.R., et al., 2021. Prostaglandin E2 receptor PTGER4-expressing macrophages promote intestinal epithelial barrier regeneration upon inflammation. Gut.

Nagashima, R., Maeda, K., Imai, Y., Takahashi, T., 1996. Lamina propria macrophages in the human gastrointestinal mucosa: their distribution, immunohistological phenotype, and function. J. Histochem. Cytochem. 44 (7), 721–731.

Naito, Y., Takagi, T., Handa, O., Ishikawa, T., Nakagawa, S., Yamaguchi, T., et al., 2003. Enhanced intestinal inflammation induced by dextran sulfate sodium in tumor necrosis factor-alpha deficient mice. J. Gastroenterol. Hepatol. 18 (5), 560–569.

Nayar, S., Morrison, J.K., Giri, M., Gettler, K., Chuang, L.S., Walker, L.A., et al., 2021. A myeloid-stromal niche and gp130 rescue in NOD2-driven Crohn's disease. Nature 593 (7858), 275–281.

Niess, J.H., Brand, S., Gu, X., Landsman, L., Jung, S., McCormick, B.A., et al., 2005. CX3CR1-mediated dendritic cell access to the intestinal lumen and bacterial clearance. Science 307 (5707), 254–258.

Ogino, T., Nishimura, J., Barman, S., Kayama, H., Uematsu, S., Okuzaki, D., et al., 2013. Increased Th17-inducing activity of CD14+ CD163 low myeloid cells in intestinal lamina propria of patients with Crohn's disease. Gastroenterology 145 (6), 1380–1391 (e1).

Ogura, Y., Bonen, D.K., Inohara, N., Nicolae, D.L., Chen, F.F., Ramos, R., et al., 2001. A frameshift mutation in NOD2 associated with susceptibility to Crohn's disease. Nature 411 (6837), 603–606.

Panea, C., Farkas, A.M., Goto, Y., Abdollahi-Roodsaz, S., Lee, C., Koscso, B., et al., 2015. Intestinal monocyte-derived macrophages control commensal-specific Th17 responses. Cell Rep. 12 (8), 1314–1324.

Papadakis, K.A., Krempski, J., Svingen, P., Xiong, Y., Sarmento, O.F., Lomberk, G.A., et al., 2015. Kruppel-like factor KLF10 deficiency predisposes to colitis through colonic macrophage dysregulation. Am. J. Physiol. Gastrointest. Liver Physiol. 309 (11), G900–G909.

Papamichael, K., Lin, S., Moore, M., Papaioannou, G., Sattler, L., Cheifetz, A.S., 2019. Infliximab in inflammatory bowel disease. Ther. Adv. Chronic Dis. 10 (20406223 19838443).

Persson, E.K., Scott, C.L., Mowat, A.M., Agace, W.W., 2013. Dendritic cell subsets in the intestinal lamina propria: ontogeny and function. Eur. J. Immunol. 43 (12), 3098–3107.

Plevy, S., Silverberg, M.S., Lockton, S., Stockfisch, T., Croner, L., Stachelski, J., et al., 2013. Combined serological, genetic, and inflammatory markers differentiate non-IBD, Crohn's disease, and ulcerative colitis patients. Inflamm. Bowel Dis. 19 (6), 1139–1148.

Prins, M.M., Verstockt, B., Ferrante, M., Vermeire, S., Wildenberg, M.E., Koelink, P.J., 2021. Monocyte TREM-1 levels associate with anti-TNF responsiveness in IBD through autophagy and Fcgamma-receptor signaling pathways. Front. Immunol. 12, 627535.

Pull, S.L., Doherty, J.M., Mills, J.C., Gordon, J.I., Stappenbeck, T.S., 2005. Activated macrophages are an adaptive element of the colonic epithelial progenitor niche necessary for regenerative responses to injury. Proc. Natl. Acad. Sci. U. S. A. 102 (1), 99–104.

Qualls, J.E., Kaplan, A.M., van Rooijen, N., Cohen, D.A., 2006. Suppression of experimental colitis by intestinal mononuclear phagocytes. J. Leukoc. Biol. 80 (4), 802–815.

Quiros, M., Nishio, H., Neumann, P.A., Siuda, D., Brazil, J.C., Azcutia, V., et al., 2017. Macrophage-derived IL-10 mediates mucosal repair by epithelial WISP-1 signaling. J. Clin. Invest. 127 (9), 3510–3520.

Rani, R., Smulian, A.G., Greaves, D.R., Hogan, S.P., Herbert, D.R., 2011. TGF-beta limits IL-33 production and promotes the resolution of colitis through regulation of macrophage function. Eur. J. Immunol. 41 (7), 2000–2009.

Rescigno, M., Urbano, M., Valzasina, B., Francolini, M., Rotta, G., Bonasio, R., et al., 2001. Dendritic cells express tight junction proteins and penetrate gut epithelial monolayers to sample bacteria. Nat. Immunol. 2 (4), 361–367.

Rivollier, A., He, J., Kole, A., Valatas, V., Kelsall, B.L., 2012. Inflammation switches the differentiation program of Ly6Chi monocytes from antiinflammatory macrophages to inflammatory dendritic cells in the colon. J. Exp. Med. 209 (1), 139–155.

Rogler, G., Hausmann, M., Vogl, D., Aschenbrenner, E., Andus, T., Falk, W., et al., 1998. Isolation and phenotypic characterization of colonic macrophages. Clin. Exp. Immunol. 112 (2), 205–215.

Rubtsov, Y.P., Rasmussen, J.P., Chi, E.Y., Fontenot, J., Castelli, L., Ye, X., et al., 2008. Regulatory T cell-derived interleukin-10 limits inflammation at environmental interfaces. Immunity 28 (4), 546–558.

Rugtveit, J., Brandtzaeg, P., Halstensen, T.S., Fausa, O., Scott, H., 1994. Increased macrophage subset in inflammatory bowel disease: apparent recruitment from peripheral blood monocytes. Gut 35 (5), 669–674.

Rugtveit, J., Bakka, A., Brandtzaeg, P., 1997. Differential distribution of B7.1 (CD80) and B7.2 (CD86) costimulatory molecules on mucosal macrophage subsets in human inflammatory bowel disease (IBD). Clin. Exp. Immunol. 110 (1), 104–113.

Ryan, P., Kelly, R.G., Lee, G., Collins, J.K., O'Sullivan, G.C., O'Connell, J., et al., 2004. Bacterial DNA within granulomas of patients with Crohn's disease- -detection by laser capture microdissection and PCR. Am. J. Gastroenterol. 99 (8), 1539–1543.

Sainathan, S.K., Hanna, E.M., Gong, Q., Bishnupuri, K.S., Luo, Q., Colonna, M., et al., 2008. Granulocyte macrophage colony-stimulating factor ameliorates DSS-induced experimental colitis. Inflamm. Bowel Dis. 14 (1), 88–99.

Sanders, T.J., McCarthy, N.E., Giles, E.M., Davidson, K.L., Haltalli, M.L., Hazell, S., et al., 2014. Increased production of retinoic acid by intestinal macrophages contributes to their inflammatory phenotype in patients with Crohn's disease. Gastroenterology 146 (5), 1278–1288 (e1-2).

Sanos, S.L., Bui, V.L., Mortha, A., Oberle, K., Heners, C., Johner, C., et al., 2009. RORgammat and commensal microflora are required for the differentiation of mucosal interleukin 22-producing NKp46+ cells. Nat. Immunol. 10 (1), 83–91.

Sartor, R.B., 2006. Mechanisms of disease: pathogenesis of Crohn's disease and ulcerative colitis. Nat. Clin. Pract. Gastroenterol. Hepatol. 3 (7), 390–407.

Schreiber, H.A., Loschko, J., Karssemeijer, R.A., Escolano, A., Meredith, M.M., Mucida, D., et al., 2013. Intestinal monocytes and macrophages are required for T cell polarization in response to Citrobacter rodentium. J. Exp. Med. 210 (10), 2025–2039.

Schridde, A., Bain, C.C., Mayer, J.U., Montgomery, J., Pollet, E., Denecke, B., et al., 2017. Tissue-specific differentiation of colonic macrophages requires TGFbeta receptor-mediated signaling. Mucosal Immunol. 10 (6), 1387–1399.

Schulz, O., Jaensson, E., Persson, E.K., Liu, X., Worbs, T., Agace, W.W., et al., 2009. Intestinal CD103+, but not CX3CR1+, antigen sampling cells migrate in lymph and serve classical dendritic cell functions. J. Exp. Med. 206 (13), 3101–3114.

Seno, H., Miyoshi, H., Brown, S.L., Geske, M.J., Colonna, M., Stappenbeck, T.S., 2009. Efficient colonic mucosal wound repair requires Trem2 signaling. Proc. Natl. Acad. Sci. U. S. A. 106 (1), 256–261.

Serbina, N.V., Pamer, E.G., 2006. Monocyte emigration from bone marrow during bacterial infection requires signals mediated by chemokine receptor CCR2. Nat. Immunol. 7 (3), 311–317.

Shaw, M.H., Kamada, N., Kim, Y.G., Nunez, G., 2012. Microbiota-induced IL-1beta, but not IL-6, is critical for the development of steady-state TH17 cells in the intestine. J. Exp. Med. 209 (2), 251–258.

Shaw, T.N., Houston, S.A., Wemyss, K., Bridgeman, H.M., Barbera, T.A., Zangerle-Murray, T., et al., 2018. Tissue-resident macrophages in the intestine are long lived and defined by Tim-4 and CD4 expression. J. Exp. Med. 215 (6), 1507–1518.

Shou, Y., Koroleva, E., Spencer, C.M., Shein, S.A., Korchagina, A.A., Yusoof, K.A., et al., 2021. Redefining the role of lymphotoxin Beta receptor in the maintenance of lymphoid organs and immune cell homeostasis in adulthood. Front. Immunol. 12, 712632.

Shouval, D.S., Biswas, A., Goettel, J.A., McCann, K., Conaway, E., Redhu, N.S., et al., 2014. Interleukin-10 receptor signaling in innate immune cells regulates mucosal immune tolerance and anti-inflammatory macrophage function. Immunity 40 (5), 706–719.

Silverberg, M.S., Satsangi, J., Ahmad, T., Arnott, I.D., Bernstein, C.N., Brant, S.R., et al., 2005. Toward an integrated clinical, molecular and serological classification of inflammatory bowel disease: report of a working party of the 2005 Montreal world congress of gastroenterology. Can. J. Gastroenterol. 19 (Suppl A:5A-36A).

Smith, P.D., Smythies, L.E., Mosteller-Barnum, M., Sibley, D.A., Russell, M.W., Merger, M., et al., 2001. Intestinal macrophages lack CD14 and CD89 and consequently are down-regulated for LPS- and IgA-mediated activities. J. Immunol. 167 (5), 2651–2656.

Smith, A.M., Rahman, F.Z., Hayee, B., Graham, S.J., Marks, D.J., Sewell, G.W., et al., 2009. Disordered macrophage cytokine secretion underlies impaired acute inflammation and bacterial clearance in Crohn's disease. J. Exp. Med. 206 (9), 1883–1897.

Smythies, L.E., Sellers, M., Clements, R.H., Mosteller-Barnum, M., Meng, G., Benjamin, W.H., et al., 2005. Human intestinal macrophages display profound inflammatory anergy despite avid phagocytic and bacteriocidal activity. J. Clin. Invest. 115 (1), 66–75.

Smythies, L.E., Shen, R., Bimczok, D., Novak, L., Clements, R.H., Eckhoff, D.E., et al., 2010. Inflammation anergy in human intestinal macrophages is due to Smad-induced IkappaBalpha expression and NF-kappaB inactivation. J. Biol. Chem. 285 (25), 19593–19604.

Sokol, H., Landman, C., Seksik, P., Berard, L., Montil, M., Nion-Larmurier, I., et al., 2020. Fecal microbiota transplantation to maintain remission in Crohn's disease: a pilot randomized controlled study. Microbiome. 8 (1), 12.

Song, C., Lee, J.S., Gilfillan, S., Robinette, M.L., Newberry, R.D., Stappenbeck, T.S., et al., 2015. Unique and redundant functions of NKp46 + ILC3s in models of intestinal inflammation. J. Exp. Med. 212 (11), 1869–1882.

Tait Wojno, E.D., Hunter, C.A., Stumhofer, J.S., 2019. The immunobiology of the Interleukin-12 family: room for discovery. Immunity 50 (4), 851–870.

Takeda, K., Clausen, B.E., Kaisho, T., Tsujimura, T., Terada, N., Forster, I., et al., 1999. Enhanced Th1 activity and development of chronic enterocolitis in mice devoid of Stat3 in macrophages and neutrophils. Immunity 10 (1), 39–49.

Tamoutounour, S., Henri, S., Lelouard, H., de Bovis, B., de Haar, C., van der Woude, C.J., et al., 2012. CD64 distinguishes macrophages from dendritic cells in the gut and reveals the Th1-inducing role of mesenteric lymph node macrophages during colitis. Eur. J. Immunol. 42 (12), 3150–3166.

Torres, J., Mehandru, S., Colombel, J.F., Peyrin-Biroulet, L., 2017. Crohn's disease. Lancet 389 (10080), 1741–1755.

Ueda, Y., Kayama, H., Jeon, S.G., Kusu, T., Isaka, Y., Rakugi, H., et al., 2010. Commensal microbiota induce LPS hyporesponsiveness in colonic macrophages via the production of IL-10. Int. Immunol. 22 (12), 953–962.

Verstockt, B., Verstockt, S., Dehairs, J., Ballet, V., Blevi, H., Wollants, W.J., et al., 2019. Low TREM1 expression in whole blood predicts anti-TNF response in inflammatory bowel disease. EBioMedicine 40, 733–742.
Viola, M.F., Boeckxstaens, G., 2020. Intestinal resident macrophages: multitaskers of the gut. Neurogastroenterol. Motil. 32 (8), e13843.
Wang, X., Allen, C., Ballow, M., 2007. Retinoic acid enhances the production of IL-10 while reducing the synthesis of IL-12 and TNF-alpha from LPS-stimulated monocytes/macrophages. J. Clin. Immunol. 27 (2), 193–200.
Watanabe, N., Ikuta, K., Okazaki, K., Nakase, H., Tabata, Y., Matsuura, M., et al., 2003. Elimination of local macrophages in intestine prevents chronic colitis in interleukin-10-deficient mice. Dig. Dis. Sci. 48 (2), 408–414.
Weaver, C.T., Elson, C.O., Fouser, L.A., Kolls, J.K., 2013. The Th17 pathway and inflammatory diseases of the intestines, lungs, and skin. Annu. Rev. Pathol. 8, 477–512.
Wilson, N.J., Boniface, K., Chan, J.R., McKenzie, B.S., Blumenschein, W.M., Mattson, J.D., et al., 2007. Development, cytokine profile and function of human interleukin 17-producing helper T cells. Nat. Immunol. 8 (9), 950–957.
Wynn, T.A., Vannella, K.M., 2016. Macrophages in tissue repair, regeneration, and fibrosis. Immunity 44 (3), 450–462.
Yao, K., Iwashita, A., Yao, T., Takemura, S., Furukawa, K., Matsui, T., et al., 1996. Increased numbers of macrophages in noninflamed gastroduodenal mucosa of patients with Crohn's disease. Dig. Dis. Sci. 41 (11), 2260–2267.
Yao, K., Yao, T., Iwashita, A., Matsui, T., Kamachi, S., 2000. Microaggregate of immunostained macrophages in noninflamed gastroduodenal mucosa: a new useful histological marker for differentiating Crohn's colitis from ulcerative colitis. Am. J. Gastroenterol. 95 (8), 1967–1973.
Yona, S., Kim, K.W., Wolf, Y., Mildner, A., Varol, D., Breker, M., et al., 2013. Fate mapping reveals origins and dynamics of monocytes and tissue macrophages under homeostasis. Immunity 38 (1), 79–91.
Zhang, H., Zheng, L., Chen, J., Fukata, M., Ichikawa, R., Shih, D.Q., et al., 2017. The protection role of Atg16l1 in CD11c(+)dendritic cells in murine colitis. Immunobiology 222 (7), 831–841.
Zigmond, E., Jung, S., 2013. Intestinal macrophages: well educated exceptions from the rule. Trends Immunol. 34 (4), 162–168.
Zigmond, E., Varol, C., Farache, J., Elmaliah, E., Satpathy, A.T., Friedlander, G., et al., 2012. Ly6C hi monocytes in the inflamed colon give rise to proinflammatory effector cells and migratory antigen-presenting cells. Immunity 37 (6), 1076–1090.
Zigmond, E., Bernshtein, B., Friedlander, G., Walker, C.R., Yona, S., Kim, K.W., et al., 2014. Macrophage-restricted interleukin-10 receptor deficiency, but not IL-10 deficiency, causes severe spontaneous colitis. Immunity 40 (5), 720–733.

CHAPTER THREE

Metabolism in tumor-associated macrophages

Jie Li[a,b], Gina M. DeNicola[c], and Brian Ruffell[a,d,*]

[a]Department of Immunology, H. Lee Moffitt Cancer Center and Research Institute, Tampa, FL, United States
[b]Cancer Biology PhD Program, University of South Florida, Tampa, FL, United States
[c]Department of Cancer Physiology, Lee Moffitt Cancer Center and Research Institute, Tampa, FL, United States
[d]Department of Breast Oncology, H. Lee Moffitt Cancer Center and Research Institute, Tampa, FL, United States
*Corresponding author: e-mail address: brian.ruffell@moffitt.org

Contents

1. Introduction 66
2. Glucose metabolism 69
   2.1 Regulation of glucose metabolism in TAMs 70
   2.2 Glucose metabolism in TAM function 75
3. Amino acid metabolism 76
   3.1 Regulation of amino acid metabolism 78
   3.2 Amino acid metabolism in TAM function 79
4. Lipid metabolism in TAM 81
   4.1 Fatty acid metabolism 82
   4.2 Arachidonic acid metabolism 83
   4.3 Cholesterol metabolism 84
5. Targeting TAM metabolism for therapy 85
6. Conclusion 90
References 90

Abstract

Macrophages functionally adapt to a diverse set of signals, a process that is critical for their role in maintaining or restoring tissue homeostasis. This process extends to cancer, where macrophages respond to a series of inflammatory and metabolic cues that direct a maladaptive healing response. Tumor-associated macrophages (TAMs) have altered glucose, amino acid, and lipid metabolic profiles, and interfering with this metabolic shift can blunt the ability of macrophages to promote tumor growth, metastasis, and the creation of an immunosuppressive microenvironment. Here we will review changes in metabolites and metabolic pathways in TAMs and link these with the phenotypic and functional properties of the cells. We will also discuss current strategies targeting TAM metabolism as a therapeutic intervention in cancer.

International Review of Cell and Molecular Biology, Volume 367
ISSN 1937-6448
https://doi.org/10.1016/bs.ircmb.2022.01.004

1. Introduction

The tumor microenvironment (TME) is a complex system of mesenchymal cells, endothelium, lymphatics, and immune cells, with an array of phenotypes and distinct lineages found within these populations (Gentles et al., 2015). The TME is also comprised of an acellular component that includes soluble proteins, the extracellular matrix, and nutrients such as glucose and amino acids. Tumors have a relatively unique metabolic profile due to poor tissue perfusion, nutrient competition, and the release of metabolic products, all of which act to create an environment that can limit cell survival and proliferation. Neoplastic cells are also well known to favor glycolysis over oxidative phosphorylation (OXPHOS) even in the presence of sufficient oxygen (i.e., the Warburg effect), resulting in a TME that is replete with lactic acid. How these various factors in the TME impact tumor cell metabolism, proliferation, and progression have been reviewed elsewhere (Elia and Haigis, 2021; Pavlova and Thompson, 2016).

With the approval of immune checkpoint blockade (ICB) and chimeric antigen receptor T cells (CAR-T) there has been a renewed interest in understanding how the challenges of a harsh TME may restrict the ability of T lymphocytes to invade, identify, and kill tumor cells (Ho and Liu, 2016; O'Neill et al., 2016; Pearce et al., 2013). For example, the competition for nutrients (e.g., glucose, amino acids) between T cells and cancer cells leads to nutrient deprivation and acidity that suppress T cell infiltration and antitumor function (Chang et al., 2015; Kedia-Mehta and Finlay, 2019; Ottensmeier et al., 2016; Pilon-Thomas et al., 2016; Reina-Campos et al., 2021). Other factors, such as high levels of potassium in the tumor interstitium, can impair uptake of glycolytic intermediates and essential amino acids within T cells and dampen antitumor effector function, while simultaneously acting to retain the stemness of $CD8^+$ T cell (Vodnala et al., 2019). Thus, T cell effector function and fate are highly influenced by metabolism and the ability of the TME to alter metabolic programs within the cells.

Beyond T cells, it is poorly understood how metabolic crosstalk between immune cells, stromal cells, and cancer cells regulates the TME (Ho and Liu, 2016; O'Neill et al., 2016; Pearce et al., 2013). Tumor-associated macrophages (TAMs) comprise a large proportion of the immune infiltrate in solid malignancies and are associated with all stages of neoplastic progression in patients, from initiation through to metastasis (Ma et al., 2020; Obeid et al., 2013; Vinogradov et al., 2014; Zheng et al., 2020a). Macrophages are critical for maintaining homeostasis within healthy tissues and are intertwined with the metabolic state of the organs (den Bossche and Saraber, 2018). How these same metabolic pathways regulate tumor progression, growth, and response to therapy are only beginning to be elucidated. In addition to direct interactions with tumor cells, both tissue-resident and monocyte-derived macrophages have established roles in suppressing T cell responses. For example, peritoneal-resident macrophages limit T cell activity and therapeutic response to antibodies against programmed cell death protein-1 (PD-1) (Chow et al., 2021), lung-resident macrophages promote the recruitment of regulatory T cells during early tumor progression (Casanova-Acebes et al., 2021), and monocyte-derived TAMs suppress dendritic cell and T cell function through expression of interleukin (IL)-10 and PD-L1, respectively (Ma et al., 2020; Nakasone et al., 2012; Ruffell et al., 2014). Understanding the metabolism of TAMs should therefore provide insight into how these cells regulate tumor progression and may offer new avenues for therapeutic intervention.

Before discussing the metabolism of TAMs, it is important to note that much of our understanding derives from in vitro polarization and the binary M1–M2 macrophage polarization system. For example, classical activation with interferon-γ (IFN-γ) and lipopolysaccharide (LPS), which results in M1 polarization, increases aerobic glycolysis. In contrast, M2 macrophages (generated with IL-4 or IL-10) prioritize ATP production through the tricarboxylic acid cycle (TCA) and OXPHOS. This is thought to reflect a functional divergence between efficient energy utilization and the production of metabolites important for immune suppression and tissue repair (M2) versus the rapid energy production and cytokine production necessary for combating infection (M1) (Eming et al., 2021). However, in vivo, TAMs integrate a diverse array of signals that may drive conflicting metabolic programs (Fig. 1) and individual cells occupy distinct microenvironments that produce heterogenous populations not conforming to M1 or M2 categories (Garrido-Martin et al., 2020; Ochocka et al., 2021; Zhang et al., 2021). Conceptually, TAMs may best be thought of as occupying a spectrum of inflammatory and repair phenotypes reflecting their presence in "wounds that do not heal," with individual cells occupying unique states depending on their localization within the tumor, the local inflammatory milieu, and their ontogeny (DeNardo and Ruffell, 2019). In this review we discuss how macrophage metabolism is altered by factors in the TME, as well as how metabolism controls the biological activity of the cells within tumors. We conclude by discussing the current strategies for targeting TAMs and the challenges associated with targeting metabolism as a therapeutic strategy.

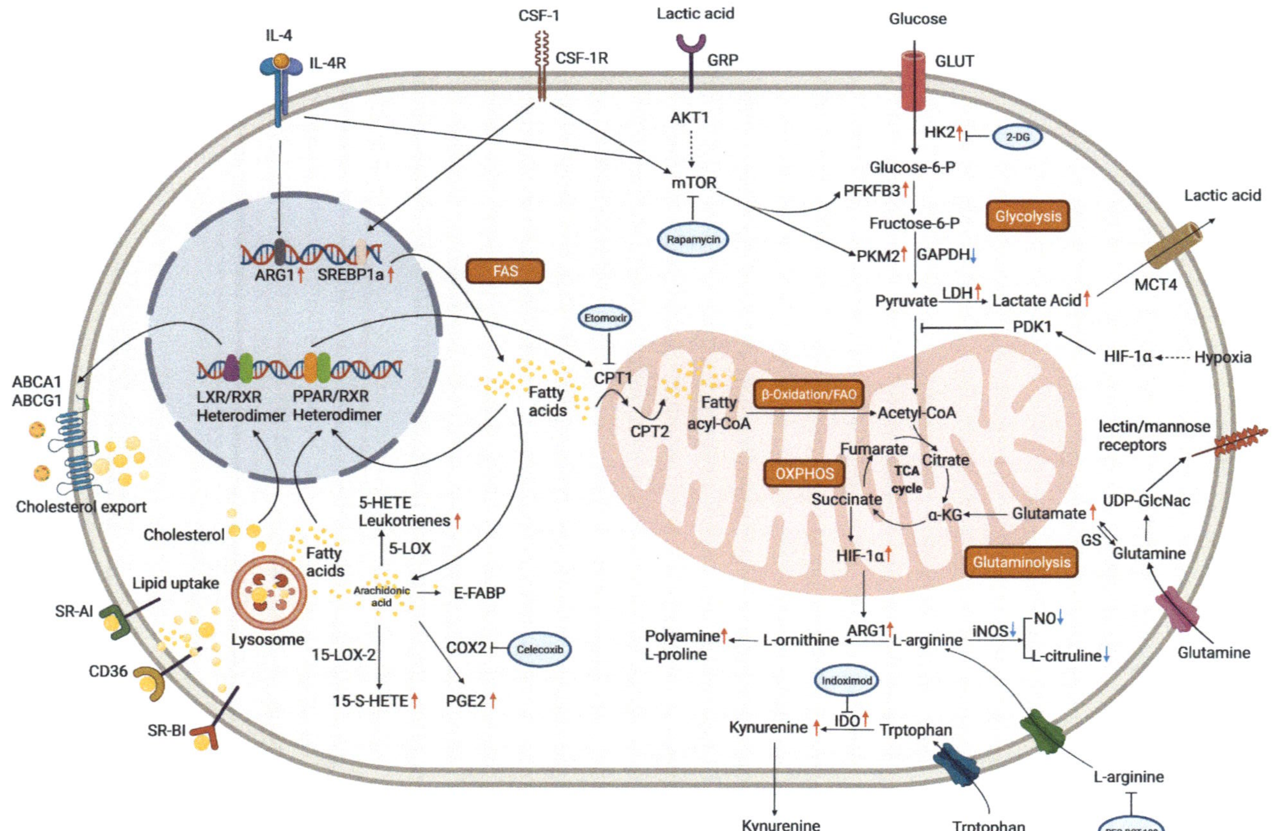

Fig. 1 See figure legend on opposite page.

2. Glucose metabolism

As mentioned above, in vitro polarization of M1 macrophages favors glucose uptake, aerobic glycolysis, and lactate production, whereas M2 macrophages favor lipoprotein uptake, fatty acid oxidation (FAO), and mitochondrial activity (Geeraerts et al., 2017). Do TAMs, which are normally thought to be M2-like, follow this dichotomy in vivo? A recent study evaluating glucose utilization through uptake of ^{18}F-fluorodeoxyglucose (FDG) found that TAMs displayed higher glucose uptake than other immune populations, including approximately twofold more than tumor cells, with a high degree of consistency across different murine cancer models (Reinfeld et al., 2021). Isolated TAMs showed a higher rate of both oxygen consumption (OCR) and extracellular acidification (ECAR), indicative of glucose flux occurring through both aerobic glycolysis and oxidative metabolism (Reinfeld et al., 2021). Similar observations have been made with human monocytes incubated with the condition medium from a melanoma cell line. These monocyte-derived TAMs displayed higher glucose uptake and lactate production than IL-4-polarized monocytes, while retaining a similar OCR (de Brito et al., 2020). Several additional studies with in vitro-derived TAMs have also noted higher ECAR or increased expression of proteins associated with glycolysis, such as hexokinase-2 (Arts et al., 2016; Liu et al., 2016; Penny et al., 2016). More importantly, a detailed study with human colon carcinoma tissue sections found that the percentage of TAMs with detectable lactate dehydrogenase (LDH) activity was much higher than the percentage of cells with isocitrate dehydrogenase (IDH) or succinate dehydrogenase (SDH) activity, although enzymatic activity overall was lower than observed in healthy tissue (Miller et al., 2017).

Fig. 1 Overview of TAM metabolic pathways. TAMs undergo metabolic reprograming in response to different cues in the TME, such as hypoxia and CSF-1. This metabolic reprograming includes activation of glycolysis, increased utilization of amino acids, and higher uptake of fatty acids and cholesterol, with a retained capacity for FAO. Depending on their activation state, some TAMs may have a disrupted TCA cycle that is fueled through glutaminolysis, with succinate accumulation promoting HIF-1α stabilization. These changes contribute to the functional reprogramming of TAMs and lead to the release of cytokines and metabolites that directly and indirectly contribute to tumor growth. This includes production of polyamines and L-proline from L-arginine and the byproducts of arachidonic acid that regulate inflammation. Drugs targeting select pathways are shown in blue ellipses.

The available data therefore indicate that TAMs shift towards glycolysis, while still retaining their capacity for OXPHOS (Wenes et al., 2016).

However, this description of a bulk population of cells masks clear differences within the various TAM subsets that exist within tumors (Garrido-Martin et al., 2020; Ochocka et al., 2021; Zhang et al., 2021). A simple interrogation of genes associated with glycolysis (e.g., *SLC2A1, SLC2A3, HK2*), FAO (e.g., *CD36, LRP1, LIPA*), amino acid metabolism (e.g., *IDO1, NOS3, FASN*), and mitochondrial function (e.g., *DLAT, SDHA, IDH1*) shows differential expression within human lung adenocarcinoma macrophage populations identified through scRNAseq (Fig. 2). Thus, there is a spectrum of metabolic states occurring within TAMs that appears to correlate with their polarization, in addition to reflecting the unique environmental niches occupied by individual cells (Hobson-Gutierrez and Carmona-Fontaine, 2018). Below we will discuss how macrophage activation and the TME act to regulate glucose metabolism within cells. We will then discuss how glucose metabolism may regulate the various functions of TAMs.

2.1 Regulation of glucose metabolism in TAMs

The classical macrophage polarization cytokines (IFN-γ, IL-4 and IL-10) are present within tumors and regulate gene expression within macrophages. Given the impact of these cytokines on glucose metabolism *in vitro* (de Brito et al., 2020; Ip et al., 2017), are they also important factors in vivo? At least in terms of glucose uptake there does not appear to be a dominant role for T cell-derived cytokines, as there was no change in FDG uptake by $CD11b^+$ tumor myeloid cells in *Rag1*-deficient mice lacking T and B cells (Reinfeld et al., 2021). The ability of tumor cell conditioned medium to promote glycolysis in macrophages also points towards a limited role for IL-4 or IFN-γ. That said, there may be differences in glucose utilization or changes in subpopulations of TAMs that have not been captured. For example, increased production of IFN-γ during immunotherapy regulates TAM phenotype and metabolism (Liu et al., 2019a), and the truncated TCA cycle observed in M1-polarized macrophages generates intermediates that could regulate TAM function (Box 1). An analysis of metabolic changes in TAMs following neutralization of specific cytokines remains to be performed. Whether tumor-derived cytokines regulate glycolysis has not been extensively evaluated, but IL-6 or IL-10 are prime candidates. For example, by limiting the production of nitric oxide (NO), IL-10 can increase the capacity of LPS-stimulated macrophages to

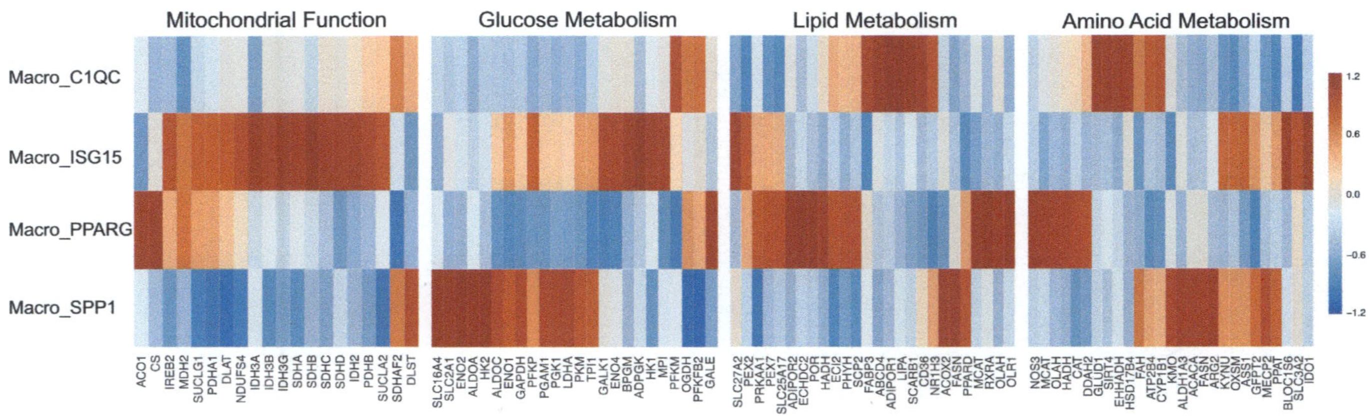

Fig. 2 Expression of metabolic genes in TAMs. Heatmap displaying expression of genes associated with mitochondrial function and glucose, lipid, and amino acid metabolism in human lung adenocarcinoma macrophage populations identified through scRNAseq. Figure was generated using http://panmyeloid.cancer-pku.cn/

BOX 1 TCA cycle metabolites

M1-polarized macrophages have a truncated TCA cycle and increasingly rely on glutamine as a carbon source (Palmieri et al., 2020), resulting in higher levels of citrate, itaconate, and α-ketoglutarate found within cells (Cordes et al., 2016; Michelucci et al., 2013; Tallam et al., 2016). This metabolic shift is driven by iNOS-dependent production of nitric oxide (Palmieri et al., 2020); thus, it is possible that iNOS-expressing TAMs will display a similar metabolic profile. Itaconate was first described as antimicrobial metabolite (McFadden and Purohit, 1977), but it has also been shown to have intrinsic anti-inflammatory properties by inhibiting expression of IL-1β, IL-6, IL-12, and IL-18 (Lampropoulou et al., 2016). This occurs through inhibition of succinate dehydrogenase oxidation of succinate (Lampropoulou et al., 2016) and through indirect promotion of nuclear factor erythroid 2-related factor 2 (Nrf2) activity (Mills et al., 2018). In a peritoneal cavity tumor model, inhibiting expression of *Irg1*, which encodes the enzyme necessary for itaconate synthesis, significantly reduce tumor growth (Weiss et al., 2018). Production of α-ketoglutarate via glutaminolysis also suppresses inflammatory cytokine expression through increased FAO and Jumonji domain-containing protein 3 (Jmjd3)-dependent epigenic regulation of genes associated with M2 polarization (*Arg1, Ym1, Retnla* and *Mrc1*) (Liu et al., 2017). Thus, iNOS-driven metabolic changes act as a negative feedback mechanism to limit expression of inflammatory cytokines, likely with context-dependent effects on tumor growth and response to immunotherapies.

engage in FAO (Baseler et al., 2016). Other tumor-derived factors that promote glycolysis in TAMs include hyaluronan fragments (Chen et al., 2019a) and exosomes (Morrissey et al., 2021).

Regions of hypoxia begin ~50 μm from functional capillaries, with the abnormal vascular structure and poor perfusion of tumors contributing to the formation of large regions with limited oxygen availability (Finger and Giaccia, 2010; Trédan et al., 2007; Vaupel, 2004). Macrophages do not shy away from these regions, but rather are specifically recruited by high levels of hypoxia-induced Semaphorin 3A through the receptor Neuropilin-1 (Casazza et al., 2013). TAMs within hypoxic regions express lower levels of MHCII and other genes associated with a T cell response, while displaying increased capacity for immunosuppression *ex vivo* (Laoui et al., 2014; Movahedi et al., 2010; Ruffell et al., 2014). This infiltration may be dependent upon increased expression of glucose transporter 1 (GLUT1) and anaerobic glycolysis (Freemerman et al., 2014; Fukuzumi et al., 1996) as motility requires rapid ATP consumption (Pantaloni et al., 2001; Semba et al., 2016) in these regions with limited oxygen availability. Indeed, inhibition of glycolysis reprograming by dichloroacetate significantly impairs migration of TAMs in a lung cancer model (Semba et al., 2016). Depending on tumor size and the state of tumor progression, hypoxia-associated TAM can comprise 50% or more of the entire TAM population.

Hypoxia induces glycolytic programs through stabilization of hypoxia-inducible factor (HIF)-1α in most cells (Kierans and Taylor, 2021), including TAMs (Wenes et al., 2016). HIF-1α is also the major driver of the immunosuppressive and angiogenic phenotype of TAMs (Corzo et al., 2010; Doedens et al., 2010), indicating that TAM metabolic activity is intricately connected with their function. While hypoxia alone promotes glycolysis in macrophages in vitro, the phenotype is accentuated by the presence of tumor cell-conditioned medium (Wenes et al., 2016). This may relate to lower glucose availability, lower pH, and higher levels of lactic acid, all of which are linked with tumor hypoxia and can potentially regulate the metabolism of TAMs (Fig. 3). Specifically, lactic acid stabilizes HIF-1α and maintains the hypoxia-induced state of TAMs (Colegio, 2015; Colegio et al., 2014). Similarly, succinate and low pH are associated with ischemia and can stabilize HIF-1α (Chouchani et al., 2014; Mills et al., 2016; Tannahill et al., 2013) and both regulate the phenotype of inflammatory macrophages (Liu et al., 2017; Park et al., 2012), as well as TAMs (Colegio et al., 2014; Wu et al., 2020a). Other pathways distinct from HIF-1α stabilization may also be involved in the response to ischemic conditions within tumors. For example, G protein-coupled receptors on TAMs respond to the acidic environment generated by lactate-producing tumor cells (Bohn et al., 2018; Chen et al., 2017).

Thus, hypoxia, catabolic byproducts, cytokines, and other tumor-derived factors all regulate glucose metabolism in TAMs. Hypoxic TAMs show the most pronounced shift towards glycolysis, while still utilizing FAO for half of their total ATP production (Wenes et al., 2016). How the interplay between hypoxia and the other components of the TME control metabolic programs in TAMs remains to be fully elucidated. For example, it is not clear the degree to which ischemic conditions modulate response to cytokines, or how either of these inputs is regulated by macrophage ontogeny. At least one study has already shown that monocyte-derived TAMs in human glioblastomas favor oxidative metabolism compared to tumor-associated microglial cells (Müller et al., 2017).

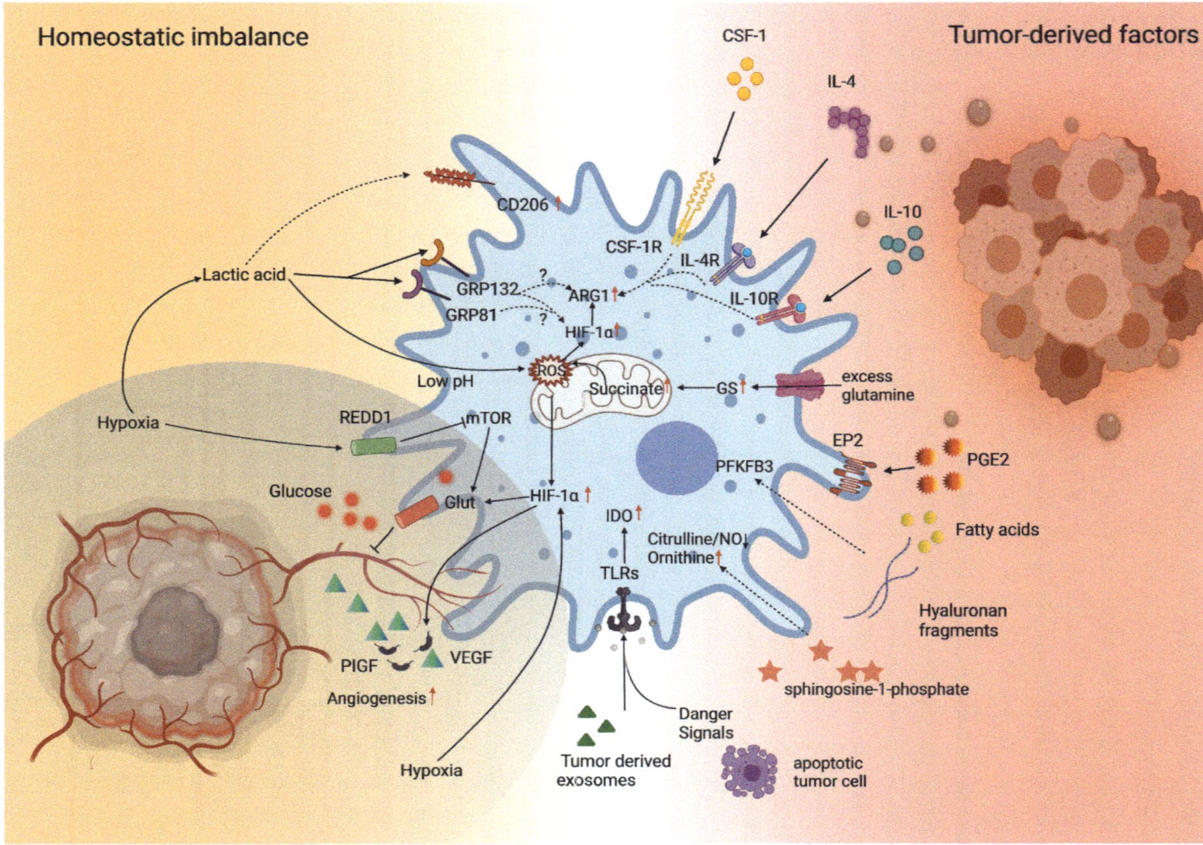

Fig. 3 See figure legend on opposite page.

2.2 Glucose metabolism in TAM function

In addition to promoting neoplastic cell survival, invasion, and extravasation, TAMs are established mediators of angiogenesis and immunosuppression. As mentioned above, hypoxia is an established driver of the angiogenic and immunosuppressive phenotype of TAMs, along with being critical for their glycolytic phenotype. Is the glycolytic shift therefore necessary for the pro-tumor function of TAMs? This is an extremely challenging question to address experimentally as it is difficult to specifically target glycolysis without altering other metabolic pathways in cells. For example, although 2-Deoxy-d-glucose (2-DG) inhibits glycolysis, it also impairs glutaminolysis and the ability of carbon to enter the TCA cycle (Wang et al., 2018). Likewise, using dichloroacetate to promote OXPHOS impairs TAM migration (Semba et al., 2016), impacting the bioactivity of the cells independent of any specific effect on their functional properties. Glycolysis may also be dispensable for macrophage polarization provided that OXPHOS remains intact (Wang et al., 2018), further complicating interpretation of results. Despite these challenges, there is direct and indirect evidence that glucose utilization controls some aspects of TAM function.

Both vascular endothelial growth factor A (VEGFA) and the VEGFA homologue placental growth factor (PlGF) (Du et al., 2008; Rolny et al., 2011) are important mediators of angiogenesis by TAMs. Macrophage-specific depletion of VEGFA limits the amount of abnormal vasculature improves tumor perfusion (Stockmann et al., 2008). As ischemic conditions are major regulators of VEGF-A expression through HIF-1α and HIF-2α (Geeraerts et al., 2017; Werno et al., 2010), and the ability of macrophages to promote angiogenesis during in vitro assays is suppressed by the addition

Fig. 3 Metabolic reprogramming of macrophages in the TME. Factors that regulate metabolism in TAMs can be divided into those related to tissue stress versus those produced by tumor cells or stromal populations. On the left, homoeostatic imbalance includes hypoxia and pseudohypoxia (e.g., lactate, low pH), which act primarily by stabilizing HIF-1α and promoting the expression of target genes such as *Glut1* and *Arg1*. Hypoxia also increases expression of REDD1, an endogenous inhibitor of mTOR, resulting in limitations on glucose consumption that would otherwise result in competition with endothelial cells. On the right, TAM reprogramming is mediated by tumor-produced factors including, but not limited to, cytokines (e.g., IL-4, IL-10, CSF-1), metabolites (e.g., PGE2, lipid sphingosine-1-phosphate, exosomes) and danger signals (e.g., hyaluronan fragments). These factors may be produced by the neoplastic cells or other immune and stromal populations found within the tumor.

of 2-DG (Penny et al., 2016) or rapamycin (Aslam et al., 2017; Chen et al., 2012), an inhibitor of mechanistic target of rapamycin (mTOR), it suggests that a hypoxia-induced glycolytic switch in macrophages is important for VEGF expression. However, this is complicated by the finding that increasing glycolysis in TAMs results in vascular normalization through increased glucose competition with endothelial cells (Wenes et al., 2016). Inhibiting mTOR with the small molecule Torin2 increases abnormal vascularization, tumor hypoxia, and metastasis, whereas the reverse is observed when a negative mTOR regulator is specifically knocked out of macrophages (Wenes et al., 2016). This may reflect the existence of a sweet spot for glucose utilization versus cytokine production in terms of TAM-mediated angiogenesis.

The impact of the glycolytic shift on other TAM functions is even less clear. Cytokine and chemokine production are certainly regulated by metabolic changes in the cells, but whether any of these changes are sufficient to impact tumor growth is unclear. Similarly, while glucose consumption by TAMs could foreseeably support an immunosuppressive TME within hypoxic regions (Thommen and Schumacher, 2018), this remains to be demonstrated. Interestingly, several studies have described that TAMs support aerobic glycolysis by neoplastic cells. Select cytokines promote the synthesis of pro-glycolytic factors such as pyruvate kinase M1/2 (PKM) and lactate dehydrogenase A (LDHA) to promote glycolysis in cancer cells (Jeong et al., 2019; Lin et al., 2014), while TAM-derived extracellular vesicles containing a HIF-1α-stabilizing lncRNA have a similar effect (Chen et al., 2019b). These reciprocal interactions between TAMs and cancer cells promote therapeutic resistance in vitro and in vivo.

3. Amino acid metabolism

Amino acids are not only central building blocks for proteins and nucleotides, but some also serve as precursors for metabolic intermediates essential for several macrophage functions. Indeed, the classification of M1 and M2 macrophages is based upon differential usage of L-arginine via arginase-1 (ARG1) or inducible nitric oxide synthase (iNOS). In addition to the function of specific metabolites, L-arginine utilization by macrophages is thought to compete with uptake by T cells, contributing to reduced anti-tumor activity (Geiger et al., 2016) (Fig. 4). In contrast, little is known about the importance of glutamine and tryptophan metabolism specifically in TAMs (Orhan et al., 2016), despite the kynurenine pathway

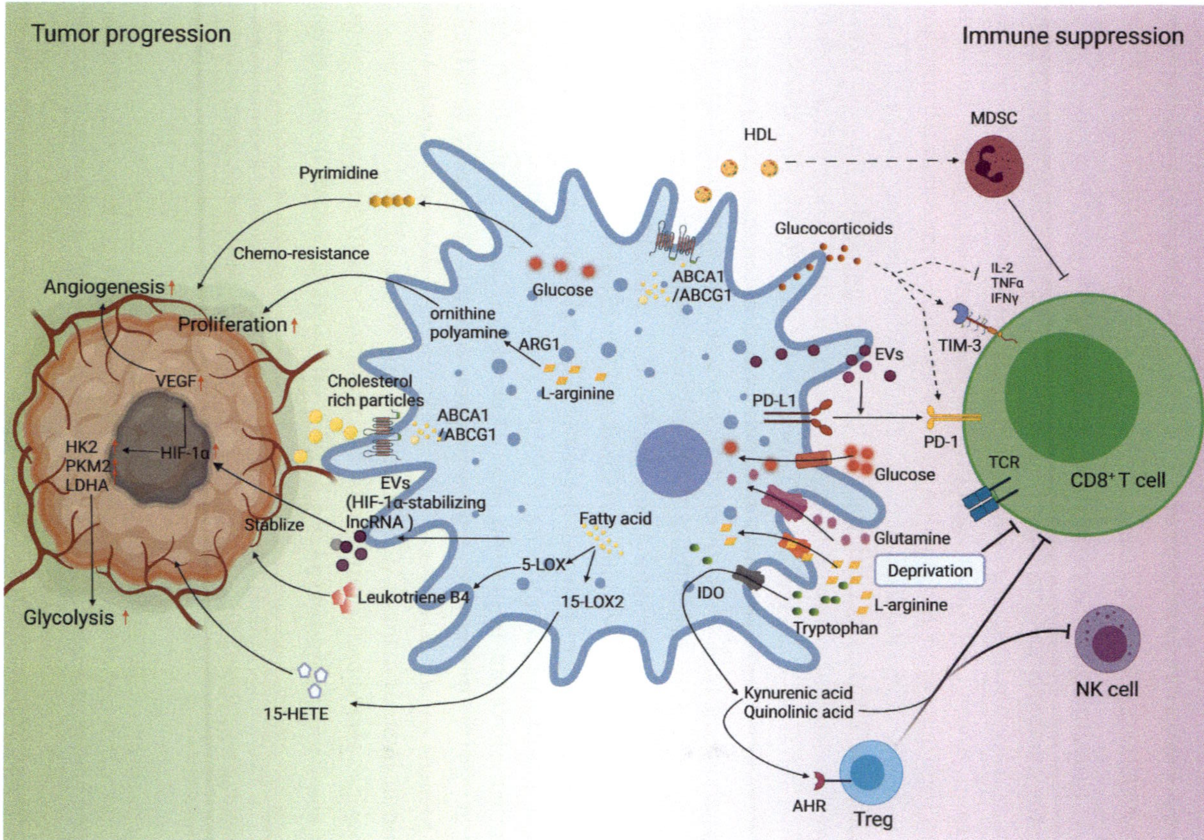

Fig. 4 See figure legend on next page.

having long been a therapeutic target of interest in cancer. Branched-chain amino acids such as leucine also participate in macrophage activation in inflammatory diseases, but their role in TAMs is unknown (Nishiyama et al., 2010; Papathanassiu et al., 2017).

3.1 Regulation of amino acid metabolism

As the quintessential genes marking macrophage polarization status in mice, much is known regarding how cytokines, damage-associated molecular patterns (DAMPs), and other signs of stress regulate L-arginine metabolism in macrophages through induction of *Arg1* and *Nos2* expression (Bronte and Zanovello, 2005; Rodriguez et al., 2017). While the importance of these factors for TAMs is less well understood, it does appear that multiple signals synergize and/or compete to generate a spectrum of macrophage phenotypes (Hobson-Gutierrez and Carmona-Fontaine, 2018). The absence of *Il4ra* reduces *Arg1* expression in TAMs by twofold in vivo, consistent with the ability of IL-4 to promote *Arg1* expression in vitro (DeNardo et al., 2009). However, neutralizing IL-4 does not produce the same effect, possibly due to an overlapping role for IL-13 in the MMTV-PyMT model system of mammary carcinoma (Shiao et al., 2015) or limited antibody penetration into ischemic areas of the tumor (Lee and Tannock, 2010). Blocking or inhibiting the receptors for IL-10 or colony-stimulating factor (CSF)-1 similarly have no impact on *Arg1* expression (Ruffell et al., 2014; Zhu et al., 2014). In contrast, *Arg1* expression is highly responsive to HIF-1α (Doedens et al., 2010) and ARG1-expressing TAMs are mostly found within hypoxic regions (Carmona-Fontaine et al., 2017). Lactic acid and low pH also directly impact *Arg1* expression by TAMs *in vivo* (Colegio et al., 2014; El-Kenawi et al., 2019), with lactic acid further augmenting

Fig. 4 TAM metabolism shapes the TME. On the left, amino acid and lipid metabolites, along with extracellular vesicles produced by TAMs, regulate tumor proliferation and angiogenesis. Overexpression of ARG1 in TAMs enhances ornithine and polyamine production, which promote tumor cell growth. Extracellular vesicle-packaged HIF-1α-stabilizing lncRNA from TAMs promotes glycolysis and angiogenesis of cancer cells by upregulating glycolytic factors (HK2, PKM2, LDHA) and VEGF. On the right, TAM metabolic activity can suppress T cell function through several distinct mechanisms. Through the uptake of glucose and amino acids TAMs are capable of depriving T cells of nutrients necessary for their proliferation, while the synthesis of glucocorticoids directly suppresses T cell function. TAMs can also regulate the immunosuppressive TME by promoting the activity of other immune populations, such as the conversion of inducible regulatory T cells (Treg) through IDO-dependent production of kynurenic acid.

Arg1 expression under hypoxic conditions (Carmona-Fontaine et al., 2017). Factors that promote expression of *Arg1* often simultaneously limit expression of *Nos2*, including IL-4 and an acidic pH, consistent with the in vitro polarization dichotomy. However, since IFN-γ and CD40 ligand are major drivers of *Nos2* expression (Marigo et al., 2016), it is possible that the in vivo relevance of these factors is due to secondary effects of a suppressed T cell response. It has also been shown that hypoxia promotes *Nos2* expression *in vitro* (Carmona-Fontaine et al., 2017). Thus, in addition to increased iNOS expression around tumor vasculature due to interactions with IFN-γ expressing leukocytes, there exists the potential for dual $iNOS^+ARG1^+$ expressing TAMs within hypoxic regions.

Glutamine is the most abundant amino acid in the blood and is preferentially taken up by cancer cells compared to immune cells within the TME (Reinfeld et al., 2021). Hypoxic TAMs do not utilize glutamine for ATP generation (Wenes et al., 2016); instead, glutamine contributes to UDP-*N*-acetylglucosamine (UDP-GlcNAc) synthesis, which is required for the initiation of N-glycosylation in the endoplasmic reticulum and branching reactions in the Golgi to support protein folding and trafficking (Dennis et al., 2009; Kornfeld and Kornfeld, 1985). This includes the glycosylation moieties found on the lectin and mannose receptors typically used as markers for protumor TAMs (Sica and Mantovani, 2012). Consistent with the importance of glutamine for this protumor polarization state, expression of glutamine synthase (GS) is observed during stimulation with IL-4 or IL-10, and inhibiting GS promotes expression of *Tnfa*, *Nos2*, *Cxcl9*, and *Cxcl10* (Palmieri et al., 2017). The effect of GS inhibition is mediated through glutamate shuttling into the TCA cycle and succinate accumulation, which can stabilize HIF-1α (Palmieri et al., 2017). Importantly, loss of GS in *Csf1r*-expressing cells promotes vascular normalization and reduces metastasis (Palmieri et al., 2017). These findings indicate that interruptions in amino acid metabolism can shift the polarization state of TAMs.

3.2 Amino acid metabolism in TAM function

Expression of ARG1 has the potential to regulate different aspects of tumor development through the production of ornithine and polyamine at expense of potential antitumor signals (i.e., nitric oxide or citrulline) (Carmona-Fontaine et al., 2017; Chang et al., 2001); however, it is the immunosuppressive effects of L-arginine depletion that have been the most extensively characterized. L-arginine is required for expression of the T cell receptor complex, proliferation, and memory establishment (Geiger et al., 2016; Popovic et al., 2007; Rath et al., 2014). Inhibition of ARG1 reduces tumor growth and macrophage-specific *Arg1*-deficiency improves response to T cell adoptive transfer (Marigo et al., 2016; Rodriguez et al., 2004). Similarly, increasing L-arginine levels induces a metabolic shift toward OXPHOS in activated T cells and promotes the generation and survival of central memory-like T cells in B16-OVA melanoma tumors (Geiger et al., 2016). Although ARG1 is not critical for the suppressive function of TAMs *ex vivo* (Marigo et al., 2016) this may simply reflect the high concentration of L-arginine normally found in cell culture medium (Rodriguez et al., 2004). That said, the role of ARG1 in humans is not clear, as it is primarily expressed by granulocytes, not macrophages, and inhibiting ARG1 does not reduce TAM suppressive activity *ex vivo* (Kryczek et al., 2006; Munder et al., 2005; Zea et al., 2005).

L-arginine metabolism through iNOS can also suppress T cell function, as iNOS inhibition or *Nos2*-deficiency significantly limits the suppressive capacity of TAMs *ex vivo* (Marigo et al., 2016; Movahedi et al., 2010). In addition to L-arginine depletion, nitric oxide can inhibit signal transduction in T cells (Albina and Henry, 1991; Bingisser et al., 1998), or produce peroxynitrites that prevent the interaction of the T cell receptor with MHC (Lu et al., 2011; Nagaraj et al., 2007). However, a suppressive role for iNOS is at odds with the general concept of M1-like macrophages promoting an anti-tumoral immune response. Indeed, *Nos2*-deficiency in myeloid cells diminishes the efficacy of cellular therapy (Marigo et al., 2016) and $iNOS^+$ TAMs promote a $CD8^+$ T cell response in vivo following stimulation with IFN-γ and monophosphoryl lipid A (Sun et al., 2021). These differences could reflect a context specific role for nitric oxide production. For example, during low-dose irradiation, iNOS-expressing macrophages induce vascular cell adhesion protein (VCAM)-1 expression on endothelial cells, leading to enhanced recruitment of $CD8^+$ T cells (Klug et al., 2013). Nitric oxide is also necessary for promoting aerobic glycolysis and glutamine anaplerosis in macrophages upon LPS stimulation (Palmieri et al., 2020). As a result, *Nos2*-deficient macrophages show increased expression of cytokines that can promote tumor progression, including IL-6, IL-1β and CCL2 (Palmieri et al., 2020). Thus, it may be that expression of iNOS is central for limiting the pro-tumoral phenotype of TAMs, especially in hypoxic populations where ARG1 and iNOS may be co-expressed.

Indoleamine 2,3-dioxygenase (IDO)-1 is the enzyme responsible for converting tryptophan to kynurenine, and expression of IDO-1 by tumor cells has long been linked to a immunosuppressive phenotype

(Meireson et al., 2020). Although the IDO-1 inhibitor epacadostat failed in a recent clinical trial in melanoma, it has been suggested this is due to a limitation of the trial design or a lack of activity against tryptophan 2,3-dioxygenase (TDO), which catalyzes the same reaction (Van Eynde et al., 2019). Alternatively, IDO-1-mediated depletion of tryptophan has been shown to cause ribosomal frameshifting events that contribute to the production of aberrant peptides, thereby actually promoting anti-tumor immunity (Bartok et al., 2021). IDO-1 expression has also been described in several different myeloid populations in cancer, including macrophages (Munn et al., 2005; Platten et al., 2015; Wang et al., 2014; Zhao et al., 2012), with expression induced through IFN-γ, CD40 ligands or toll-like receptor activation (Orhan et al., 2016; Platten et al., 2015; Werner et al., 1989). Tryptophan catabolism by IDO-1 in macrophages and monocytes suppress T cells in vitro, inhibiting proliferation, cytokine production, and cytotoxic function by down regulating the TCR ζ chain (Fallarino et al., 2006; Zhao et al., 2012). Several kynurenine related metabolites, such as quinolinic acid, can also act as ligands for immune modulatory receptors (Mándi and Vécsei, 2012; Orhan et al., 2016). However, other studies have not observed *Ido1* expression in murine macrophages, with only limited expression in dendritic cells (Ruffell et al., 2014). Human scRNAseq data also shows *IDO1* expression primarily in tumor dendritic cells. This may reflect differences in the extent of IFN-γ exposure among the different populations, as macrophages with an interferon response signature show detectable *IDO1* expression (Cheng et al., 2021). These data hint that a subset of IDO1-expressing macrophages may be involved in regulating the anti-tumor response, but definitive evidence for this remains lacking.

4. Lipid metabolism in TAM

Macrophages play a central role in local and systemic lipid homeostasis, as well as the pathology of lipid-associated diseases such as atherosclerosis. Macrophages acquire exogenous lipids through phagocytosis, micropinocytosis, or scavenger receptors such as CD36 that recognize lipid particles (e.g., VLDL, LDL, oxidized lipoproteins) (Arandjelovic and Ravichandran, 2015; Remmerie and Scott, 2018; Tabas, 2010). Lipids are then degraded in the lysosome into free cholesterol and fatty acids. Fatty acids can enter the TCA cycle following oxidation, with in vitro studies showing that M2 macrophages favor FAO, while M1 macrophages engage in increased fatty acid synthesis (FAS) (Odegaard et al., 2007; Vats et al., 2006). In contrast, cholesterol is either converted into cholesterol esters for storage as lipid droplets, or effluxed from the plasma membrane through diffusion or active export via the ATP-binding cassette subfamily A member 1 (ABCA1) and ABCG1 transporters. ABCA1 and ABCG1 transfer cholesterol to lipid poor apolipoproteins, contributing to HDL formation and reverse cholesterol transport (RCT) to the liver for lipid recycling and clearance (Arandjelovic and Ravichandran, 2015; Tabas, 2010). Critically, TAMs have substantially higher levels of intracellular lipids than macrophages in normal tissue and share some morphological properties with the foamy macrophages found in atherosclerotic plaques (Conza et al., 2021; Donadon et al., 2020; El-Kenawi et al., 2021; Su et al., 2020; Wu et al., 2019). Cholesterol and fatty acids also active nuclear receptors that have demonstrated roles in the protumoral phenotype of TAMs, as well as serve as precursors for the generation of steroid hormones and eicosanoids, respectively (Chawla, 2010; Huang et al., 2014; Kratz et al., 2014).

4.1 Fatty acid metabolism

The primary regulator of lipid metabolism in macrophages is the accumulation of the lipids themselves, reflecting a feedback mechanism for macrophages to efficiently respond to lipid buildup. The family of peroxisome proliferator-activated receptors (PPARs) recognize fatty acids and form heterodimers with the retinoid X receptors (RXRs) to regulate the expression of genes that control lipid uptake, metabolism, and efflux. Many of these genes are also regulated by CCAAT enhancer binding proteins (C/EBPs) and sterol regulatory element binding proteins (SREBPs) (Park et al., 2018; Remmerie and Scott, 2018; Schulman, 2017). In particular, SREBP is induced by CSF1 and is critical for normal macrophage differentiation through promotion of de novo fatty acid synthesis (Ecker et al., 2010). Active $CD8^+$ T cell responses reduce immune suppression in the TME via IFN-γ suppression of SREBP in TAMs, which reduces FAS, mitochondrial integrity, and cell survival (Liu et al., 2019a). Increased FAS (Liu et al., 2019a), reduced FAO (Xiang et al., 2018), and lipid scavenging through the CD36 receptor (Al-Khami et al., 2017; Su et al., 2020) have all been shown to significantly contribute to lipid accumulation in TAMs. This phenotype is likely driven at least in part by hypoxic conditions in the TME, similar to findings in atherosclerosis (Crucet et al., 2013). Interfering with any of these pathways normalizes the lipid burden in TAMs and reduces their pro-tumorigenic functions, while also increasing their potential to promote anti-tumor immunity (Goossens et al., 2019; Liu et al., 2019b; Rabold et al., 2020).

Although lipid accumulation is a hallmark of TAMs, and interfering with this accumulation reduces tumor growth, the mechanisms through which this occurs are poorly defined. PPARα, PPARβ/δ, and PPARγ, all of which have been reported to be expressed in TAMs (Schumann et al., 2015; Wu et al., 2020b), regulate overlapping and unique metabolic programs in response to fatty acids (Cariello et al., 2021). For example, PPARα promotes FAO, PPARγ increases FAS, while both enhance lipid uptake through CD36 expression. PPAR activation could thus perpetuate the "foamy" phenotype of TAMs unable to restore homeostasis in the TME. TAM functionality can be directly regulated by PPAR-induced gene expression, with CSF-1 promoting FAS, leading to PPARβ activation and expression of IL-10 (Park et al., 2015). Increased FAO or reduced expression of cytokines such as TNF-α could also potentially support pro-tumor activity. In support of the former, administration of the CPT1a inhibitor etomoxir reduces TAM infiltration and ARG1 expression (Niu et al., 2017; Wu et al., 2020b). However, since etomoxir does not mimic the genetic loss of the FAO pathway, this may reflect an artifact of this specific inhibitor (Nomura et al., 2016). In addition, cleavage of PPARγ by caspase-1 in TAMs has been shown to both inhibit and promote FAO, with both results leading to increased tumor growth (Niu et al., 2017; Wu et al., 2020b). Part of this complexity may arise because—although the PPAR family members have different ligand binding properties—knocking out a single family member would invariably impact the activity of the others. Specific fatty acids or their related metabolites could also have unique effects on TAM function, with arachidonic acid, linoleic acid and palmitic acid all shown to promote a pro-tumor phenotype (Linton et al., 2018). Beyond potentially activating PPARs, altered lipid composition in the endoplasmic reticulum of TAMs has recently been shown to induce an XBP1-mediated stress response that supports their survival (Conza et al., 2021).

4.2 Arachidonic acid metabolism

In addition to lipid accumulation altering the phenotype and survival of TAMs, arachidonic acid is the key precursor for eicosanoids, such as prostaglandins E_2 (PGE2) and leukotrienes, as well as other products of lipoxygenases (LOX) and cyclooxygenases (COXs) (Zheng et al., 2020b). High levels of arachidonic acid are detected in tumors, with the expression of genes related to arachidonic acid and cholesterol metabolism correlated with immune infiltration, suggesting the potential for arachidonic acid metabolism to regulate crosstalk between cells in the TME (Hao et al., 2019).

In a lung metastasis model, alveolar macrophages were found to express 5-LOX and produce leukotriene B4, with either 5-LOX inhibition or clodronate liposome depletion of macrophages reducing the proliferation of metastatic cells (Nosaka et al., 2018). Similarly, TAMs isolated from renal cell carcinoma highly express 15-LOX2 and produce 15(*S*)-hydroxyeicosatetraenoic acid (15-HETE). Although the in vivo relevance of this is not known, a pan-LOX inhibitor reduced TAM expression of IL-10 and CCL2 *ex vivo* (Daurkin et al., 2011), both of which promote an immunosuppressive TME.

Tumor production of PGE2, formed from arachidonic acid by COXs and PGE2 synthases, can regulate cancer cell proliferation, chemoresistance, angiogenesis, and immune suppression (Liu et al., 2015). For example, PGE2 interferes with the ability of NK cells to recruit type I conventional dendritic cells (Böttcher et al., 2018) and generates an immunoregulatory phenotype in TAMs (Olesch et al., 2015; Prima et al., 2017). COX-2 expression is also observed in macrophages, with expression induced by cancer cells and COX-2-expressing TAMs found in both human and murine tumors (Bianchini et al., 2007; Gan et al., 2016; Li et al., 2015). In vitro functional assays suggest that COX-2 expressing TAMs can regulate a number of tumorigenic properties, including angiogenesis, invasion and survival (Bianchini et al., 2007; Gan et al., 2016; Li et al., 2015); however, *Ptgs2* depletion in myeloid cells does not impact UVB-induced skin cancer (Jiao et al., 2014). TAM-derived extracellular vesicles also show no change in COX2 or PGE2, but rather higher levels of the eicosanoid thromboxane B_2 (Cianciaruso et al., 2019). Thus, whether COX2 expression by TAMs has a role in advanced tumors remains unclear.

4.3 Cholesterol metabolism

Although macrophages are critical for normal cholesterol metabolism, the ability of TAMs to regulate cholesterol and steroid levels in tumors is only beginning to emerge. The loss of either the ABCA1 or ABCG1 transporters in myeloid cells significantly reduces the growth of subcutaneous B16 and MB49 tumors (Sag et al., 2015; Zamanian-Daryoush et al., 2014). Importantly, this phenotype is not observed in mice on a normal diet, but rather emerges when mice are fed a Western-like diet, or when the mice are aged to the point of spontaneously developing cholesterol-laden macrophages (Baldán et al., 2006). ABCA1/ABCG1-deficient macrophages show enhanced sensitivity to IFN-γ—and reduced sensitivity to IL-4—due to the buildup of lipid rafts that differentially regulate the activity of the corresponding receptors (Goossens et al., 2019; Sag et al., 2015). Conversely, tumor-derived factors, such as hyaluronic acid, induce membrane cholesterol efflux by macrophages, reinforcing the expression of genes associated with M2 polarization (Goossens et al., 2019). In addition to altering cytokine receptor activity, cholesterol accumulation directly regulates gene expression through activation of the Liver X Receptors (LXRs), which recognize oxygenated derivatives of cholesterol and regulate cholesterol metabolism in conjunction with RXRs (Wang and Tontonoz, 2018). In support of a functional role for this pathway, the LXR agonist TO901317 reduces expression of IRF4-dependent gene expression in TAMs and limits recruitment of regulatory T cells (Carbó et al., 2021), while the LXR agonist GW3965 reduces the survival of TAMs (Carbó et al., 2021) and myeloid-derived suppressor cells *in vivo* (Tavazoie et al., 2018). Thus, excessive cholesterol accumulation regulates gene expression and apoptosis in TAMs, leading to the accumulation of cytotoxic lymphocytes (Goossens et al., 2019; Sag et al., 2015).

Beyond regulating the phenotype of TAMs, cholesterol is a precursor for glucocorticoid production, which promotes T cell dysfunction and resistance to immunotherapy in the MC38 colon carcinoma model (Acharya et al., 2020). TAMs may also act as a source of cholesterol to "feed" cancer cells (Cianciaruso et al., 2019; Peres et al., 1997, 2003; Rabold et al., 2020) similar to the recent description of neutrophils providing triglycerides and their derivatives to metastatic breast cancer cells (Li et al., 2020a). Consistent with this, we have recently shown that macrophages have the capacity to transfer cholesterol, an anabolic precursor for androgens, to prostate cancer cells. Depleting TAMs therefore reduces the level of intratumoral dihydrotestosterone and enhances response to androgen-deprivation therapy (El-Kenawi et al., 2021). It remains to be determined if there are specific targets in TAMs that can be used to regulate cholesterol metabolism in this population, or if there is a viable therapeutic window for molecules such as the LXR agonists that impact multiple organs and cell types.

5. Targeting TAM metabolism for therapy

Due to the prominent role of TAMs in generating the immunosuppressive TME, in addition to the ability of TAMs to regulate angiogenesis and cancer cell survival across a range of tumor models, there remains an intense interest in developing therapeutics that either deplete or repolarize TAMs (Coussens et al., 2013; DeNardo and Ruffell, 2019; Guerriero et al., 2017;

Mazzieri et al., 2011; Palucka and Coussens, 2016; Sanford et al., 2013; Varner et al., 2017; Vonderheide, 2018). Interfering with the CSF1-CSF1R, CCL2-CCR2, or CXCL12-CXCR4 axes have been shown to functionally deplete TAMs by targeting their recruitment and survival within tumors (Bonapace et al., 2014; DeNardo et al., 2011; Li et al., 2019; Papadopoulos et al., 2017; Raschioni et al., 2018). Agents that block the pro-tumor polarization state of TAMs, such as histone deacetylase (HDAC) inhibitors (Bonapace et al., 2014; Papadopoulos et al., 2017) and phosphoinositide 3-kinase (PI3K)-γ inhibitors (Kaneda et al., 2016), or that directly stimulate TAMs, such as CD40 agonists (Vonderheide, 2018), have also demonstrated pre-clinical efficacy. It has been hypothesized that harnessing the potential of TAMs to phagocytose tumors cells or promote a cytotoxic lymphocyte response will prove more efficacious than simply depleting TAMs. That said, combinatorial approaches such as chemotherapy and immune checkpoint blockade are usually required for macrophage-targeted agents to demonstrate efficacy, and it remains to be determined which of these approaches will prove clinically viable.

Since TAM metabolism is inextricably connected to the functionality of the cells, metabolic reprogramming might be an elegant way to repolarize TAMs toward an anti-tumoral phenotype. It is possible that the metabolic programs in hypoxic TAMs would allow for agents with broadly expressed targets to act selectively within TAMs. Thus, agents that target glucose metabolism by inhibiting glycolysis (e.g., 2-DG, Rapamycin, Temsirolimus, Monocarboxylate transporter 1 inhibitors) (Arts et al., 2016; Aslam et al., 2017; Chen et al., 2012; Li et al., 2020b; Penny et al., 2016; Wang et al., 2018; Wenes et al., 2016; Zhang and Li, 2020) or OXPHOS (e.g., Metformin, IACS-010759) (Ding et al., 2015; Incio et al., 2015; Uehara et al., 2019; Umar et al., 2021); modulate amino acids by reducing cycling arginine (e.g., PEG-BCT-100, AEB-1102) (Diaz et al., 2021; Mussai et al., 2015; Santo et al., 2018) or inhibit tryptophan catabolism (e.g., Epacodostat, Indoximod) (Munn et al., 2005; Platten et al., 2015; Wainwright et al., 2012; Wang et al., 2014; Zhao et al., 2012); or reprogram lipid metabolism by blocking FAO (e.g., Etomoxir) (Su et al., 2020; Wu et al., 2019) or inhibiting PGE2 synthesis (e.g., Celecoxib) (Bianchini et al., 2007; Poczobutt et al., 2016), could function at least in part through regulation of TAM metabolism.

Alternatively, since utilizing pathways shared by most cell types would make preferential targeting of TAMs extremely challenging, the specialized role for macrophages in maintaining tissue homeostasis, such as with cholesterol, may offer unique opportunities for selective targeting (Table 1).

Table 1 Agents with the potential to target TAM metabolism.

Target	Drugs	Effects on TAMs	Diseases	Clinical trials	References
Glucose metabolism					
Hexokinase-2	2-Deoxyglucose (DG)	Inhibit glycolysis and protumor function in TAMs	Cancer	NCT00633087 NCT00096707	Penny et al. (2016), Wang et al. (2018)
AMPK	Metformin	Inhibit OXPHOS and reprogram TAMs	Type II diabetes Solid tumors	Approved for type II diabetes NCT02437656 NCT04758000	Ding et al. (2015), Incio et al. (2015), Uehara et al. (2019)
mTOR	Rapamycin Temsirolimus	Inhibit glycolysis and repolarize TAMs	Solid tumors	NCT00919035 NCT00411788 NCT02642094	Arts et al. (2016), Aslam et al. (2017), Chen et al. (2012), Wenes et al. (2016)
MCT1	AZD3965	Inhibit lactate transport and repolarize TAMs	Advance tumor	NCT01791595	Li et al. (2020b), Zhang and Li (2020)
Mitochondrial complex I	IACS-010759	Inhibit OXPHOS and reduce cytokines production in macrophage	Advanced solid and hematopoietic tumor	NCT03291938	Umar et al. (2021)
Amino acid metabolism					
Arginine	PEG-BCT-100 (ADI-PEG20), AEB-1102	Reduce cycling arginine and inhibit arginine uptake by TAMs	Acute myeloid leukemia Myelodysplastic syndrome Hyperargininemia Solid tumors	NCT02561234 NCT02488044 NCT01092091 NCT03455140	Diaz et al. (2021), Mussai et al. (2015), Santo et al. (2018)

Continued

Table 1 Agents with the potential to target TAM metabolism.—cont'd

Target	Drugs	Effects on TAMs	Diseases	Clinical trials	References
IDO	Indoximod Epacadostat BMS-986205	Inhibit tryptophan catabolism in TAMs	Solid tumors	NCT04049669 NCT03493945 NCT03463161 NCT03459222 NCT03854032	Munn et al. (2005) Platten et al. (2015), Wainwright et al. (2012), Wang et al. (2014), Zhao et al. (2012)
Lipid metabolism					
TREM2	PY314	Lipid metabolic reprogramming and TAM depletion	Alzheimers disease Advanced solid tumors	NCT04691375	Katzenelenbogen et al. (2020), Molgora et al. (2020), Nakamura and Smyth (2020)
PARP1/2	Olaparib	TAM reprogramming via the SREBP1	Solid tumors	Approved for breast and ovarian cancer NCT03786796 NCT02338622	Mehta et al. (2020)
CD36	VT1021	Block uptake of oxidized lipids in TAMs	Advanced solid tumors	Active, not recruiting NCT03364400	Huang et al. (2014), Su et al. (2020)
PARP14	RBN012759	Reduce TAMs response to FAO and repolarize TAMs	Non-small cell lung cancer		Schenkel et al. (2021)
CPT-1	Etomoxir	Inhibit FAO and repolarize TAMs	Type II diabetes heart failure	Terminated due to severe hepatotoxicity	Su et al. (2020), Wu et al. (2019)

COX-2	Celecoxib	Inhibit PGE2 synthesis and cytokines production in TAMs	Osteoarthritis, Rheumatoid arthritis Acute pain Melanoma	Approved for Osteoarthritis, Rheumatoid arthritis, Acute pain NCT02429427 NCT02885974	Bianchini et al. (2007), Poczobutt et al. (2016)
HMG-CoA reductase	Simvastatin	Reduce systemic LDL cholesterol; inhibit LXR in TAMs	Non-small cell lung cancer Pancreatic cancer	Approved for reducing low-density lipoprotein cholesterol NCT01099085 NCT03324425	Jin et al. (2019), Mo et al. (2018)
Cathelicidin	Vitamin D LL-37	Enhance cytotoxicity of TAMs	High-grade B cell lymphoma Multiple myeloma	NCT00953225 NCT01631526 NCT02603757	Bruns et al. (2015), Busch et al. (2018)

Red, terminated; Green, recruiting; Black, completed.

For example, triggering receptor expressed on myeloid cells 1/2 (TREM1/2) (Katzenelenbogen et al., 2020; Molgora et al., 2020; Nakamura and Smyth, 2020) and the scavenger receptor CD36 (Huang et al., 2014; Su et al., 2020) regulate lipid metabolism in TAMs and can be targeted to regulate their activation state, although CD36 also promotes oxidized lipid accumulation and dysfunction in T cells (Xu et al., 2021). Other receptors on macrophages may control their functional state by directly stimulating the expression of metabolic genes. This has been shown for v-set immunoglobulin-domain-containing 4 (VSIG4), which enhances PDK2 expression in macrophages and reduces pyruvate metabolism (Huang et al., 2019; Kim et al., 2020; Li et al., 2017; Liao et al., 2014). Thus, any agent that profoundly alters macrophage function would also be expected to modify cellular metabolism, with the impact of the PARP inhibitor Olaparib on glucose and lipid metabolism in TAMs being one recent example (Mehta et al., 2020).

6. Conclusion

Advances in high-dimensional single-cell analyses and metabolic flux have improved our understanding of immunometabolism and brought the potential to identify new therapeutic strategies for cancer (Artyomov and den Bossche, 2020; Zhou et al., 2020). How best to translate these findings remains an open question, especially given the emerging state of the field. There are also several unknowns regarding our basic understanding of macrophage metabolism. Is there a difference in the metabolic state of resident and monocyte-derived macrophage populations? Does adaption to specific TMEs explain the metabolic differences between specific subpopulations of TAMs? How do macrophages metabolically adapt to an array of signals in vivo, and is this adaptation driven by one or more dominant pathways? Finally, are there systemic changes in the body that regulate the function of TAMs? With the overlapping and powerful impact of diet, the microbiome, obesity, and age on cancer incidence, progression, and diagnosis, it seems likely that macrophages are responding to these cues and impacting patient outcomes.

References

Acharya, N., et al., 2020. Endogenous glucocorticoid signaling regulates CD8+ T cell differentiation and development of dysfunction in the tumor microenvironment. Immunity 53, 658–671.e6.

Albina, J.E., Henry, W.L., 1991. Suppression of lymphocyte proliferation through the nitric oxide synthesizing pathway. J. Surg. Res. 50, 403–409.

Al-Khami, A.A., et al., 2017. Exogenous lipid uptake induces metabolic and functional reprogramming of tumor-associated myeloid-derived suppressor cells. Onco. Targets. Ther. 6, e1344804.

Arandjelovic, S., Ravichandran, K.S., 2015. Phagocytosis of apoptotic cells in homeostasis. Nat. Immunol. 16, 907–917.

Arts, R.J.W., et al., 2016. Transcriptional and metabolic reprogramming induce an inflammatory phenotype in non-medullary thyroid carcinoma-induced macrophages. Onco. Targets. Ther. 5, e1229725.

Artyomov, M.N., den Bossche, J.V., 2020. Immunometabolism in the single-cell era. Cell Metab. 32, 710–725.

Aslam, M., et al., 2017. FRI-121 rapamycin and zoledronic acid synergistically repolarize tumour associated macrophages towards anti-HCC responses. J. Hepatol. 66, S465.

Baldán, Á., et al., 2006. Deletion of the transmembrane transporter ABCG1 results in progressive pulmonary lipidosis*. J. Biol. Chem. 281, 29401–29410.

Bartok, O., et al., 2021. Anti-tumour immunity induces aberrant peptide presentation in melanoma. Nature 590, 332–337.

Baseler, W.A., et al., 2016. Autocrine IL-10 functions as a rheostat for M1 macrophage glycolytic commitment by tuning nitric oxide production. Redox Biol. 10, 12–23.

Bianchini, F., et al., 2007. Expression of cyclo-oxygenase-2 in macrophages associated with cutaneous melanoma at different stages of progression. Prostag. Oth. Lipid M. 83, 320–328.

Bingisser, R.M., Tilbrook, P.A., Holt, P.G., Kees, U.R., 1998. Macrophage-derived nitric oxide regulates T cell activation via reversible disruption of the Jak3/STAT5 signaling pathway. J. Immunol. 1950 (160), 5729–5734.

Bohn, T., et al., 2018. Tumor immunoevasion via acidosis-dependent induction of regulatory tumor-associated macrophages. Nat. Immunol. 19, 1319–1329.

Bonapace, L., et al., 2014. Cessation of CCL2 inhibition accelerates breast cancer metastasis by promoting angiogenesis. Nature 515, 130–133.

Böttcher, J.P., et al., 2018. NK cells stimulate recruitment of cDC1 into the tumor microenvironment promoting cancer immune control. Cell 172, 1022–1037.e14.

Bronte, V., Zanovello, P., 2005. Regulation of immune responses by L-arginine metabolism. Nat. Rev. Immunol. 5, 641–654.

Bruns, H., et al., 2015. Vitamin D–dependent induction of cathelicidin in human macrophages results in cytotoxicity against high-grade B cell lymphoma. Sci. Transl. Med. 7, 282ra47.

Busch, L., et al., 2018. Lenalidomide enhances MOR202-dependent macrophage-mediated effector functions via the vitamin D pathway. Leukemia 32, 2445–2458.

Carbó, J.M., et al., 2021. Pharmacologic activation of LXR alters the expression profile of tumor-associated macrophages and the abundance of regulatory T cells in the tumor microenvironment. Cancer Res. 81, 968–985.

Cariello, M., Piccinin, E., Moschetta, A., 2021. Transcriptional regulation of metabolic pathways via lipid-sensing nuclear receptors. Cell. Mol. Gastroenterol. Hepatol. 11, 1519–1539.

Carmona-Fontaine, C., et al., 2017. Metabolic origins of spatial organization in the tumor microenvironment. Proc. Natl. Acad. Sci. U. S. A. 114, 2934–2939.

Casanova-Acebes, M., et al., 2021. Tissue-resident macrophages provide a pro-tumorigenic niche to early NSCLC cells. Nature 1–7. https://doi.org/10.1038/s41586-021-03651-8.

Casazza, A., et al., 2013. Impeding macrophage entry into hypoxic tumor areas by Sema3A/Nrp1 signaling blockade inhibits angiogenesis and restores antitumor immunity. Cancer Cell 24, 695–709.

Chang, C.I., Liao, J.C., Kuo, L., 2001. Macrophage arginase promotes tumor cell growth and suppresses nitric oxide-mediated tumor cytotoxicity. Cancer Res. 61, 1100–1106.

Chang, C.-H., et al., 2015. Metabolic competition in the tumor microenvironment is a driver of cancer progression. Cell 162, 1229–1241.
Chawla, A., 2010. Control of macrophage activation and function by PPARs. Circ. Res. 106, 1559–1569.
Chen, W., et al., 2012. Macrophage-induced tumor angiogenesis is regulated by the TSC2–mTOR pathway. Cancer Res. 72, 1363–1372.
Chen, P., et al., 2017. Gpr132 sensing of lactate mediates tumor–macrophage interplay to promote breast cancer metastasis. Proc. Natl. Acad. Sci. U. S. A. 114, 580–585.
Chen, D.-P., et al., 2019a. Glycolytic activation of peritumoral monocytes fosters immune privilege via the PFKFB3-PD-L1 axis in human hepatocellular carcinoma. J. Hepatol. 71, 333–343.
Chen, F., et al., 2019b. Extracellular vesicle-packaged HIF-1α-stabilizing lncRNA from tumour-associated macrophages regulates aerobic glycolysis of breast cancer cells. Nat. Cell Biol. 21, 498–510.
Cheng, S., et al., 2021. A pan-cancer single-cell transcriptional atlas of tumor infiltrating myeloid cells. Cell 184, 792–809.e23.
Chouchani, E.T., et al., 2014. Ischaemic accumulation of succinate controls reperfusion injury through mitochondrial ROS. Nature 515, 431–435.
Chow, A., et al., 2021. Tim-4+ cavity-resident macrophages impair anti-tumor CD8+ T cell immunity. Cancer Cell. https://doi.org/10.1016/j.ccell.2021.05.006.
Cianciaruso, C., et al., 2019. Molecular profiling and functional analysis of macrophage-derived tumor extracellular vesicles. Cell Rep. 27, 3062–3080.e11.
Colegio, O.R., 2015. Lactic acid polarizes macrophages to a tumor-promoting state. Onco. Targets. Ther. 5, e1014774.
Colegio, O.R., et al., 2014. Functional polarization of tumour-associated macrophages by tumour-derived lactic acid. Nature 513, 559–563.
Conza, G.D., et al., 2021. Tumor-induced reshuffling of lipid composition on the endoplasmic reticulum membrane sustains macrophage survival and pro-tumorigenic activity. Nat. Immunol. 22, 1403–1415.
Cordes, T., et al., 2016. Immunoresponsive gene 1 and itaconate inhibit succinate dehydrogenase to modulate intracellular succinate levels. J. Biol. Chem. 291, 14274–14284.
Corzo, C.A., et al., 2010. HIF-1α regulates function and differentiation of myeloid-derived suppressor cells in the tumor microenvironment. J. Exp. Med. 207, 2439–2453.
Coussens, L.M., Zitvogel, L., Palucka, A.K., 2013. Neutralizing tumor-promoting chronic inflammation: a magic bullet? Science 339, 286–291.
Crucet, M., et al., 2013. Hypoxia enhances lipid uptake in macrophages: role of the scavenger receptors Lox1, SRA, and CD36. Atherosclerosis 229, 110–117.
Daurkin, I., et al., 2011. Tumor-associated macrophages mediate immunosuppression in the renal cancer microenvironment by activating the 15-Lipoxygenase-2 pathway. Cancer Res. 71, 6400–6409.
de Brito, N.M., et al., 2020. Aerobic glycolysis is a metabolic requirement to maintain the M2-like polarization of tumor-associated macrophages. Biochim. Biophys. Acta Mol. Cell Res. 1867, 118604.
den Bossche, J.V., Saraber, D.L., 2018. Metabolic regulation of macrophages in tissues. Cell. Immunol. 330, 54–59.
DeNardo, D.G., Ruffell, B., 2019. Macrophages as regulators of tumour immunity and immunotherapy. Nat. Rev. Immunol. 19, 369–382.
DeNardo, D.G., et al., 2009. CD4+ T cells regulate pulmonary metastasis of mammary carcinomas by enhancing protumor properties of macrophages. Cancer Cell 16, 91–102.
DeNardo, D.G., et al., 2011. Leukocyte complexity predicts breast cancer survival and functionally regulates response to chemotherapy. Cancer Discov. 1, 54–67.

Dennis, J.W., Nabi, I.R., Demetriou, M., 2009. Metabolism, cell surface organization, and disease. Cell 139, 1229–1241.

Diaz, G.A., et al., 2021. Clinical effect and safety profile of pegzilarginase in patients with arginase 1 deficiency. J. Inherit. Metab. Dis. 44, 847–856.

Ding, L., et al., 2015. Metformin prevents cancer metastasis by inhibiting M2-like polarization of tumor associated macrophages. Oncotarget 6, 36441–36455.

Doedens, A.L., et al., 2010. Macrophage expression of hypoxia-inducible factor-1α suppresses T-cell function and promotes tumor progression. Cancer Res. 70, 7465–7475.

Donadon, M., et al., 2020. Macrophage morphology correlates with single-cell diversity and prognosis in colorectal liver metastasis. J. Exp. Med. 217.

Du, R., et al., 2008. HIF1α induces the recruitment of bone marrow-derived vascular modulatory cells to regulate tumor angiogenesis and invasion. Cancer Cell 13, 206–220.

Ecker, J., et al., 2010. Induction of fatty acid synthesis is a key requirement for phagocytic differentiation of human monocytes. Proc. Natl. Acad. Sci. U. S. A. 107, 7817–7822.

Elia, I., Haigis, M.C., 2021. Metabolites and the tumour microenvironment: from cellular mechanisms to systemic metabolism. Nat. Metab. 3, 21–32.

El-Kenawi, A., et al., 2019. Acidity promotes tumour progression by altering macrophage phenotype in prostate cancer. Br. J. Cancer 121, 556–566.

El-Kenawi, A., et al., 2021. Macrophage-derived cholesterol contributes to therapeutic resistance in prostate cancer. Cancer Res. https://doi.org/10.1158/0008-5472.can-20-4028.

Eming, S.A., Murray, P.J., Pearce, E.J., 2021. Metabolic orchestration of the wound healing response. Cell Metab. 33, 1726–1743.

Fallarino, F., et al., 2006. The combined effects of tryptophan starvation and tryptophan catabolites Down-regulate T cell receptor ζ-chain and induce a regulatory phenotype in naive T cells. J. Immunol. 176, 6752–6761.

Finger, E.C., Giaccia, A.J., 2010. Hypoxia, inflammation, and the tumor microenvironment in metastatic disease. Cancer Metastasis Rev. 29, 285–293.

Freemerman, A.J., et al., 2014. Metabolic reprogramming of macrophages glucose transporter 1 (GLUT1)-mediated glucose metabolism drives a proinflammatory phenotype. J. Biol. Chem. 289, 7884–7896.

Fukuzumi, M., Shinomiya, H., Shimizu, Y., Ohishi, K., Utsumi, S., 1996. Endotoxin-induced enhancement of glucose influx into murine peritoneal macrophages via GLUT1. Infect. Immun. 64, 108–112.

Gan, L., et al., 2016. Cyclooxygenase-2 in tumor-associated macrophages promotes metastatic potential of breast cancer cells through Akt pathway. Int. J. Biol. Sci. 12, 1533–1543.

Garrido-Martin, E.M., et al., 2020. M1hot tumor-associated macrophages boost tissue-resident memory T cells infiltration and survival in human lung cancer. J. Immunother. Cancer 8, e000778.

Geeraerts, X., Bolli, E., Fendt, S.-M., Ginderachter, J.A.V., 2017. Macrophage metabolism as therapeutic target for cancer, atherosclerosis, and obesity. Front. Immunol. 8, 289.

Geiger, R., et al., 2016. L-arginine modulates T cell metabolism and enhances survival and anti-tumor activity. Cell 167, 829–842.e13.

Gentles, A.J., et al., 2015. The prognostic landscape of genes and infiltrating immune cells across human cancers. Nat. Med. 21, 938–945.

Goossens, P., et al., 2019. Membrane cholesterol efflux drives tumor-associated macrophage reprogramming and tumor progression. Cell Metab. 29, 1376–1389.e4.

Guerriero, J.L., et al., 2017. Class IIa HDAC inhibition reduces breast tumours and metastases through anti-tumour macrophages. Nature 543, 428–432.

Hao, Y., et al., 2019. Investigation of lipid metabolism dysregulation and the effects on immune microenvironments in pan-cancer using multiple omics data. BMC Bioinform. 20, 195.

Ho, P.-C., Liu, P.-S., 2016. Metabolic communication in tumors: a new layer of immunoregulation for immune evasion. J. Immunother. Cancer 4, 4.
Hobson-Gutierrez, S.A., Carmona-Fontaine, C., 2018. The metabolic axis of macrophage and immune cell polarization. Dis. Model. Mech. 11, dmm034462.
Huang, S.C.-C., et al., 2014. Cell-intrinsic lysosomal lipolysis is essential for alternative activation of macrophages. Nat. Immunol. 15, 846–855.
Huang, X., et al., 2019. VSIG4 mediates transcriptional inhibition of Nlrp3 and Il-1β in macrophages. Sci. Adv. 5, eaau7426.
Incio, J., et al., 2015. Metformin reduces desmoplasia in pancreatic cancer by reprogramming stellate cells and tumor-associated macrophages. PLos One 10, e0141392.
Ip, W.K.E., Hoshi, N., Shouval, D.S., Snapper, S., Medzhitov, R., 2017. Anti-inflammatory effect of IL-10 mediated by metabolic reprogramming of macrophages. Science 356, 513–519.
Jeong, H., et al., 2019. Tumor-associated macrophages enhance tumor hypoxia and aerobic glycolysis. Cancer Res. 79. canres.2545.2018.
Jiao, J., et al., 2014. Cell-type-specific roles for COX-2 in UVB-induced skin cancer. Carcinogenesis 35, 1310–1319.
Jin, H., et al., 2019. Targeting lipid metabolism to overcome EMT-associated drug resistance via integrin β3/FAK pathway and tumor-associated macrophage repolarization using legumain-activatable delivery. Theranostics 9, 265–278.
Kaneda, M.M., et al., 2016. PI3Kγ is a molecular switch that controls immune suppression. Nature 539, 437–442.
Katzenelenbogen, Y., et al., 2020. Coupled scRNA-seq and intracellular protein activity reveal an immunosuppressive role of TREM2 in cancer. Cell 182, 872–885.e19.
Kedia-Mehta, N., Finlay, D.K., 2019. Competition for nutrients and its role in controlling immune responses. Nat. Commun. 10, 2123.
Kierans, S.J., Taylor, C.T., 2021. Regulation of glycolysis by the hypoxia-inducible factor (HIF): implications for cellular physiology. J. Physiol. 599, 23–37.
Kim, H.J., et al., 2020. Abstract 3268: anti-tumor effect of VSIG4 in conversion of immunosuppressive type II macrophage via repolarization. Immunology, 3268. https://doi.org/10.1158/1538-7445.am2020-3268.
Klug, F., et al., 2013. Low-dose irradiation programs macrophage differentiation to an iNOS +/M1 phenotype that orchestrates effective T cell immunotherapy. Cancer Cell 24, 589–602.
Kornfeld, R., Kornfeld, S., 1985. Assembly of asparagine-linked oligosaccharides. Annu. Rev. Biochem. 54, 631–664.
Kratz, M., et al., 2014. Metabolic dysfunction drives a mechanistically distinct proinflammatory phenotype in adipose tissue macrophages. Cell Metab. 20, 614–625.
Kryczek, I., et al., 2006. B7-H4 expression identifies a novel suppressive macrophage population in human ovarian carcinoma. J. Exp. Med. 203, 871–881.
Lampropoulou, V., et al., 2016. Itaconate links inhibition of succinate dehydrogenase with macrophage metabolic remodeling and regulation of inflammation. Cell Metab. 24, 158–166.
Laoui, D., et al., 2014. Tumor hypoxia does not drive differentiation of tumor-associated macrophages but rather fine-tunes the M2-like macrophage population. Cancer Res. 74, 24–30.
Lee, C.M., Tannock, I.F., 2010. The distribution of the therapeutic monoclonal antibodies cetuximab and trastuzumab within solid tumors. BMC Cancer 10, 255.
Li, H., et al., 2015. Cyclooxygenase-2 in tumor-associated macrophages promotes breast cancer cell survival by triggering a positive-feedback loop between macrophages and cancer cells. Oncotarget 6, 29637–29650.

Li, J., et al., 2017. VSIG4 inhibits proinflammatory macrophage activation by reprogramming mitochondrial pyruvate metabolism. Nat. Commun. 8, 1322.
Li, X., et al., 2019. CXCL12/CXCR4 pathway orchestrates CSC-like properties by CAF recruited tumor associated macrophage in OSCC. Exp. Cell Res. 378, 131–138.
Li, P., et al., 2020a. Lung mesenchymal cells elicit lipid storage in neutrophils that fuel breast cancer lung metastasis. Nat. Immunol., 1–12. https://doi.org/10.1038/s41590-020-0783-5.
Li, B., et al., 2020b. Expression of monocarboxylate transporter 1 in immunosuppressive macrophages is associated with the poor prognosis in breast cancer. Front. Oncol. 10, 574787.
Liao, Y., et al., 2014. VSIG4 expression on macrophages facilitates lung cancer development. Lab. Invest. 94, 706–715.
Lin, S., et al., 2014. Lactate-activated macrophages induced aerobic glycolysis and epithelial-mesenchymal transition in breast cancer by regulation of CCL5-CCR5 axis: a positive metabolic feedback loop. Oncotarget 5, 110426–110443.
Linton, S.S., et al., 2018. Tumor-promoting effects of pancreatic cancer cell exosomes on THP-1-derived macrophages. PLos One 13, e0206759.
Liu, B., Qu, L., Yan, S., 2015. Cyclooxygenase-2 promotes tumor growth and suppresses tumor immunity. Cancer Cell Int. 15, 106.
Liu, D., et al., 2016. Comprehensive proteomics analysis reveals metabolic reprogramming of tumor-associated macrophages stimulated by the tumor microenvironment. J. Proteome Res. 16, 288–297.
Liu, P.-S., et al., 2017. α-Ketoglutarate orchestrates macrophage activation through metabolic and epigenetic reprogramming. Nat. Immunol. 18, 985–994.
Liu, C., et al., 2019a. Treg cells promote the SREBP1-dependent metabolic fitness of tumor-promoting macrophages via repression of CD8+ T cell-derived interferon-γ. Immunity 51, 381–397.e6.
Liu, M., et al., 2019b. Metabolic rewiring of macrophages by CpG potentiates clearance of cancer cells and overcomes tumor-expressed CD47-mediated 'don't-eat-me' signal. Nat. Immunol. 20, 265–275.
Lu, T., et al., 2011. Tumor-infiltrating myeloid cells induce tumor cell resistance to cytotoxic T cells in mice. J. Clin. Invest. 121, 4015–4029.
Ma, R.-Y., et al., 2020. Monocyte-derived macrophages promote breast cancer bone metastasis outgrowth. J. Exp. Med. 217.
Mándi, Y., Vécsei, L., 2012. The kynurenine system and immunoregulation. J. Neural Transm. 119, 197–209.
Marigo, I., et al., 2016. T cell cancer therapy requires CD40-CD40L activation of tumor necrosis factor and inducible nitric-oxide-synthase-producing dendritic cells. Cancer Cell 30, 377–390.
Mazzieri, R., et al., 2011. Targeting the ANG2/TIE2 axis inhibits tumor growth and metastasis by impairing angiogenesis and disabling rebounds of proangiogenic myeloid cells. Cancer Cell 19, 512–526.
McFadden, B.A., Purohit, S., 1977. Itaconate, an isocitrate lyase-directed inhibitor in *Pseudomonas indigofera*. J. Bacteriol. 131, 136–144.
Mehta, A.K., et al., 2020. Targeting immunosuppressive macrophages overcomes PARP inhibitor resistance in BRCA1-associated triple-negative breast cancer. Nat. Cancer, 1–17. https://doi.org/10.1038/s43018-020-00148-7.
Meireson, A., Devos, M., Brochez, L., 2020. IDO expression in cancer: different compartment, different functionality? Front. Immunol. 11, 531491.
Michelucci, A., et al., 2013. Immune-responsive gene 1 protein links metabolism to immunity by catalyzing itaconic acid production. Proc. Natl. Acad. Sci. U. S. A. 110, 7820–7825.

Miller, A., et al., 2017. Exploring metabolic configurations of single cells within complex tissue microenvironments. Cell Metab. 26, 788–800.e6.

Mills, E.L., et al., 2016. Succinate dehydrogenase supports metabolic repurposing of mitochondria to drive inflammatory macrophages. Cell 167, 457–470.e13.

Mills, E.L., et al., 2018. Itaconate is an anti-inflammatory metabolite that activates Nrf2 via alkylation of KEAP1. Nature 556, 113–117.

Mo, X., et al., 2018. Antiglioma via regulating oxidative stress and remodeling tumor-associated macrophage using lactoferrin-mediated biomimetic codelivery of simvastatin/fenretinide. J. Control. Release 287, 12–23.

Molgora, M., et al., 2020. TREM2 modulation remodels the tumor myeloid landscape enhancing anti-PD-1 immunotherapy. Cell 182, 886–900.e17.

Morrissey, S.M., et al., 2021. Tumor-derived exosomes drive immunosuppressive macrophages in a pre-metastatic niche through glycolytic dominant metabolic reprogramming. Cell Metab. https://doi.org/10.1016/j.cmet.2021.09.002.

Movahedi, K., et al., 2010. Different tumor microenvironments contain functionally distinct subsets of macrophages derived from Ly6C(high) monocytes. Cancer Res. 70, 5728–5739.

Müller, S., et al., 2017. Single-cell profiling of human gliomas reveals macrophage ontogeny as a basis for regional differences in macrophage activation in the tumor microenvironment. Genome Biol. 18, 234.

Munder, M., et al., 2005. Arginase I is constitutively expressed in human granulocytes and participates in fungicidal activity. Blood 105, 2549–2556.

Munn, D.H., et al., 2005. GCN2 kinase in T cells mediates proliferative arrest and anergy induction in response to indoleamine 2,3-dioxygenase. Immunity 22, 633–642.

Mussai, F., et al., 2015. Arginine dependence of acute myeloid leukemia blast proliferation: a novel therapeutic target. Blood 125, 2386–2396.

Nagaraj, S., et al., 2007. Altered recognition of antigen is a mechanism of CD8+ T cell tolerance in cancer. Nat. Med. 13, 828–835.

Nakamura, K., Smyth, M.J., 2020. TREM2 marks tumor-associated macrophages. Signal Transduct. Target. Ther. 5, 233.

Nakasone, E.S., et al., 2012. Imaging tumor-stroma interactions during chemotherapy reveals contributions of the microenvironment to resistance. Cancer Cell 21, 488–503.

Nishiyama, A., Yokote, Y., Sakagami, H., 2010. Changes in amino acid metabolism during activation of mouse macrophage-like cell lines. In Vivo (Athens Greece) 24, 857–860.

Niu, Z., et al., 2017. Caspase-1 cleaves PPARγ for potentiating the pro-tumor action of TAMs. Nat. Commun. 8, 766.

Nomura, M., et al., 2016. Fatty acid oxidation in macrophage polarization. Nat. Immunol. 17, 216–217.

Nosaka, T., et al., 2018. Alveolar macrophages drive hepatocellular carcinoma lung metastasis by generating leukotriene B 4. J. Immunol. 200, ji1700544.

Obeid, E., Nanda, R., Fu, Y.-X., Olopade, O.I., 2013. The role of tumor-associated macrophages in breast cancer progression. Int. J. Oncol. 43, 5–12.

Ochocka, N., et al., 2021. Single-cell RNA sequencing reveals functional heterogeneity of glioma-associated brain macrophages. Nat. Commun. 12, 1151.

Odegaard, J.I., et al., 2007. Macrophage-specific PPARγ controls alternative activation and improves insulin resistance. Nature 447, 1116–1120.

Olesch, C., et al., 2015. MPGES-1-derived PGE2 suppresses CD80 expression on tumor-associated phagocytes to inhibit anti-tumor immune responses in breast cancer. Oncotarget 6, 10284–10296.

O'Neill, L.A.J., Kishton, R.J., Rathmell, J., 2016. A guide to immunometabolism for immunologists. Nat. Rev. Immunol. 16, 553–565.

Orhan, F., et al., 2016. Tryptophan metabolism along the kynurenine pathway downstream of toll-like receptor stimulation in peripheral monocytes. Scand. J. Immunol. 84, 262–271.

Ottensmeier, C.H., et al., 2016. Upregulated glucose metabolism correlates inversely with CD8+ T-cell infiltration and survival in squamous cell carcinoma. Cancer Res. 76, 4136–4148.

Palmieri, E.M., et al., 2017. Pharmacologic or genetic targeting of glutamine synthetase skews macrophages toward an M1-like phenotype and inhibits tumor metastasis. Cell Rep. 20, 1654–1666.

Palmieri, E.M., et al., 2020. Nitric oxide orchestrates metabolic rewiring in M1 macrophages by targeting aconitase 2 and pyruvate dehydrogenase. Nat. Commun. 11, 698.

Palucka, A.K., Coussens, L.M., 2016. The basis of oncoimmunology. Cell 164, 1233–1247.

Pantaloni, D., Clainche, C.L., Carlier, M.-F., 2001. Mechanism of actin-based motility. Science 292, 1502–1506.

Papadopoulos, K.P., et al., 2017. First-in-human study of AMG 820, a monoclonal anti-Colony-stimulating factor 1 receptor antibody, in patients with advanced solid tumors. Clin. Cancer Res. 23, 5703–5710.

Papathanassiu, A.E., et al., 2017. BCAT1 controls metabolic reprogramming in activated human macrophages and is associated with inflammatory diseases. Nat. Commun. 8, 16040.

Park, S.-Y., Bae, D.-J., Kim, M.-J., Piao, M.L., Kim, I.-S., 2012. Extracellular low pH modulates phosphatidylserine-dependent phagocytosis in macrophages by increasing stabilin-1 expression. J. Biol. Chem. 287, 11261–11271.

Park, J., et al., 2015. M-CSF from cancer cells induces fatty acid synthase and PPARβ/δ activation in tumor myeloid cells, leading to tumor progression. Cell Rep. 10, 1614–1625.

Park, H.Y., Kang, H.S., Im, S.-S., 2018. Recent insight into the correlation of SREBP-mediated lipid metabolism and innate immune response. J. Mol. Endocrinol. 61, R123–R131.

Pavlova, N.N., Thompson, C.B., 2016. The emerging hallmarks of cancer metabolism. Cell Metab. 23, 27–47.

Pearce, E.L., Poffenberger, M.C., Chang, C.-H., Jones, R.G., 2013. Fueling immunity: insights into metabolism and lymphocyte function. Science 342, 1242454.

Penny, H.L., et al., 2016. Warburg metabolism in tumor-conditioned macrophages promotes metastasis in human pancreatic ductal adenocarcinoma. Onco. Targets. Ther. 5.

Peres, C.M., de Bittencourt, P.I.H., da Costa, M., Curi, R., 1997. Transference of fatty acids from macrophages to lymphocytes in culture. Biochem. Soc. Trans. 25, 344S.

Peres, C.M., de Bittencourt, P.I.H., Mendonça, J.R., Curi, R., 2003. Evidence that macrophages transfer arachidonic acid and cholesterol to tissues in vivo. Cell Biochem. Funct. 21, 317–323.

Pilon-Thomas, S., et al., 2016. Neutralization of tumor acidity improves antitumor responses to immunotherapy. Cancer Res. 76, 1381–1390.

Platten, M., von Knebel Doeberitz, N., Oezen, I., Wick, W., Ochs, K., 2015. Cancer immunotherapy by targeting IDO1/TDO and their downstream effectors. Front. Immunol. 5, 673.

Poczobutt, J.M., et al., 2016. Expression profiling of macrophages reveals multiple populations with distinct biological roles in an immunocompetent orthotopic model of lung cancer. J. Immunol. 196, 2847–2859.

Popovic, P.J., Zeh, H.J., Ochoa, J.B., 2007. Arginine and immunity. J. Nutr. 137, 1681S–1686S.

Prima, V., Kaliberova, L.N., Kaliberov, S., Curiel, D.T., Kusmartsev, S., 2017. COX2/mPGES1/PGE2 pathway regulates PD-L1 expression in tumor-associated macrophages and myeloid-derived suppressor cells. Proc. Natl. Acad. Sci. U. S. A. 114, 1117–1122.

Rabold, K., et al., 2020. Enhanced lipid biosynthesis in human tumor-induced macrophages contributes to their protumoral characteristics. J. Immunother. Cancer 8, e000638.
Raschioni, C., et al., 2018. CXCR4/CXCL12 signaling and protumor macrophages in primary tumors and sentinel lymph nodes are involved in luminal B breast cancer progression. Dis. Markers 2018, 5018671.
Rath, M., Müller, I., Kropf, P., Closs, E.I., Munder, M., 2014. Metabolism via arginase or nitric oxide synthase: two competing arginine pathways in macrophages. Front. Immunol. 5, 532.
Reina-Campos, M., Scharping, N.E., Goldrath, A.W., 2021. CD8+ T cell metabolism in infection and cancer. Nat. Rev. Immunol., 1–21. https://doi.org/10.1038/s41577-021-00537-8.
Reinfeld, B.I., et al., 2021. Cell-programmed nutrient partitioning in the tumour microenvironment. Nature, 1–7. https://doi.org/10.1038/s41586-021-03442-1.
Remmerie, A., Scott, C.L., 2018. Macrophages and lipid metabolism. Cell. Immunol. 330, 27–42.
Rodriguez, P.C., et al., 2004. Arginase I production in the tumor microenvironment by mature myeloid cells inhibits T-cell receptor expression and antigen-specific T-cell responses. Cancer Res. 64, 5839–5849.
Rodriguez, P.C., Ochoa, A.C., Al-Khami, A.A., 2017. Arginine metabolism in myeloid cells shapes innate and adaptive immunity. Front. Immunol. 8, 93.
Rolny, C., et al., 2011. HRG inhibits tumor growth and metastasis by inducing macrophage polarization and vessel normalization through downregulation of PlGF. Cancer Cell 19, 31–44.
Ruffell, B., et al., 2014. Macrophage IL-10 blocks CD8+ T cell-dependent responses to chemotherapy by suppressing IL-12 expression in intratumoral dendritic cells. Cancer Cell 26, 623–637.
Sag, D., Cekic, C., Wu, R., Linden, J., Hedrick, C.C., 2015. The cholesterol transporter ABCG1 links cholesterol homeostasis and tumour immunity. Nat. Commun. 6, 6354.
Sanford, D.E., et al., 2013. Inflammatory monocyte mobilization decreases patient survival in pancreatic cancer: a role for targeting the CCL2/CCR2 axis. Clin. Cancer Res. 19, 3404–3415.
Santo, C.D., et al., 2018. Metabolic therapy with PEG-arginase induces a sustained complete remission in immunotherapy-resistant melanoma. J. Hematol. Oncol. 11, 68.
Schenkel, L.B., et al., 2021. A potent and selective PARP14 inhibitor decreases protumor macrophage gene expression and elicits inflammatory responses in tumor explants. Cell Chem. Biol. https://doi.org/10.1016/j.chembiol.2021.02.010.
Schulman, I.G., 2017. Liver X receptors link lipid metabolism and inflammation. FEBS Lett. 591, 2978–2991.
Schumann, T., et al., 2015. Deregulation of PPARβ/δ target genes in tumor-associated macrophages by fatty acid ligands in the ovarian cancer microenvironment. Oncotarget 6, 13416–13433.
Semba, H., et al., 2016. HIF-1α-PDK1 axis-induced active glycolysis plays an essential role in macrophage migratory capacity. Nat. Commun. 7, 11635.
Shiao, S.L., et al., 2015. TH2-polarized CD4+ T cells and macrophages limit efficacy of radiotherapy. Cancer Immunol. Res. 3, 518–525.
Sica, A., Mantovani, A., 2012. Macrophage plasticity and polarization: in vivo veritas. J. Clin. Invest. 122, 787–795.
Stockmann, C., et al., 2008. Deletion of vascular endothelial growth factor in myeloid cells accelerates tumorigenesis. Nature 456, 814–818.
Su, P., et al., 2020. Enhanced lipid accumulation and metabolism are required for the differentiation and activation of tumor-associated macrophages. Cancer Res. 80, 1438–1450.

Sun, L., et al., 2021. Activating a collaborative innate-adaptive immune response to control metastasis. Cancer Cell 39, 1361–1374.e9.

Tabas, I., 2010. Macrophage death and defective inflammation resolution in atherosclerosis. Nat. Rev. Immunol. 10, 36–46.

Tallam, A., et al., 2016. Gene regulatory network inference of immunoresponsive gene 1 (IRG1) identifies interferon regulatory factor 1 (IRF1) as its transcriptional regulator in mammalian macrophages. PLos One 11, e0149050.

Tannahill, G.M., et al., 2013. Succinate is an inflammatory signal that induces IL-1β through HIF-1α. Nature 496, 238–242.

Tavazoie, M.F., et al., 2018. LXR/ApoE activation restricts innate immune suppression in cancer. Cell 172, 825–840.e18.

Thommen, D.S., Schumacher, T.N., 2018. T cell dysfunction in cancer. Cancer Cell 33, 547–562.

Trédan, O., Galmarini, C.M., Patel, K., Tannock, I.F., 2007. Drug resistance and the solid tumor microenvironment. J. Natl. Cancer Inst. 99, 1441–1454.

Uehara, T., et al., 2019. Metformin induces CD11b+-cell-mediated growth inhibition of an osteosarcoma: implications for metabolic reprogramming of myeloid cells and anti-tumor effects. Int. Immunol. 31, 187–198.

Umar, S., et al., 2021. Metabolic regulation of RA macrophages is distinct from RA fibroblasts and blockade of glycolysis alleviates inflammatory phenotype in both cell types. Cell. Mol. Life Sci., 1–15. https://doi.org/10.1007/s00018-021-03978-5.

Van den Eynde, B.J., van Baren, N., Baurain, J.-F., 2019. Is there a clinical future for IDO1 inhibitors after the failure of epacadostat in melanoma? Annu. Rev. Cancer Biol. 4, 1–16.

Varner, J.A., Foubert, P., Kaneda, M.M., 2017. PI3Kgamma activates integrin alpha4 and promotes immune suppressive myeloid cell polarization during tumor progression. *Cancer*. Immunol. Res. 5. canimm.0143.2017.

Vats, D., et al., 2006. Oxidative metabolism and PGC-1β attenuate macrophage-mediated inflammation. Cell Metab. 4, 13–24.

Vaupel, P., 2004. Tumor microenvironmental physiology and its implications for radiation oncology. Semin. Radiat. Oncol. 14, 198–206.

Vinogradov, S., Warren, G., Wei, X., 2014. Macrophages associated with tumors as potential targets and therapeutic intermediates. Nanomedicine 9, 695–707.

Vodnala, S.K., et al., 2019. T cell stemness and dysfunction in tumors are triggered by a common mechanism. Science 363, eaau0135.

Vonderheide, R.H., 2018. The immune revolution: a case for priming, not checkpoint. Cancer Cell 33, 563–569.

Wainwright, D.A., et al., 2012. IDO expression in brain tumors increases the recruitment of regulatory T cells and negatively impacts survival. Clin. Cancer Res. 18, 6110–6121.

Wang, B., Tontonoz, P., 2018. Liver X receptors in lipid signalling and membrane homeostasis. Nat. Rev. Endocrinol. 14, 452–463.

Wang, X.-F., et al., 2014. The role of indoleamine 2,3-dioxygenase (IDO) in immune tolerance: focus on macrophage polarization of THP-1 cells. Cell. Immunol. 289, 42–48.

Wang, F., et al., 2018. Glycolytic stimulation is not a requirement for M2 macrophage differentiation. Cell Metab. 28, 463–475.e4.

Weiss, J.M., et al., 2018. Itaconic acid mediates crosstalk between macrophage metabolism and peritoneal tumors. J. Clin. Invest. 128, 3794–3805.

Wenes, M., et al., 2016. Macrophage metabolism controls tumor blood vessel morphogenesis and metastasis. Cell Metab. 24, 701–715.

Werner, E.R., et al., 1989. Parallel induction of tetrahydrobiopterin biosynthesis and indoleamine 2,3-dioxygenase activity in human cells and cell lines by interferon-γ. Biochem. J. 262, 861–866.

Werno, C., et al., 2010. Knockout of HIF-1α in tumor-associated macrophages enhances M2 polarization and attenuates their pro-angiogenic responses. Carcinogenesis 31, 1863–1872.

Wu, H., et al., 2019. Lipid droplet-dependent fatty acid metabolism controls the immune suppressive phenotype of tumor-associated macrophages. EMBO Mol. Med. 11.

Wu, J.-Y., et al., 2020a. Cancer-derived succinate promotes macrophage polarization and cancer metastasis via succinate receptor. Mol. Cell 77, 213–227.e5.

Wu, L., et al., 2020b. RIPK3 orchestrates fatty acid metabolism in tumor-associated macrophages and hepatocarcinogenesis. Cancer Immunol. Res. 8, 710–721.

Xiang, W., et al., 2018. Monoacylglycerol lipase regulates cannabinoid receptor 2-dependent macrophage activation and cancer progression. Nat. Commun. 9, 2574.

Xu, S., et al., 2021. Uptake of oxidized lipids by the scavenger receptor CD36 promotes lipid peroxidation and dysfunction in CD8+ T cells in tumors. Immunity. https://doi.org/10.1016/j.immuni.2021.05.003.

Zamanian-Daryoush, M., et al., 2014. Myeloid-specific genetic ablation of ATP-binding cassette transporter ABCA1 is protective against cancer. Oncotarget 5, 71965–71980.

Zea, A.H., et al., 2005. Arginase-producing myeloid suppressor cells in renal cell carcinoma patients: a mechanism of tumor evasion. Cancer Res. 65, 3044–3048.

Zhang, L., Li, S., 2020. Lactic acid promotes macrophage polarization through MCT-HIF1α signaling in gastric cancer. Exp. Cell Res. 388, 111846.

Zhang, Y., et al., 2021. Single-cell analyses reveal key immune cell subsets associated with response to PD-L1 blockade in triple-negative breast cancer. Cancer Cell. https://doi.org/10.1016/j.ccell.2021.09.010.

Zhao, Q., et al., 2012. Activated CD69+ T cells Foster immune privilege by regulating IDO expression in tumor-associated macrophages. J. Immunol. 188, 1117–1124.

Zheng, X., et al., 2020a. Spatial density and distribution of tumor-associated macrophages predict survival in non–small cell lung carcinoma. Cancer Res. 80, 4414–4425.

Zheng, X., et al., 2020b. Metabolism in tumour-associated macrophages: a quid pro quo with the tumour microenvironment. Eur. Respir. Rev. 29, 200134.

Zhou, B., et al., 2020. The angiocrine Rspondin3 instructs interstitial macrophage transition via metabolic–epigenetic reprogramming and resolves inflammatory injury. Nat. Immunol. 21, 1430–1443.

Zhu, Y., et al., 2014. CSF1/CSF1R blockade reprograms tumor-infiltrating macrophages and improves response to T-cell checkpoint immunotherapy in pancreatic cancer models. Cancer Res. 74, 5057–5069.

CHAPTER FOUR

Macrophages in the kidney in health, injury and repair

Rachel M.B. Bell and Bryan R. Conway*
Centre for Cardiovascular Science, University of Edinburgh, Edinburgh, United Kingdom
*Corresponding author: e-mail address: bryan.conway@ed.ac.uk

Contents

1. Introduction 102
2. Identification of renal macrophages 103
3. Ontology of renal macrophage and their role in the developing kidney 107
   3.1 Developmental origin of kidney macrophages 107
   3.2 Roles of macrophages in kidney development 112
4. Macrophage phenotypic heterogeneity 112
   4.1 Macrophage activation states 112
   4.2 Mononuclear phagocyte phenotyping in the era of single cell RNA sequencing 114
   4.3 Single cell RNA sequencing in the healthy kidney 115
   4.4 Single cell RNA sequencing studies in kidney disease 116
5. Role of macrophages in renal immunosurveillance 119
6. Macrophages in renal injury 120
   6.1 Pro-inflammatory macrophages predominate during acute kidney injury 120
   6.2 Macrophage activation in chronic kidney disease 124
   6.3 Macrophage to myofibroblast transition 126
7. Macrophages in renal repair 128
8. Therapeutic prospects 130
9. Conclusions 134
References 134

Abstract

Macrophages are a key component of the renal mononuclear phagocyte system, playing a major role in defense against infection, renal injury and repair. Yolk sac macrophage precursors seed the early embryonic kidney and are important for renal development. Later, renal macrophages are derived from hematopoietic stem cells and in adult life, there is a significant contribution from circulating monocytes, which is enhanced in response to infection or injury. Macrophages are highly plastic and can alter their phenotype in response to cues from parenchymal renal cells. Danger-associated molecules released from injured kidney cells may activate macrophages toward a pro-inflammatory phenotype, mediating further recruitment of inflammatory

International Review of Cell and Molecular Biology, Volume 367
ISSN 1937-6448
https://doi.org/10.1016/bs.ircmb.2022.01.005

cells, exacerbating renal injury and activating renal fibroblasts to promote scarring. In acute kidney injury, once the injury stimulus has abated, macrophages may adopt a more reparative phenotype, dampening the immune response and promoting repair of renal tissue. However, in chronic kidney disease ongoing activation of pro-inflammatory monocytes and persistence of reparative macrophages leads to glomerulosclerosis and tubulointerstitial fibrosis, the hallmarks of end-stage kidney disease. Several strategies to inhibit the recruitment, activation and secretory products of pro-inflammatory macrophages have proven beneficial in pre-clinical models and are now undergoing clinical trials in patients with kidney disease. In addition, macrophages may be utilized in cell therapy as a "Trojan Horse" to deliver targeted therapies to the kidney. Single-cell RNA sequencing has identified a previously unappreciated spectrum of macrophage phenotypes, which may be selectively present in injury or repair, and ongoing functional analyses of these subsets may identify more specific targets for therapeutic intervention.

1. Introduction

Macrophages are one component of the renal mononuclear phagocyte system (MPS), which also comprises monocytes and dendritic cells (DCs) (Van Furth and Cohn, 1968). While these are often thought of as discrete cell types, it is now appreciated that there is considerable phenotypic and functional overlap (Nelson et al., 2012). Renal macrophages form a dense reticular network that permeates the tubulointerstitial compartment of both the cortex and medullary regions of the kidney. In addition, macrophages are located within and surround the glomerulus.

Macrophages have multiple functions within the kidney. They seed the kidney early in embryogenesis and may play an important role in the development of the kidney. In adult life, renal macrophages have key roles in organ homeostasis and immune surveillance, defending against ascending urinary infections, as well as being pivotal to the response to renal injury. They are professional phagocytes, clearing pathogens and endogenous cell debris during infection or following tissue injury. While DCs are considered to be the key antigen-presenting cell type, macrophages also express the major histocompatibility complex and play an important part in the activation of the adaptive immune response.

Macrophages exhibit considerable heterogeneity according to their location within the kidney, their interaction with endogenous renal cells and their function. Furthermore, they are very plastic cells, changing their morphology, cell surface markers and function in response to environmental cues. For example, in the context of acute injury, they may adopt a pro-inflammatory phenotype and exacerbate tissue damage, whereas once the injurious stimulus is removed, they may be important in mediating repair.

Macrophages have traditionally been defined by the cell surface markers they express; however, there is substantial overlap in expression of these markers with other members of the MPS. No one marker is specific for macrophages in the kidney, therefore, a panel of antibodies is required. For example, in the mouse kidney, both monocytes and macrophages express CD64 and CD11b, with macrophages being differentiated from monocytes by the high levels of F4/80 expression in the former. An additional complexity is that macrophage cell surface markers are not consistent between organs and species, for example, F4/80 is not expressed in human macrophages. This lack of consistent and specific cell markers has led to confusion in nomenclature and to difficulty in comparing the results from different groups of investigators.

New technologies such as single-cell RNA sequencing (scRNA-Seq) have identified additional macrophage heterogeneity that would not be adequately captured by using standard antibody panels in flow cytometry. They have also enabled assessment of the molecular pathways associated with each macrophage phenotype, leading to a greater understanding of the mechanisms that promote plasticity and confer specialized functions.

These recent advances offer the possibility of developing therapies that could target macrophage subsets more precisely. Identification of cell surface markers specific to injurious phenotypes may enable their targeted depletion while sparing those with a reparative phenotype. Alternatively, understanding the transcription factors that promote phenotype switches may enable reprogramming of injurious macrophages toward a more reparative phenotype. In addition, knowledge of the molecular pathways activated in macrophage subsets may enable the development of small molecule inhibitors of injurious pathways or activators of anti-inflammatory pathways.

In this chapter, we will summarize what is known about the renal macrophage heterogeneity, phenotype and function including recent advances using scRNA-Seq and suggest ways by which this knowledge may be harnessed to ameliorate kidney injury and promote repair.

2. Identification of renal macrophages

Renal macrophages constitute the predominant cell type in the renal MPS. One of the challenges in the field has been the lack of specific markers that can be used to definitively denote a renal macrophage and distinguish them from other immune cells (Nelson et al., 2012). Typically renal macrophages have been identified by the use of cell surface markers on flow cytometry or immunohistochemistry. However, many markers are shared by other members of the MPS, such as monocytes or DCs or by other leucocytes such as neutrophils or cells of the adaptive immune system (Viehmann et al., 2018). Furthermore, cell surface expression of these markers may be altered in different macrophage activation states. Hence, combinatorial panels of markers are required to identify the renal macrophage through positive or negative selection across sequential gating strategies in flow cytometry or by using multiplex immunofluorescence on tissue sections.

Flow cytometry is widely utilized for the identification and quantification of renal macrophages following enzymatic digestion of the kidney into single cell suspension. In common with all leucocytes, renal macrophages express CD45, therefore, initial immune lineage gating on CD45 is useful to discriminate them from endogenous renal cells. This is typically followed by negative selection for immune cells that are not part of the MPS. In mice, this can be achieved using antibodies against CD3 (T-cell), CD19 (B-cell), Ly6G (neutrophil) and Siglec-F (eosinophils) (Conway et al., 2020). Human neutrophils do not express Ly6G, however, CD15 can be used to exclude both neutrophils and eosinophils in humans (Pillay et al., 2013).

Following the exclusion of other immune cell types, the majority of $CD45^+$ cells in the kidney will be members of the MPS, however, the nomenclature used to define different MPS cell phenotypes is challenging, none more so than when differentiating macrophages and DCs. This has caused significant controversy in the field as various strategies have been employed by research groups to characterize macrophages and DCs, such that it can be difficult to interpret the findings. The difficulty in distinguishing macrophage and DCs may arise from their significant functional overlap, with both cell types being capable of phagocytosis and antigen presentation, hence they will both express cell surface markers relating to these shared functions. In addition, both macrophages and DCs may be further subdivided into subsets, with not all subsets expressing the same cell surface markers. For example, in both humans and mice, conventional DCs are comprised of type 1 (cDC1) and type 2 (cDC2) subsets. cDC1 cells express CLEC9A and CD103 and activate cytotoxic T-cells, while cDC2 cells prime $CD4^+$ cells and express CD209, but importantly also CD11b, making it difficult to differentiate them from monocytes/macrophages (Kurts et al., 2020). An additional DC subset, plasmacytoid DCs, can be identified by expression of SIGLEC-H (mouse) or CD303 (human) (Musumeci et al., 2019). Finally, both macrophages and DCs exhibit marked phenotypic plasticity in response to environmental cues (Gottschalk and Kurts, 2015), for example, macrophages programmed with IL-4 and IL-13 toward an alternatively activated phenotype express higher levels of mannose receptor on their cell surface (Gordon, 2003).

While CD11b and CD11c were traditionally used to tag macrophages and DCs, respectively, it is now known that these markers may be expressed by both cell types (Kawakami et al., 2013; Kruger et al., 2004). This has implications for studies that have used transgenic mice to tag or deplete MPS cells such as CD11b-DTR mice (Machida et al., 2010) or CD11c-DTR mice (Snelgrove et al., 2012) for macrophages and DCs, respectively. Furthermore, markers that define resident macrophages in other organs such as Tim-4 (Shaw et al., 2018) or CD64 (Tamoutounour et al., 2012) are either not expressed by macrophages in the adult kidney (Tim-4) (Salei et al., 2020) or are additionally expressed by renal dendritic cells (CD64) (Schraml et al., 2013). Hence, no one marker can effectively discriminate between renal macrophages and other MPS cells, however, markers that are more highly expressed in renal macrophages than other MPS cells include CD64, the Fc gamma receptor 1, which is in keeping with their role in phagocytosis of opsonized micro-organisms and F4/80, a G-protein coupled transmembrane receptor (Waddell et al., 2018). Recently, researchers have taken advantage of transcription factors that are specifically expressed in DC subsets to differentiate them from macrophages using transgenic mice. In Zbtb46-GFP mice both cDC1 and cDC2 cells are labeled with GFP, while Snx22-GFP may be used to specifically label cDC1 cells (Brahler et al., 2018). Intriguingly, this approach revealed that in the kidney the dense network of $CD11c^{+}$ cells with dendritic processes were macrophages and not DCs as previously thought. Conversely, the $Zbtb46^{+}$ and $Snx22^{+}$ DCs were more rounded in shape and less abundant, localizing to the peri-vascular region. These studies exemplify the evolution in our understanding of macrophages and DCs and the need for more specific markers.

Renal macrophages may be distinguished from circulating monocytes in mice using a classic flow cytometry gating strategy, with monocytes being $CD11b^{+}F4/80^{Lo}$ and macrophages being $CD11b^{+}F4/80^{Hi}$ (Kawakami et al., 2013) (Fig. 1A). Importantly, as cDC2 cells also express CD11b, they may also appear in the $CD11b^{+}F4/80^{Lo}$ gate (Fig. 1B), however, purer macrophage populations may be obtained by first removing $CD11c^{Hi}CD64^{-}$

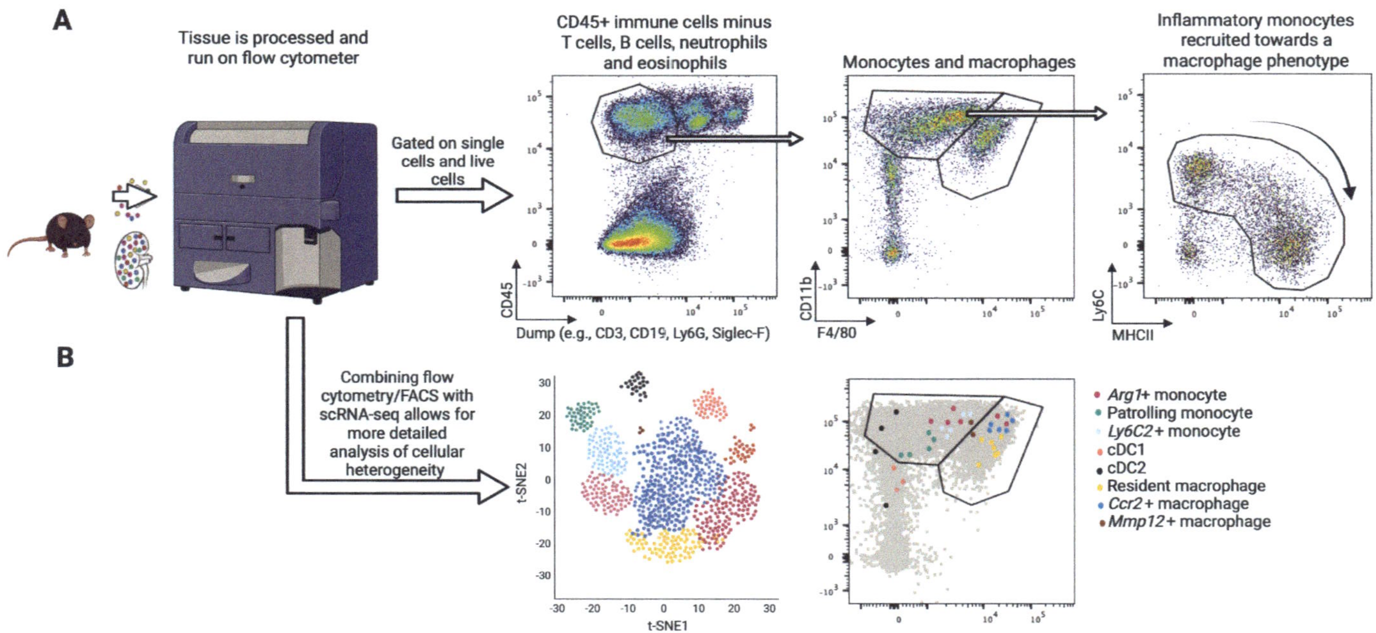

Fig. 1 (A) Typical flow cytometry gating strategy to characterize monocytes and macrophages in the kidney. Initial gating on live CD45+ cells that do not express other leukocyte markers selects for MPS cells. Macrophages are $CD11b^{+} F4/80^{Hi}$, whereas monocytes are $CD11b^{+} F4/80^{Lo}$ and in this example from a kidney 2 days after ureteric obstruction, there is an influx of monocytes, reversing the typical 2:1 ratio of macrophages to monocytes. As Ly6CHi inflammatory monocytes are recruited to the kidney, they differentiate toward a more macrophage phenotype, losing cell surface Ly6C expression and upregulating Class II MHC, giving the impression of a "waterfall effect." (B) Single-cell RNA-seq of the kidney in injury and repair highlights MPS subsets and SMART-Seq2 technology enables these to be mapped back to the monocyte and macrophage gates, revealing the true of extent of the MPS heterogeneity that would not be readily identified by standard flow cytometry.

DCs (Tamoutounour et al., 2012). Monocytes may be further subdivided into $Ly6C^{+}CCR2^{+}Cx3cr1^{-}$ inflammatory monocytes and $Ly6C^{-}CCR2^{-}$ $Cx3cr1^{+}$ patrolling monocytes (Geissmann et al., 2003). The $Ly6C^{+}$ inflammatory monocytes are recruited to the kidney during injury and differentiate toward a macrophage phenotype (Lin et al., 2009). This differentiation process can be observed on a plot of Ly6C against MHCII, with $Ly6C^{+}MHCII^{-}$ inflammatory monocytes transitioning toward a $Ly6C^{-}MHCII^{+}$ macrophage phenotype passing through an intermediate $Ly6C^{Int}/MHCII^{Int}$ state, sometimes referred to as the "monocyte waterfall" (Fig. 1A) (Tamoutounour et al., 2012).

A similar flow cytometry approach may be taken in humans to identify renal monocytes and macrophages, although this must be modified to reflect the fact that human monocytes do not express Ly6C and that the F4/80 antigen in humans is expressed in eosinophils rather than macrophages (Hamann et al., 2007). In humans, monocytes are categorized into a classical $CD14^{++}CD16^{-}$ subset, which most closely resemble mouse $Ly6C^{+}$ inflammatory monocytes; and a non-classical $CD14^{+}CD16^{+}$ subset, which share phenotypic features with $Ly6C^{-}$ patrolling monocytes and adhere to the vascular endothelium (Geissmann et al., 2003). An additional intermediate $CD14^{++}CD16^{+}$ monocyte subset is also present in humans, although whether this is a discrete subset or whether it contains additional subpopulations remains unclear (Cormican and Griffin, 2020).

A panel of markers is also required to identify macrophages in histological sections. In mice, F4/80 is a useful marker of macrophages in the tubulointerstitium, however, it does not stain all glomerular macrophages (Sung et al., 2017). CD68 has been used as a marker of macrophages in both the glomerulus and tubulointerstitium in tissue sections from human and murine kidneys, however, it is also expressed by DCs. Additional macrophage markers used in human disease include galectin-3 (Kikuchi et al., 2004), mannose receptor (CD206) and the hemoglobin-haptoglobin scavenger receptor (CD163) (Klessens et al., 2017), although these molecules are more highly expressed by alternatively activated macrophages.

3. Ontology of renal macrophage and their role in the developing kidney

3.1 Developmental origin of kidney macrophages

Recent advances in lineage tracing have improved our understanding of the ontogeny of renal macrophages (Hoeffel and Ginhoux, 2018). During

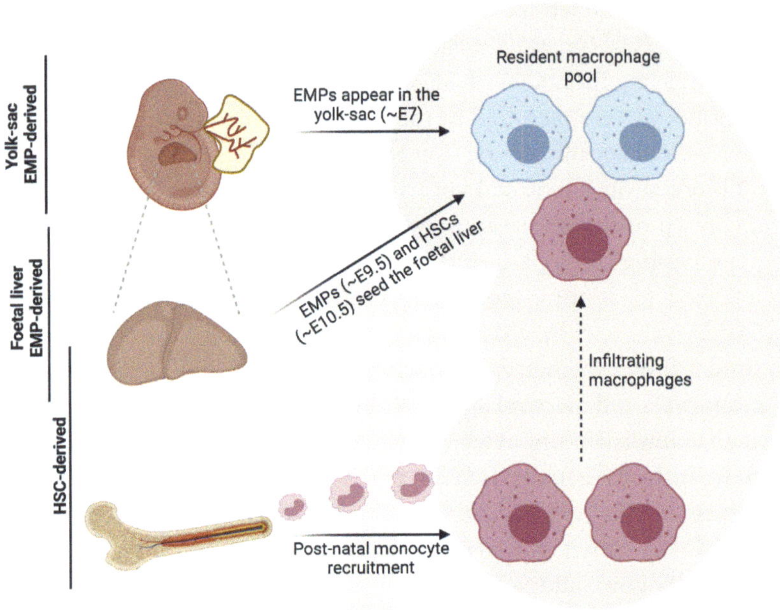

Fig. 2 Ontology of renal macrophages. During embryogenesis, macrophages colonize the kidney in several spatially and temporally regulated waves. The first wave of macrophage precursors arise at E7.5 from yolk sac-derived erythro-myeloid progenitors (EMPs). From E8.25, a second set of EMPs seed the fetal liver, where they then expand and differentiate into monocytes/pre-macrophages, which then migrate to the kidney. A second source of macrophage precursors are hematopoietic stem cells (HSCs) produced in the aorta-gonado-mesonephros (AGM). From E10.5, these HSCs colonize the fetal liver and later the bone marrow and differentiate into monocytes that seed the embryonic kidney and continue to contribute to the renal macrophage pool in adult life.

embryogenesis, macrophages colonize tissues in several spatially and temporally regulated waves. Across the lifespan, kidney resident macrophages are derived from all the progenitor waves, however, the relative proportion from each wave varies during kidney development, aging and injury/disease (Fig. 2) (Ginhoux and Guilliams, 2016; Mass and Gentek, 2021). In mice, the first wave of macrophage precursors develops in the extra-embryonic yolk-sac at day 7 of embryonic development (E7.5), when erythro-myeloid progenitors (EMPs) appear from cadherin-5$^+$ cells of the blood islands and capillary endothelia. These cells form independently of the transcription factor c-Myb and bypass the intermediate monocyte phase, directly acquiring a core macrophage transcriptional program and differentiating into pre-macrophages. When the yolk-sac and embryonic vasculature connect at E8.5, these yolk-sac macrophages migrate into fetal tissues. From E8.25, a second set of EMPs, which are c-Myb-dependent, are generated from hemogenic endothelia in the yolk-sac (Hoeffel et al., 2015). Many of these EMPs travel through the embryonic vasculature to seed the fetal liver, where they then expand and differentiate into monocytic intermediates and/or pre-macrophages. Upon exiting the fetal liver, monocytes/pre-macrophages travel through the circulation to all embryonic tissues, except the brain, as the blood–brain barrier is closed at this time-point. A second source of macrophage precursors are hematopoietic stem cells (HSCs), which are produced in the hemogenic epithelia, primarily in the aorta-gonado-mesonephros (AGM). From E10.5, these HSCs colonize the fetal liver, expand, and differentiate into monocytic intermediates before going on to populate tissues and mature into tissue-resident macrophages. HSCs further seed the bone marrow and spleen during embryogenesis and are maintained postnatally as progenitors that can generate a constant supply of monocytes that can be released into the circulation (Liu et al., 2019).

To gain a full understanding of the origins of the kidney-resident macrophages, various fate-mapping mouse models have been used that employ injection of 4-hydroxytamoxifen to irreversibly label progenitors and their progeny using inducible Cre recombinase-mediated reporter labeling. Gentek exploited the fact that cadherin 5 (Cdh5) is expressed by all endothelial cells including hematopoietic precursors, but is down-regulated upon hematopoietic transition (Gentek et al., 2018). They employed tamoxifen-inducible $Cdh5^{\mathrm{CreERT/tdT}}$ mice to specifically label yolk-sac derived EMPs or AGM-derived HSCs by administering tamoxifen at either E7.5 or E10.5, respectively. Following labeling of yolk sac progenitors with tamoxifen at E7.5, approximately 70% of kidney macrophages and brain microglia were labeled at E14.5. However, while a similar proportion of microglia retained the label at E16.5, only ~5% of kidney macrophages remained labeled, indicating that the yolk sac-derived renal macrophages were being diluted by recruitment of new cells. Conversely, after administration of tamoxifen at E10.5 to label HSCs, few renal macrophages were labeled by E14.5, however, the labeling increased progressively in the later embryonic stages and adulthood, reaching ~50% by postnatal week 6. This suggests that a very transient wave of yolk sac-derived macrophages populate the kidney during development but are replaced by HSC-derived macrophages from E14.5 onwards. Similar results were obtained using lineage tracing models employing runt-related transcription factor 1 (*Runx1*), which is expressed in yolk-sac macrophage progenitors as they bud from the hemogenic endothelium (Hoeffel et al., 2015). Administration of tamoxifen at E7.5 to *Runx1*$^{CreERT/eYFP}$ mice labeled ~25% of renal macrophages at E13.5, which had markedly reduced by E16.5; conversely following labeling of precursors with tamoxifen at E8.5, there was a progressive increase in the proportion of labeled renal macrophages in late embryogenesis, with approximately 30% retaining label through 6 weeks postnatally.

Further studies have been performed using inducible reporter mice that express Cre recombinase under inducible promoters of Colony-stimulating factor 1 receptor (CSF-1R) or Cx3cr1, which are both expressed by all yolk-sac macrophage progenitors. In *Csf1r*$^{CreERT/GFP}$ or Cx3cr1$^{CreERT+/-}$/ihCD59$^{+/-}$ mice, yolk sac-derived macrophage precursors labeled by administration of tamoxifen at E8.5 make a minimal contribution to the pool of kidney macrophages at postnatal week 10 (Epelman et al., 2014; Liu et al., 2020). Yolk sac-derived macrophages in the late embryonic kidney are MERTK^{+}TIM4^{+}MHCII^{-} and are replaced by MERTK^{-} TIM4^{-}MHCII^{+} HSC-derived macrophages early postnatally (Salei et al., 2020). As with microglia and gut macrophages, adult renal macrophages continue to express Cx3cr1 (Liu et al., 2019) and administration of tamoxifen to Cx3cr1$^{CreERT+/-}$/ihCD59$^{+/-}$ mice labels all renal macrophages at birth. The proportion of labeled macrophages then decreases rapidly in the first month post-natal reaching a steady-state of 60% by week 13, suggesting that they may be replaced by unlabeled cells infiltrating from the bloodstream.

To assess the contribution of adult kidney macrophages from bone marrow-derived precursors, Liu et al. utilized Ms4a3-Cre mice. The *Ms4a3* gene is specifically expressed by granulocyte-monocyte progenitors (GMPs) but not by tissue-resident macrophages or DCs. As such, GMPs, which are derived from HSCs and give rise to various leukocytes including macrophages, may be traced based on *Ms4a3* expression. In the kidneys of *Ms4a3*$^{Cre/tdT}$ mice, there was no labelling at birth, but the percentage of labeling increased to reach a steady-state at postnatal week 12 suggesting that monocytes contribute ~40% of kidney macrophages (Liu et al., 2019), complementing the 60% derived from embryonic macrophages as determined in the analysis of Cx3cr1$^{CreERT+/-}$/ihCD59$^{+/-}$ mice as above. There was a sex difference in labeling, with a greater proportion of renal macrophage derived from bone marrow precursors at week 20 in male mice compared to females, as has previously been reported for peritoneal macrophages (Bain et al., 2020).

Intriguingly, to determine whether $CD64^+F4/80^{Hi}CD11b^{Lo}$ resident macrophages could be derived from DC precursors in the bone marrow, Salei et al. employed reporter mice under the *Clec9a* promoter, thought to be a specific marker of the common DC precursor (CDP) (Salei et al., 2020). As expected, the vast majority of DCs in the kidneys of adult Clec9a-$Cre^{+/-}$RosaTdTomato were labeled with the reporter, however, in addition ~60% of the $CD64^+F4/80^{Hi}CD11b^+$ cells were labeled, suggesting that cells classically described as macrophages were derived from CDPs. These cells were transcriptionally similar to $CD64^+F4/80^{Hi}CD11b^+$ cells that did not express the promoter, suggesting a similar macrophage phenotype despite different ontogeny. This further draws into question the utility of describing cells as macrophages or DCs given that cells which express similar surface markers may be ontologically diverse.

Taken together, these studies elegantly demonstrate that in the early embryo, kidney macrophages are initially derived from yolk sac EMPs and are $MERTK^+TIM4^+MHCII^-$, however, by birth these are replaced by HSC-derived cells from the AGM. The HSC-derived cells persist in the kidney postnatally, but are gradually diluted by the arrival of bone marrow derived monocytes, which contribute approximately 40% of renal macrophages by 3 months of age. Hence, the rate of replacement of embryonic-derived kidney macrophages with bone marrow-derived monocytes is much slower than for the gut (Bain et al., 2014), but much more extensive than the brain, where embryonic microglia self-renew and persist (Liu et al., 2019, 2020).

The origin of human kidney macrophages is less well understood. Stewart et al. performed scRNA-Seq on human embryonic and adult kidneys to compare the MPS subsets at different stages of kidney development (Stewart et al., 2019). A single monocyte cluster was observed in the embryonic kidney, which most closely resembled the $CD14^{++}CD16^-$ subset in adult kidneys. In early development, the fetal monocytes exhibited lower expression of inflammatory and phagocytosis genes than in the adult kidney, in keeping with the sterile environment of the amniotic sac. Two clusters of fetal macrophages were observed, with the second cluster expressing more inflammatory genes and increasing proportionally in later development. Tissue macrophages in the adult kidney expressed similar markers to the fetal macrophages including *CD163*, *MRC1* and *C1QC* and also expressed *RUNX1*, the transcription factor used to trace yolk sac-derived cells as above, suggesting that they may originate from fetal macrophages rather than infiltrating monocytes.

3.2 Roles of macrophages in kidney development

The function of macrophages in kidney development remains to be fully elucidated, although it is becoming clear that they orchestrate tissue remodeling and contribute to the morphogenesis of the developing kidney (Munro and Hughes, 2017). A recent study demonstrated that macrophages in the embryonic kidney align with newly forming cortical interstitial blood vessels and are enriched for mRNAs linked to vascular development (Munro et al., 2019). Furthermore, depletion of macrophages in cultured kidney explants by blocking the CSF-1/CSF1R pathway resulted in a greater proportion of isolated, discontinuous endothelial structures compared to controls which had characteristic net-like vascular plexuses. This is consistent with macrophages having the capacity to promote vascular anastomoses in other settings, for example, in response to vascular rupture (Liu et al., 2016). Macrophages were also shown to engulf erythrocytes and apoptotic endothelial cells in the kidney, with cell clearance being one mechanism by which these phagocytic cells may remodel developing tissues (Munro et al., 2019). In agreement with this, treating cultured kidney explants with CSF-1 accelerates renal growth and ureteric bud branching in conjunction with increased macrophage numbers and an alternatively activated profile (Rae et al., 2007). Treatment of neonatal mice with CSF-1 promotes development of kidneys of higher mass (Alikhan et al., 2011), whereas blockade of the CSF1/CSF1R pathway in adult mice results in reduced kidney to bodyweight ratio (Sauter et al., 2014). This body of evidence supports the concept that embryonic macrophages are essential for normal kidney development; however, more research is needed to get a clear picture of the mechanisms by which this is mediated as this may identify strategies to promote renal repair.

4. Macrophage phenotypic heterogeneity

4.1 Macrophage activation states

Macrophages are very plastic and may acquire a range of activation states in response to physiological or pathological stimuli. Historically, these have been broadly classified into an M1 or pro-inflammatory phenotype and an M2 or reparative phenotype (Tang et al., 2019a). However, many of these activation states have been generated from *in vitro* studies and it is clear that the milieu of the injured or repairing kidney is much more complex and that discrete phenotypes are likely to be replaced by a spectrum of activation

(Mosser and Edwards, 2008). Furthermore, macrophage activation status may change over time depending on the balance of injurious and pro-reparative stimuli in the surrounding environment (Lee et al., 2011; Ricardo et al., 2008). Despite these limitations, the paradigm of M1 and M2 activation may be useful as a broad concept and hence will be discussed below.

The M1 or classical pro-inflammatory activation status may occur in response to exogenous stimuli such as products derived from pathogens or endogenous stimuli released in response to kidney injury. Classical exogenous stimuli include lipopolysaccharide (LPS), a constituent of bacteria cell walls, which induces macrophage secretion of noxious substances including reactive oxygen species (ROS) and nitric oxide, pro-inflammatory cytokines including IL-1, IL-6, IL-12 and tumor necrosis factor (TNF)-α and cell surface expression of MHCII and CD80 and CD86. This classical activation may be further accentuated in the presence of pro-inflammatory cytokines such as γ-interferon (IFN-γ) or TNF-α released by adjacent macrophages or T-cells. This phenotype is consistent with the role of inflammatory macrophages in directly killing pathogens, recruiting inflammatory cells and activating Th1 cells to trigger an adaptive immune response. Endogenous stimuli, called danger-associated molecular patterns (DAMPs), may be released from injured kidney cells to promote an inflammatory macrophage phenotype. Kidney DAMPs include heat shock proteins (Lang et al., 2005), high mobility box protein-1 (Wu et al., 2010), and biglycan (Schaefer et al., 2005). Binding of DAMPs to pattern recognition receptors on macrophages including toll-like receptors (TLRs) (Moreth et al., 2014) or Mincle (Lv et al., 2017; Tanaka et al., 2020) promotes release of chemokines to recruit inflammatory cells and mediate further tissue injury. The M1 phenotype tends to predominate early after kidney injury, therefore depletion of macrophages in the early injury phase is beneficial (Lee et al., 2011).

The M2 alternative activation status comprises a range of macrophage phenotypes each with their own activation stimulus, but broadly characterized by a reparative phenotype and the ability to activate Th2 cells (Chen et al., 2019). The best known M2 activation stimulus employs a combination of IL-4 and IL-13 cytokines, which are characteristically produced by Th2 cells in response to parasitic infections such as helminths. Activation of the IL-4/IL-13 receptor on macrophages results in an M2a phenotype, which is characterized by cell surface expression of class II MHC for antigen presentation and scavenger receptors such as mannose receptor and CD163 for phagocytosis. (Gordon, 2003). It is also associated with the release of CCL17 and CCL22 to attract lymphocytes including Th2 cells, and increased expression of arginase-1 and Ym1. The M2a process may be mediated by galectin-3 (MacKinnon et al., 2008) and the resultant phenotype is immunomodulatory and counteracts an M1 response, thereby dampening inflammation. However, if the M2a activation state is sustained, this may be detrimental as it promotes fibrosis (Murray and Wynn, 2011). For example, the M2a enzyme arginase-1 converts arginine to ornithine, a precursor for collagen synthesis and the alternative activation marker galectin-3 may drive hepatic and renal fibrosis (Henderson et al., 2006). M2b macrophages are induced by combinatorial activation with immune complexes and either LPS or IL-1. While M2b macrophages, like M1 macrophages, express high levels of MHCII and CD80, they are immunosuppressive in part through release of IL-10. M2c macrophages are induced by IL-10, TGF-β, glucocorticoids or phagocytosis of apoptotic cells. They share some markers of M2a phenotype including expression of mannose receptor and expression of arginase-1, but they are strongly anti-inflammatory *via* secretion of IL-10 and TGF-β.

4.2 Mononuclear phagocyte phenotyping in the era of single cell RNA sequencing

As described above, until recently MNP cells have been classified primarily on the basis of cell surface markers or activation states determined from *in vitro* studies. However, classification by cell surface markers is biased toward our existing knowledge and the markers chosen may not be the most optimal for discriminating between MNP subsets. Furthermore, the reductionist stimuli used in *in vitro* studies do not capture the complex milieu of the renal parenchyma and hence the macrophage phenotypes that emerge are unlikely to be representative of the diverse range of activation states that are present in the kidney in health and disease. The emergence of scRNA-Seq technology has the potential to revolutionize our understanding of the MNP heterogeneity in the kidney. It enables the transcriptomes of thousands of individual cells in the kidney to be captured simultaneously. Bioinformatic approaches enable cells with similar transcriptomes to be clustered, highlighting novel cell phenotypes and quantifying changes in the relative proportion of cells adopting each phenotype in health or disease. Furthermore, as the transcriptome of each cell phenotype is known, genes that are differentially expressed in each macrophage subset may highlight specific markers of each phenotype in an unbiased manner. Where differentially expressed genes encode cell surface markers, there is the potential to refine the current antibody panels used for flow cytometry in order to fully capture the MNP heterogeneity in the kidney. Additionally, pathway analysis of the differentially expressed genes may be informative regarding the functional role of the cells in each of the MNP clusters or generate hypotheses to be tested in further studies.

4.3 Single cell RNA sequencing in the healthy kidney

Park et al. performed the first large-scale scRNA-seq analysis of the healthy mouse kidney, identifying all the major known cell subtypes that comprise the nephron (Park et al., 2018). However, as immune cell types contribute a relatively small proportion of the total cells in the uninjured kidney, they were unable to separate the MNP into discrete subclusters. To enrich for immune cells, Zimmerman and colleagues initially sorted $CD45^+$ immune cells from the kidneys of healthy mice, rats, pigs and humans and then discarded lymphocytes to target MNPs for scRNA-Seq (Zimmerman et al., 2019). Bioinformatic analysis segregated the remaining immune cells into nine clusters, which were annotated using the differentially expressed genes in each cluster as neutrophils, natural killer (NK) cells, innate lymphoid cells, $Ly6C^{Hi}$ and $Ly6C^{Lo}$ monocytes, macrophages, conventional DCs (cDC1 and cDC2) and an unidentified cluster. Not all of these clusters were present in other species, indicating the difficulty of translating pre-clinical studies in mice to other species. However, a large cluster of cells which expressed *C1q* and other markers of resident macrophages was generic to all species. Other differentially expressed genes in this cluster were *Cd74* and *Cd81*, which encoded cell surface proteins and flow cytometry validated that $CD74^+CD81^+$ cells were more abundant in resident $CD11b^+F4/80^{Hi}$ macrophages compared to infiltrating $CD11b^+F4/80^{Lo}$ monocytes in mice. $CD74^+CD81^+$ cells were also observed in flow cytometry in rats and human kidneys, suggesting that they may be useful pan-species markers of resident renal macrophages. Other scRNA-seq analyses of whole kidney tissue from mice (Conway et al., 2020; Park et al., 2018) and humans (Wu et al., 2018) have also identified this *C1q*-expressing macrophage cluster. Conversely, genes encoding proteins often used to tag macrophages in immunohistochemistry analysis of the kidney, such as *Cd68* and *Cd163,* were not specific to the macrophage cluster in scRNA-Seq in the human kidney, calling into question their specificity to identify macrophages in biopsies from patients with kidney disease.

scRNA-Seq has also been performed in the healthy human adult kidney, identifying four MNP cell clusters, which based on their transcriptome, were classified as $CD14^+$ classical monocytes, $CD16^+$ non-classical monocytes, DCs and a tissue-macrophage population that was skewed toward an anti-inflammatory phenotype expressing M2 markers such as CD206 (Stewart et al., 2019). As droplet-based scRNA-Seq is agnostic of the location of the cells within the kidney, the researchers cross-referenced their transcriptomic data to publicly available bulk RNA-seq data revealed generated from specific regions of the kidney. This determined that the monocytes were preferentially located in the medulla and pelvis, and hence are ideally situated to sense and counteract ascending infection. The localization of immune cells to specific regions was absent in fetal kidneys, as were signatures indicative of roles in inflammation and immune defense, consistent with lack of activation in the relatively sterile environment of the amniotic sac.

4.4 Single cell RNA sequencing studies in kidney disease

Several scRNA-Seq studies have now been employed to characterize how MNP phenotypes are altered in pre-clinical murine models and in human kidney disease. Kirita et al. employed single nuclear RNA sequencing (snRNA-seq) to characterize cell states across the time course of injury and repair in a murine model of bilateral ischemia reperfusion injury (Kirita et al., 2020). Six leukocyte clusters were identified including three macrophage subtypes, cDC1 and cDC2 dendritic cells, T and B cells. As both endogenous renal cells and immune cells were captured in this study, analysis of the signals that mediate cross-talk between renal parenchymal cells and immune cells could be explored through ligand-receptor analysis. This suggested that CCL2 released from endogenous kidney cells signaled to CCR2 on leukocytes to recruit them to the kidney. There was a temporal evolution of this response, with the initial signaling mediated by fibroblasts and endothelial cells, followed by leukocyte-leukocyte signaling and later CCL2 expression in proximal tubules that failed to repair at 2 days and 6 weeks post-injury, which may sustain the inflammatory response. The same group performed snRNA-Seq on fibrotic mouse kidney 14 days after ureteric obstruction (Wu et al., 2019). Only one macrophage cluster was identified, though researchers identified a novel "de-differentiated" proximal tubule cluster that expressed numerous secreted pro-inflammatory cytokines that may recruit inflammatory cells including Ccl2, macrophage proliferative cytokine Il-34, and neutrophil chemo-attractants Cxcl1 and Cxcl2.

scRNA-seq analysis has also been performed in chronic models of kidney disease including folic acid nephropathy model (Dhillon et al., 2021). In chronic kidney disease there was an increase in immune cell heterogeneity, with 14 immune clusters identified compared with only 5 clusters in the healthy kidney (Park et al., 2018). Clusters included monocytes, macrophages, cDC1 and cDC2 cells, with a greater proportion of cells in these MNP clusters in chronic kidney disease compared with the healthy kidney.

To characterize the temporal evolution of MNP cell phenotypes during renal injury and repair, our group employed scRNA-seq in the murine model of reversible unilateral ureteric obstruction (R-UUO), in which there is a wave of inflammation driving tubulointerstitial fibrosis during the obstructive phase, which regresses following reversal of obstruction (Conway et al., 2020). In total, we identified 12 myeloid cell subsets, with the proportion of cells in each subset differing markedly between normal kidney tissue, early or late injury (day 2 or day 7 post obstruction) and during repair 2 weeks following reversal of obstruction. We observed clusters consistent with Ly6C^{Hi} and Ly6C^{Lo} monocytes in the healthy kidney in keeping with previous studies (Zimmerman et al., 2019), however, an additional cluster that was transcriptomically similar to Ly6C^{Hi} monocytes was observed uniquely early after induction of injury and was tagged by expression of *Arg1*. A similar *Arg1*$^{+}$ cluster (Dick et al., 2019) was observed during scRNA-Seq analysis of the immune response in the heart after myocardial infarction and *Arg1* expression is upregulated in monocytes recruited to the kidney in ischemia-reperfusion injury (Tanaka et al., 2020), suggesting that the *Arg1*+ cells represent a novel monocyte activation status that is generic across different injuries and multiple organs. The *Arg1*+ cells expressed genes activated in hypoxic stress, pro-inflammatory genes, and genes encoding ECM components and ECM cross-linkers. The recruitment of these cells in the acute injury period and their pro-inflammatory transcriptome suggests that they most closely resemble an M1 phenotype, however, the fact that they express high levels of *Arg1* is intriguing as it encodes arginase-1, which is traditionally considered to be a marker of the reparative (M2) activation state. Furthermore, both the Ly6C^{Hi} monocytes and Arg1+ cluster express *Chil3*, which encodes Ym1, which again is more commonly associated with a M2 phenotype and has been shown to activate fibroblasts following renal ischemia-perfusion injury (Montgomery et al., 2017). Hence, the unbiased clustering and assignment of differentially expressed genes performed in scRNA-Seq studies calls into question the relevance of macrophage activation states determined from *in vitro* studies.

A strength of scRNA-Seq technology is that pseudotime analysis can be employed to assess the trajectory of gene expression changes to infer evolution of MNP phenotypes over time. During ureteric obstruction, we observed a trajectory from Ly6c^{Hi} monocytes through Arg1^{+} cells to an additional myeloid cell cluster that expressed genes characteristic of macrophages including *Cd74* and *Cd81*, but which also expressed *Ccr2*. To test the hypothesis that the Ly6C^{Hi} monocytes gave rise to the Ccr2^{Hi} macrophage-like cells we employed paired blood exchange in CD45.1/CD45.2 mismatched pairs to track circulating monocytes recruited to the obstructed kidney, observing that by 7 days after obstruction they did indeed adopt a F4/80^{Hi} macrophage phenotype, while continuing to express CCR2.

To assess how our 12 myeloid cell subsets mapped to standard flow cytometry analysis, we integrated droplet and plate-based scRNA-seq, with the latter enabling simultaneous analysis of cell surface antibody binding and scRNA-Seq within the same cells (Fig. 1B). This demonstrated that the classical F4/80^{Hi}CD11b^{+} and F4/80^{Lo}CD11b^{+} gating strategy does not capture the full extent of myeloid cell heterogeneity in the kidney. As expected, the Ly6C^{Hi} and Ly6C^{Lo} monocytes mapped to the F4/80^{Lo}CD11b^{+} monocyte gate, whereas resident and mannose receptor^{Hi} macrophages gated to the F4/80^{Hi}CD11b^{+} gate. Interestingly, the Arg1 + cells straddled monocyte and macrophage gates, in keeping with a phenotypic evolution from monocytes to macrophages.

Single-nuclear RNAseq has also been performed in nephrectomy specimens or kidney biopsies from patients with chronic kidney disease. Nephrectomy specimens from patients with diabetic kidney disease were found to have a higher proportion of infiltrating leukocytes compared to control kidneys (Wilson et al., 2019). While monocytes, plasma cells, T and B cells were detected, a discrete macrophage cluster was not observed, despite an influx of macrophages to the kidney in diabetic nephropathy being well-documented (Tesch, 2010). This may reflect differences in the cells captured by scRNA-Seq and snRNA-Seq, with the latter capturing more podocytes and mesangial cells, while the former captures a greater proportion of immune cells, in particular myeloid cells (O'Sullivan et al., 2019).

The Accelerating Medicines Partnership in SLE consortium has performed scRNA-Seq in biopsy cores from patients with lupus nephritis and from healthy controls (Arazi et al., 2019), identifying 21 immune clusters including five subsets of myeloid cells. These included resident macrophages, DCs, pro-inflammatory monocytes, phagocytic CD16^{+} macrophages and

M2-like $CD16^+$ macrophages. Only the pro-inflammatory monocytes were found in the blood and pseudotime analysis suggested that when recruited to the kidney, these transited through a phagocytic macrophage phenotype to arrive at the M2-like state with the evolving phenotype associated with a reduction in pro-inflammatory genes and an increase in phagocytic genes, potentially important in scavenging debris and excess extracellular matrix in the injured kidney.

5. Role of macrophages in renal immunosurveillance

Macrophages form a continuous reticular network throughout the tubulointerstitium of the kidney, which may be best appreciated in images taken in macrophage reporter models such as the MacGreen mouse, which expresses eGFP under the *Csf1r* promoter (Sasmono et al., 2003). This extensive network renders them well placed to respond rapidly to ascending urinary infection or kidney injury. In the mouse, $CD11b^{Hi}Cx3cr1^{Hi}Ly6C^{Lo}$ patrolling monocytes crawl along the luminal side of the capillary wall in the kidneys through the interaction of the integrin LFA1 with ICAM1 expressed on endothelial cells (Auffray et al., 2007). They can sense viral nucleic acids *via* TLR7, recruit neutrophils to mount an inflammatory response and then phagocytose endothelial cell debris, thereby restoring tissue architecture. Hence, they have been characterized as "housekeepers" of the renal endothelium (Carlin et al., 2013). A similar function has been ascribed to $CD14^+CD16^+$ non-inflammatory monocytes in humans (Cros et al., 2010).

The kidney represents a unique environment for macrophages as they are exposed to markedly increased osmolarity in the medulla compared with the cortex. Intriguingly, this osmolar gradient may be responsible for differences in the MNP phenotypes observed between the cortex and medulla (Berry et al., 2017). High medullary sodium concentration activates NFAT5 in tubular epithelial cells to promote CCL2 release which recruits $CD14^+$ $CD16^-$ inflammatory monocytes to the medulla and primes them to produce pro-inflammatory cytokines such as IL-6 and TNF-α to counteract ascending bacterial infection. Furthermore, disruption of the medullary osmotic gradient due to treatment with tolvaptan in patients with autosomal dominant polycystic kidney disease or in the ischemic kidney in patients with sickle cell disease, was associated with an increased risk of urinary tract infection despite increased urinary flow.

6. Macrophages in renal injury

Almost all forms of acute kidney injury (AKI) and chronic kidney disease (CKD), regardless of cause, are directly or indirectly immune-mediated (Tecklenborg et al., 2018). Recent developments have examined the role of macrophages in mediating kidney injury and promoting renal fibrosis, the final common pathway by which CKD progresses to end-stage kidney disease (ESKD) (Tang et al., 2019a). Macrophages play a prominent role in CKD, indeed, biopsies from patients with CKD show that the degree of macrophage accumulation correlates with glomerulosclerosis and tubulointerstitial fibrosis, both of which result in the loss of normal renal architecture, reduction in microvascular capillary density, hypoxia, and tubular atrophy (Klessens et al., 2017).

6.1 Pro-inflammatory macrophages predominate during acute kidney injury

In response to the onset of injury, damaged renal tubular epithelial cells release DAMPs, which act as injury signals to macrophages leading to augmented phagocytosis, antigen processing and presentation, and secretion of pro-inflammatory cytokines (Tecklenborg et al., 2018). Bone marrow-derived monocytes are recruited into injured tissues and mature into monocyte-derived macrophages (Fig. 3). Recruitment and local proliferation of monocytes is known to make a significant contribution to the overall pool of activated macrophages in the acutely injured kidney in pre-clinical models of renal injury, such as ischemia-reperfusion injury (IRI), unilateral ureteral obstruction (UUO), and renal allograft rejection, with proliferation of tissue-resident macrophages having a lesser impact (Jose et al., 2003; Le Meur et al., 2002; Lin et al., 2009).

The deleterious role of circulating monocytes and tissue macrophages in early injury has been demonstrated by depleting monocyte/macrophage in pre-clinical models. Administration of diphtheria toxin to mice that express human diphtheria toxin (DT) under the CD11b promoter prior to or in early injury has been shown to attenuate injury in various kidney disease models including glomerulonephritis (Duffield et al., 2005) and ureteric obstruction (Lin et al., 2009; Machida et al., 2010). Conversely, administration of DT to CD11b-hDTR mice did not reduce kidney injury following ischemia-reperfusion injury, while administration of liposomal clodronate was beneficial (Ferenbach et al., 2012). Diptheria toxin induced global

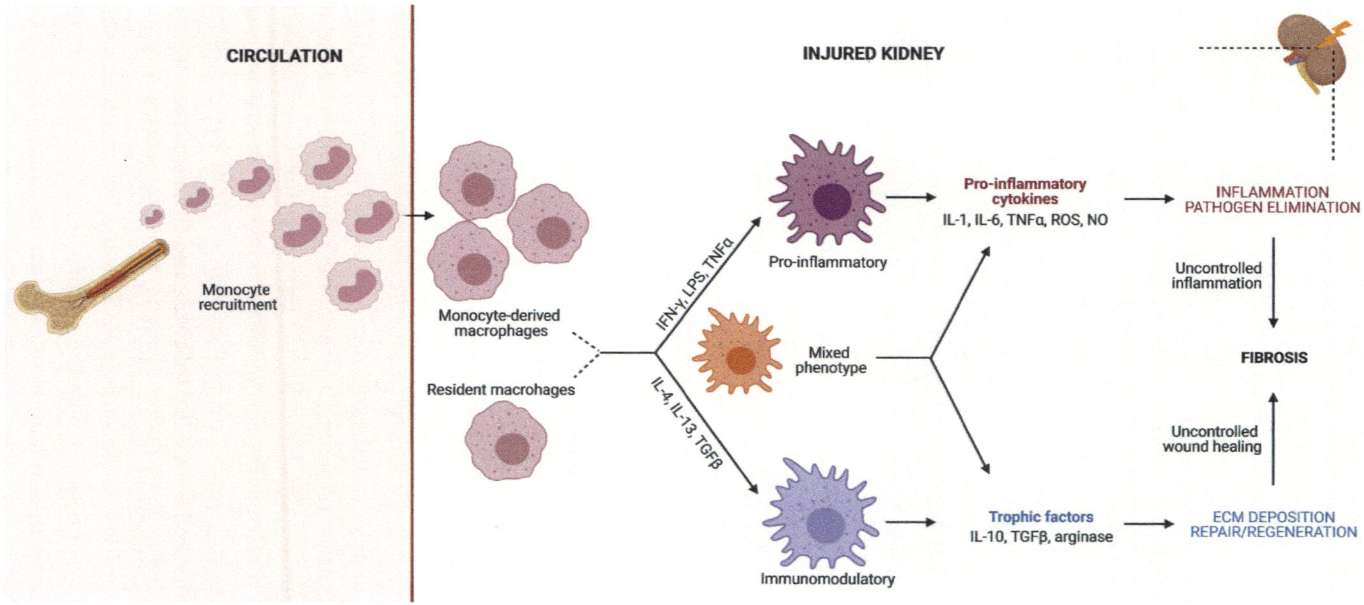

Fig. 3 Macrophage phenotypes in the injured kidney. In response to kidney injury, there is increased production and release of monocytes into the peripheral circulation from the bone marrow. Circulating monocytes enter the inflamed tissue differentiate into pro-inflammatory macrophages in early injury. In later injury, there is a switch toward a predominantly reparative, immunomodulatory phenotype to enable resolution of inflammation and repair. However, in chronic kidney disease, ongoing recruitment of pro-inflammatory monocytes and persistence of immunomodulatory macrophages in the tissues results in renal fibrosis.

macrophage depletion in the kidney, suggesting that more targeted depletion of the most injurious subsets may be the most beneficial strategy. As such, a better understanding of the mechanisms by which monocytes are recruited to the kidney, the phenotypes that they adopt and the molecular pathways by which they mediate their pathogenic effect is required to identify more specific anti-inflammatory therapies. Conversely, adoptive transfer of macrophages stimulated by LPS or IFN-γ to induce a pro-inflammatory phenotype induces more severe renal injury in a number of models of AKI (Ikezumi et al., 2003a,b; Wang et al., 2007).

$Ly6C^{Hi}CCR2^{Hi}CX3CR1^{lo}$ monocytes ($CD14^{+}CD16^{-}$ in humans) are considered pro-inflammatory, accumulating in sites of inflammation where they participate in the immune response. The role of $Ly6C^{Hi}$ monocytes in acute kidney injury has been investigated by inhibiting recruitment to the kidney by blocking the CCL2-CCR2 signaling pathway. Genetic deficiency of *Ccr2* or blockade of the CCR2 receptor inhibits tubular injury in IRI (Xu et al., 2019), ureteric obstruction (Braga et al., 2018; Li et al., 2008) and glomerulonephritis (Lloyd et al., 1997). CX3CR1, expressed in patrolling monocytes, may also be important in recruiting injurious cells to the kidney. Genetic deficiency of *Cx3cr1* inhibits monocyte infiltration and fibrosis in IRI and ureteric obstruction (Li et al., 2008; Oh et al., 2008). However, $CX3CR1^{Hi}$ monocytes migrate into tissues less frequently than $Ly6C^{Hi}$ monocytes (Auffray et al., 2007) and *Cx3cr1* is also expressed at high levels by resident renal macrophages (Liu et al., 2019), so not all of the beneficial effect may be through inhibition of monocyte recruitment. Furthermore, $Ly6C^{Hi}CX3CR1^{Lo}$ monocytes may switch to $CX3CR1^{Hi}$ macrophage-like phenotype and once this switch has occurred, CX3CR1 may mediate renal fibrosis (Peng et al., 2015).

Following acute kidney injury, tubular cells release small (<100 μM) membrane bound vesicles called exosomes and these may mediate signaling to monocytes by transfer of genetic material including mRNAs and miRs. Exosomes from injured tubular cells are enriched for *Ccl2* mRNA and when taken up by macrophages this mediates a pro-inflammatory response (Lv et al., 2020). Furthermore, urinary exosomes from patients with IgA nephropathy are enriched for *CCL2* mRNA, and this correlates with interstitial macrophage infiltration in the kidneys. Following AKI, tubular cell exosomes contain miR-19b-3p and miR-23a and when these exosomes are taken up by macrophages it promotes a pro-inflammatory phenotype, in part by neutralizing anti-inflammatory molecules such as A20 (Li et al., 2019; Lv et al., 2020).

The injured kidney cells also release DAMPs such as heat shock proteins (Lang et al., 2005), high mobility box protein-1 (Wu et al., 2010), and biglycan (Schaefer et al., 2005), which bind to pattern recognition receptors on macrophages to induce a pro-inflammatory phenotype, characterized by upregulation of cell surface markers, such as histocompatibility complex (MHC) class II, CD80 and CD86 and the IL-1b receptor. In addition, injured tubular cells release β-glucosylceramide, which binds to the Mincle receptor on macrophages, activating Syk kinase and inducing release of pro-inflammatory cytokines including TNF-α and IL-6, which mediate sustained kidney injury (Tanaka et al., 2020). Furthermore, Syk kinase is a key component of the intracellular signaling downstream of complement receptors and binding of the complement component 5a to its receptor on myeloid cells propagates further renal inflammation through production of HMGB1 and IL-1β (Zhang et al., 2020). Mice in which the myeloid cells are deficient in Syk kinase are protected from AKI, suggesting that Syk kinase inhibition may be an important therapeutic strategy to prevent activation towards pro-inflammatory macrophage phenotype (Ryan et al., 2016a,b).

Additional intracellular signaling pathways may mediate pro-inflammatory gene expression. Phosphorylation of apoptosis signal-regulating kinase 1 (ASK1/MAP3K5) in turn activates stress activated pathways such as p38 mitogen-activated protein kinase (MAPK) and the JUN N-terminal kinase 1 (JNK1) pathway in acute kidney disease (Amos et al., 2018; Ma et al., 2009). Activation of JNK is integral to adoption of a pro-inflammatory macrophage phenotype (Ikezumi et al., 2004). Many of these signaling pathways terminate in nuclear transcription factor-κβ (NF-κβ) which mediates expression of pro-inflammatory cytokines. Inhibiting the action of NF-κβ by administration of decoy oligodeoxynucleotides reduced inflammatory cytokine expression and renal injury in a rat model of glomerulonephritis (Tomita et al., 2000).

The activated macrophages secrete pro-inflammatory cytokines such as TNF-α, IL-1β, IL-12, IL-18, and IL-23 which can mediate renal cell injury and recruit additional inflammatory cells (Han et al., 2019). TNF-α, in particular, is an important mediator of inflammation in kidney disease and has been implicated in promoting podocyte or tubular cell apoptosis, thereby accelerating the renal injury (Ryu et al., 2011). Infiltrating macrophages are the primary source of TNF-α in the injured kidney, although it can also be expressed by other cells within the kidney, such as DCs and proximal tubular epithelial cells (Dong et al., 2007; Klahr and Morrissey, 1998;

Timoshanko et al., 2003). Pro-inflammatory macrophages produce additional cytotoxic factors, such as ROS and nitric oxide generated by inducible nitric oxide synthase (iNOS), both of which induce mitochondrial damage and tubular cell apoptosis, leading to tubular atrophy if the inflammation remains unresolved (Kipari et al., 2006). Macrophages activated in a pro-inflammatory environment further secrete matrix metalloproteinases (MMPs), such as MMP-1, -9 and -12, which contribute to the degradation of extracellular matrix, but may also promote ongoing injury by activating inflammatory cytokines (Tan et al., 2013).

6.2 Macrophage activation in chronic kidney disease

In CKD, the ongoing injury promotes sustained recruitment of inflammatory monocytes to the kidney and persistent activation toward a pro-inflammatory phenotype. Hence, some of the same mechanisms by which monocytes mediate renal injury in AKI may also be applicable to CKD.

Mice with genetic deficiency of *Ccr2* or intercellular adhesion molecule-1 (*Icam1*) exhibit reduced monocyte recruitment and less histological renal injury, proteinuria and decline in kidney function in models of chronic disease including diabetic nephropathy (Chow et al., 2005, 2006, 2007). In addition, the complement pathway is implicated in CKD, with increased expression of components of the complement pathway found in kidneys from patients with diabetic kidney disease (Flyvbjerg, 2017; Sheng et al., 2020; Woroniecka et al., 2011) and mice deficient for complement factor 5a are protected from diabetic nephropathy (Tan et al., 2020). The intracellular signaling pathways that are activated in AKI exhibit sustained activation in CKD including the JAK-STAT pathway, with expression of JAK-STAT pathway components correlating with the severity of tubulointerstitial fibrosis in patients with diabetic kidney disease (Berthier et al., 2009). Mice with genetic disruption of the JAK-STAT pathway exhibit reduced glomerular inflammation and matrix accumulation in streptozotocin-induced diabetes (Lu et al., 2009). There is evidence for persistence of pro-inflammatory mediators released by macrophages, with elevated TNF-α expression in kidney biopsies from patients with CKD and serum levels of soluble TNF-α receptors predicting renal outcomes in patients with CKD (Coca et al., 2017; Schrauben et al., 2021). Inhibition of TNF-α is a potential therapeutic target as this slows the progressive decline in renal function in animal models of CKD (Elmarakby et al., 2006).

scRNA-seq studies have shed light on the mechanisms by which AKI may promote ongoing inflammation in CKD. Unbiased clustering of tubular cells has identified a subpopulation of that undergo maladaptive repair and persist after AKI in pre-clinical models. This tubular cell subset expresses high levels of NFκβ-regulated genes including vascular endothelial growth factor-1 (*VCAM1*) (Kirita et al., 2020) and a similar tubular cell phenotype has been identified in the human kidney (Muto et al., 2021). Furthermore, the VCAM1+ tubular cells persisted in the kidney long after the AKI resolved and were surrounded by macrophages (Kirita et al., 2020), indicating the potential for tubular cell-macrophage cross-talk. Administration of TNF-α to tubular cells *in vitro* replicates this phenotype and the subsequent tubular cell expression of VCAM1 binds macrophages by binding β1-integrin (Wuthrich, 1992). Hence, a positive feedback loop may be generated with pro-inflammatory macrophages producing TNF-α to activate maladaptive repair in tubular cells, which in turn retains macrophages in the kidney, promoting transition to chronic inflammation in CKD.

As detailed below, after AKI the pro-inflammatory macrophages recruited to the kidney may later switch towards a reparative or M2 phenotype. If the renal injury is transient, this may be beneficial as the TGF-β they produce dampens the inflammation and promotes wound healing. However, in chronic disease, persistence of immunoregulatory macrophages may be detrimental with chronic TGF-β secretion promoting renal fibrosis (Fig. 3). In renal biopsies from patients with diabetic nephropathy, there was a correlation between the number of glomerular $CD163^+$ immunoregulatory macrophages and glomerulosclerosis, tubulointerstitial fibrosis and tubular atrophy (Klessens et al., 2017). Furthermore, studies employing protocol kidney allograft biopsies collected 1, 5 and 10 years following transplantation have reported that the number of $CD163^+$ macrophages, but not T-cells in the 1 year biopsies correlate with subsequent progression of interstitial fibrosis (Ikezumi et al., 2015).

Functional studies in pre-clinical models further support a pathogenic role for immunomodulatory M2 macrophages. Following severe ischemia-reperfusion injury, a model of progression from AKI to CKD, administration of liposomal clodronate to deplete M2 macrophages during the fibrotic phase reduced renal fibrosis (Kim et al., 2015). The benefit of clodronate depletion was abrogated by adoptive transfer of bone-marrow-derived macrophages that had been differentiated *ex vivo* toward an M2 phenotype by treatment with IL-10/TGF-β. A similar paradigm has been observed following ureteric obstruction; from day 4 onwards after obstruction, macrophages adopted an M2 phenotype and depletion of macrophages during this phase reduced renal fibrosis, whereas adoptive transfer of purified M2 macrophages promoted the accumulation of αSMA^{+} myofibroblasts (Shen et al., 2014).

In addition, inhibition of signaling pathways that promote an M2 activation phenotype may ameliorate renal fibrosis. IL-4 promotes an M2a phenotype and mice deficient in the IL4 receptor develop less renal fibrosis in a number of pre-clinical models (Liang et al., 2017). STAT6 lies downstream of IL-4 signaling in the M2 programming pathway and STAT6 deficiency in bone marrow cells inhibits fibroblast accumulation in obstructed kidneys (Yan et al., 2015). Galectin-3 is required for alternative but not classical macrophage activation (MacKinnon et al., 2008) and mice that are galectin-3 deficient are protected from renal fibrosis. Adoptive transfer of galectin-3$^{+/+}$ but not galectin-3$^{-/-}$ macrophages to galectin-3$^{-/-}$ mice restored the pro-fibrotic phenotype, highlighting the role of macrophage galectin-3 (Henderson et al., 2008). The Wnt-β-catenin signaling pathways promotes an M2 phenotype *in vitro* and macrophage specific ablation of β-catenin reduces accumulation of M2 polarized macrophages in the kidney and tubulointerstitial fibrosis following ureteric obstruction (Feng et al., 2018).

6.3 Macrophage to myofibroblast transition

It is increasingly recognized that macrophages may not only promote scarring through paracrine signaling to activate adjacent renal fibroblasts, but that they, themselves, may differentiate to a myofibroblast phenotype and contribute to extracellular matrix collagen production; however, the relative contribution that this makes to renal fibrosis is controversial. Bone marrow transplant studies using donor marrow from transgenic mice that enable fate tracing have been used to assess the relative contribution of bone marrow-derived cells that contribute to the myofibroblast pool. In studies where the donor bone marrow cells were labeled with either human placental alkaline phosphatase or eGFP up to 30% of α-SMA^{+} myofibroblasts were derived from bone marrow cells in models of ischemic-reperfusion or adriamycin nephropathy, respectively (Broekema et al., 2007; Li et al., 2007). While these studies looked at the contribution of total bone marrow-derived cells to myofibroblasts others looked specifically at the contribution of circulating monocytes by employing *Lyz2*-Cre reporter mice, where the reporter fluorochrome is specific to cells that currently or previously express the monocyte/macrophage marker gene *Lyz2* (Wang et al., 2017). Up to 35% of α-SMA^{+} myofibroblasts were labeled with the reporter, indicating a substantial contribution from macrophage to mesenchymal trans-differentiation.

However, the interpretation of these studies is difficult as macrophages and fibroblasts are adjacently located in the tubulointerstitium and their elongated processes make it difficult to distinguish them by immunofluorescence even on confocal microscopy. To circumnavigate this difficulty, other studies have employed bone marrow from donors that express fluorescent proteins under promoters of prototypical myofibroblast genes such as α-SMA (*Acta2*) or collagen (*Col1a1*). In these studies, fluorescence will only be detected in the donor bone marrow-derived cells in the kidney if they transdifferentiate to express *Acta2* or *Col1a1*. Estimates of the proportion of cells derived from bone marrow cells in these models varies widely from less than 1% up to 35% despite both reports using the ureteric obstruction model (LeBleu et al., 2013; Lin et al., 2008) More recently, parabiotic experiments involving CD45.1 and CD45.2 mismatched pairs have been employed to try to reconcile the discrepancies observed in bone marrow transplant models (Kramann et al., 2018). These demonstrated that approximately 10% of renal PDGFR-b^{+} myofibroblasts may be derived from circulating immune cells.

The contribution of immune cells *per se* to direct production of the extracellular matrix has been dissected by knocking out the *Col1a1* gene specifically in immune cells using *Cd45*Cre-*Col1a1*$^{fl/fl}$ mice (Buchtler et al., 2018). Immune cell specific *Col1a1* knockout reduced the global *Col1a1* expression in the renal cortex by 50% and total extracellular matrix deposition by 33% in a model of adenine nephropathy, without reducing the number of α-SMA^{+} myofibroblasts. Importantly, this was associated with a 33% improvement in glomerular filtration rate compared with wild-type mice, suggesting that preventing immune cell collagen production may be an important therapeutic target.

TGF-β signaling may be important in mediating the macrophages to myofibroblast transition. Macrophages express TGF-β receptors and binding of TGF-β induces pro-fibrotic gene expression through the SMAD pathway. When kidneys from wild-type mice were transplanted into SMAD3$^{-/-}$ recipients, they accumulated fewer M2-macrophages and developed less fibrosis compared with transplantation into SMAD3$^{+/+}$ controls, suggesting that the TGF-β/SMAD3 signaling pathway in immune cells is critical for the development of fibrosis (Wang et al., 2017). A similar reduction in fibrosis was observed with adoptive transfer of SMAD3$^{-/-}$ compared with SMAD3$^{+/+}$ bone marrow-derived macrophages into irradiated wild-type recipients that had undergone ureteric obstruction (Wang et al., 2016). Further studies indicated that Src kinase may mediate the TGF-β induced macrophage to myofibroblast transition (Tang et al., 2018) and that

STAT6 signaling in bone marrow-derived cells was crucial for development of a fibroblast phenotype (Yan et al., 2015).

7. Macrophages in renal repair

The mechanisms through which macrophages may promote renal repair is less well understood, in particular in human disease, as biopsies are rarely performed in patients who are improving clinically. In pre-clinical models of repair following AKI the inflammatory milieu is reversed by the infiltration of immune cells, including macrophages, which promote a reparative microenvironment by secreting anti-inflammatory cytokines.

There is a switch from a pro-inflammatory M1 phenotype in AKI to a M2-phenotype during repair, with depletion of the M2 cells inhibiting the reparative process (Lee et al., 2011). Clodronate is more selective than administering DT to CD11b-iDTR mice in depleting renal macrophages, with relative sparing of the $CD206^+$ and $CD11c^+$ cells (Ferenbach et al., 2012). Indeed, when these cells were depleted by administration of diphtheria toxin to clodronate-treated mice the protective effect of clodronate was lost, indicating the presence of a cytoprotective intrarenal population of mononuclear phagocytes sensitive to diphtheria toxin-mediated ablation. Conversely, enhancing the number of macrophages during the repair phase by administration of CSF-1 promotes repair by inducing macrophage expression of anti-inflammatory cytokines and insulin growth factor-1, which may promote tubular repair (Alikhan et al., 2011; Zhang et al., 2012). Similarly, tubular secretion of GM-CSF promoted an M2 macrophage phenotype *via* STAT5 signaling and this was important for renal repair post ischemia-reperfusion injury (Huen et al., 2015). Proliferation of resident macrophages may be important in mediating the effect of CSF-1 (Zhang et al., 2012), and further support for a protective role for resident renal macrophages is derived from the fact that selective depletion of resident $F4/80^{Hi}CD11b^+$ but not $F4/80^{Lo}CD11b^+$ infiltrating monocytes by administration of clodronate results in failure to repair from ischemia-reperfusion injury (Park et al., 2020).

The mechanisms by which reparative macrophages may promote kidney repair include suppression of inflammation, phagocytosis of apoptotic debris and excess extracellular matrix and promotion of endogenous kidney cell regeneration. However, the signaling pathways that mediate these actions are less well understood. During repair after ischemia-reperfusion injury, resident macrophages switch to a phenotype similar to that found in the embryo, including downregulation of MHCII and upregulation of the

Wnt pathway (Lever et al., 2019). Furthermore, following IRI macrophages secrete Wnt7b, which acts to restore tubular integrity by promoting tubular cell proliferation and inhibiting tubular apoptosis (Lin et al., 2010). In addition, macrophage derived sphingosine-1-phosphate may induce lipocalin-2 production to mediate tubular proliferation and repair (Sola et al., 2011).

Scavenging of apoptotic debris during repair is important both for inducing a reparative macrophage phenotype and also for restoring normal tissue homeostasis. Phagocytosis of apoptotic debris mediates a switch away from a pro-inflammatory phenotype and this may be important in limiting renal injury (Duffield et al., 2001; Hughes et al., 2004). Macrophages may also secrete apoptosis inhibitor of macrophage (AIM) protein, which can promote clearance of tubular luminal debris by mediating binding to KIM-1 on tubular cells to promote phagocytosis (Arai et al., 2016). Unlike lung, liver and brain resident macrophages, resident kidney macrophages express *Visr*, which encodes V-domain immunoglobulin suppressor of T-cell activation (VISTA). Genetic deficiency of *Visr* resulted in incomplete repair after ischemia-reperfusion injury, partly due to impaired phagocytosis of apoptotic cells but also due to failure to dampen the T-cell activation response (Park et al., 2020).

scRNA-seq studies may also highlight putative reparative macrophage phenotypes. In our model of reversible ureteric obstruction, we observed a cluster on scRNA-Seq that was comprised of cells that were present exclusively in kidneys that had undergone reversal of ureteric obstruction to induce degradation of excess scar tissue. The cells in this cluster transcriptionally resembled resident macrophages, but additionally expressed *Mmp12* and scavenger receptors, including *Gpnmb* and *Mrc1*. A similar *Mmp12*$^+$ phenotype has been observed in monocytes recruited to the liver during resolution of cirrhosis (Ramachandran et al., 2012). The transcriptome of the *Mmp12*+ cells was replicated *in vitro* by administering collagen to bone marrow-derived macrophages, suggesting that phagocytosis of scar tissue may promote this phenotype. While the *Mmp12* gene was a key marker gene in these cells, it may not be the primary driver of a reparative phenotype, as genetic deficiency of *Mmp12* had no effect on renal injury following ureteric obstruction, indeed administrtion of an anti-MMP12 neutralising antibody ameliorated glomerulonephritis (Abraham et al., 2012, 2018; Kaneko et al., 2003). Further studies will be required to determine whether these *Mmp12*$^+$ cells may promote kidney repair and if so how they might be harnessed to accelerate scar degradation.

8. Therapeutic prospects

Broadly, potential avenues for therapeutic intervention on MNP cells to ameliorate kidney disease focus on inhibiting the recruitment, proliferation or activation of pro-inflammatory monocytes or augmenting the reparative capacity of tissue macrophages (Fig. 4). A wide range of strategies have been employed including monoclonal antibodies, small molecule inhibitors, RNA interference, administration of microvesicles or using macrophages as cell-based therapy.

Blockade of pro-inflammatory monocyte recruitment to the kidney is beneficial in a number of models of acute and chronic kidney injury. As described above, administration of a CCR2 antagonist has been shown to be effective in preventing the development of fibrosis in various pre-clinical models (Kanamori et al., 2007; Sayyed, 2011; Xu et al., 2019). These encouraging results led to the first human trials of a selective CCR2 antagonist, CCX140-B, which has shown promise with a reduction in proteinuria in patients with diabetic nephropathy who were already established on renin-angiotensin blockade (de Zeeuw et al., 2015). Preventing activation of the complement pathway by administration of a human C1-inhibitor reduced macrophage infiltration and kidney injury in murine ischemia-reperfusion injury (Danobeitia et al., 2017) whereas inhibition of complement components such as factor 5a or its receptor was beneficial in pre-clinical models of diabetic nephropathy (Tan et al., 2020; Yiu et al., 2018).

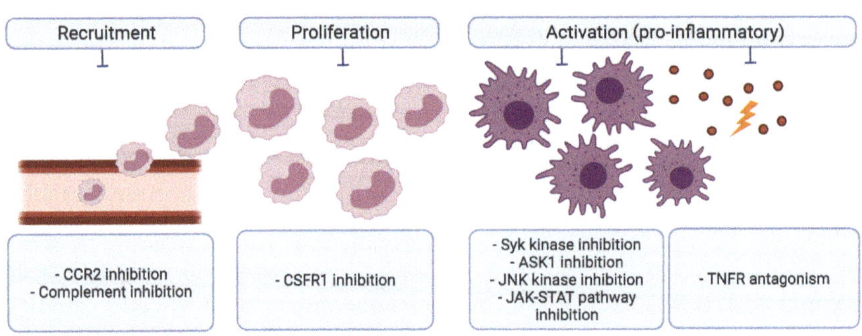

Fig. 4 Strategies targeting macrophages in kidney disease. A number of molecular targets have been identified to inhibit monocyte recruitment to the injured kidney, reduce proliferation of monocytes and macrophages or prevent activation toward a pro-inflammatory phenotype.

Preventing the robust myeloid cell proliferation that occurs early after acute kidney injury may be achieved by blocking the CSF1-CSF1R pathway. Administration of a neutralizing antibody or small molecule inhibitors targeting CSF1R (also known as c-fms) can ameliorate renal injury and fibrosis in a range of pre-clinical models of acute and chronic kidney disease, including UUO (Le Meur et al., 2002) and diabetic nephropathy (Lim et al., 2009). In addition, monoclonal antibody against c-fms reduced macrophage proliferation and tubulitis in a murine model of acute renal allograft rejection (Jose et al., 2003). However, c-fms inhibitors have not yet been studied in human kidney disease.

An alternative strategy is to inhibit the intracellular pathways *via* which monocytes recruited to the injured kidney are activated toward a pro-inflammatory phenotype. Syk kinase mediates pro-inflammatory intracellular signaling downstream from complement receptors and Mincle and activation of Syk kinase has been observed in macrophages in rapidly progressive glomerulonephritis in humans (McAdoo et al., 2015; Ryan et al., 2016b). Syk kinase inhibitors have been beneficial in a number of pre-clinical models including ischemia-reperfusion injury (Ryan et al., 2016a), ureteric obstruction (Ma et al., 2016), vasculitis, glomerulonephritis (McAdoo et al., 2014, 2020) and allograft rejection (Ramessur Chandran et al., 2015). Syk inhibitors are currently in phase 2 trials in human kidney disease. The full results of a clinical trial of the Syk kinase inhibitor, *Fostamatinib,* in patients with IgA nephropathy are awaited (NCT02112838), while the ongoing FOSTAMR Trial is examining the efficacy of fostamatinib in patients with chronic active antibody-mediated rejection (NCT03991780).

The ASK1-JNK-P38 kinase signaling cascade is another target attractive target, with ASK1 inhibitors reducing inflammation, fibrosis and decline in kidney function in pre-clinical models (Liles et al., 2018), while JNK or p38 kinase inhibitors ameliorate renal pathology in models of renal fibrosis (Ma et al., 2007), anti-GBM disease (Flanc et al., 2007) and cisplatin nephrotoxicity (Ramesh and Reeves, 2005). The ASK1 inhibitor, selonsertib, is now in phase 2 clinical trials in patients with diabetic kidney disease. In the first published trial, selonsertib had no effect on albuminuria while the effect on GFR was complicated by the inhibitor unexpectedly reducing tubular creatinine secretion, thereby apparently decreasing eGFR. Although the primary end-points were not achieved, *post hoc* analysis suggested that following the early reduction in eGFR observed in the first 4 weeks related to inhibition of creatinine secretion, there was a slower decline in eGFR over the subsequent 44 weeks (Chertow et al., 2019). A phase 2b trial of selonsertib in moderately advanced diabetic nephropathy (NCT04026165) has been completed and results are awaited. Inhibitors of the JAK-STAT pathway have been trialed in phase 2 studies in humans, with Baricitinib, an oral, reversible, pan-JAK1 and JAK2 inhibitor, reducing albuminuria in patients with type 2 diabetic nephropathy (Tuttle et al., 2018). Targeting the signaling pathways that promote macrophage to myofibroblast differentiation may also be beneficial, with administration of a small molecule inhibitor of STAT6 inhibiting renal accumulation of $CD45^+$ cells that co-express myofibroblast markers and reducing fibrosis in ureteric obstruction and folic acid nephropathy models (Jiao et al., 2021).

Antagonism of the injurious molecules by which pro-inflammatory macrophages may induce injury and death of endogenous renal cells represents a more specific therapeutic approach. Administration of monoclonal neutralizing antibodies that target the TNF receptor are beneficial in a number of pre-clinical models including Alport nephropathy (Ryu et al., 2011).

An interesting area of research is whether macrophages could be utilized in cell therapy as "Trojan horses" to deliver anti-inflammatory cytokines to modulate kidney disease. Indeed, adoptive transfer of macrophages that have been programmed *ex vivo* to adopt an anti-inflammatory phenotype dampens inflammation and ameliorates kidney injury in a number of pre-clinical models. Administration of splenic macrophages treated with IL4 and IL-13 to promote an M2a phenotype inhibited the recruitment of host inflammatory immune cells and reduced functional and histological injury in adriamycin nephropathy in SCID mice (Wang et al., 2007) and an even greater benefit was observed when macrophages were polarized toward an M2c phenotype using IL-10 and TGF-β *ex vivo* (Lu et al., 2013). Similarly macrophages programmed with IL-25, which promotes a Th2 phenotype, ameliorate adriamycin nephropathy (Cao et al., 2011) and dexamethasone pre-treated macrophages are beneficial in glomerulonephritis (Ikezumi et al., 2003a). In addition, intra-arterial injection of macrophages transduced with adenovirus expressing either IL-10 or a dominant negative κβ to deactivate NF-κβ signaling prevented recruitment of host macrophages and glomerular injury in glomerulonephritis (Wilson et al., 2002, 2005). One concern of adoptive transfer is that, given the inherent plasticity of macrophages, they may switch to adopt a pro-inflammatory phenotype once they encounter DAMPs and cytokines in the injured kidney. Indeed this has been observed with bone marrow-derived macrophages, which undergo proliferation when they enter the kidney, but not with more mature splenic macrophages (Cao et al., 2014). Clearly splenic macrophages are not readily harvestable for cell therapy in humans; however, a similar sustained anti-inflammatory phenotype is observed with more accessible peritoneal macrophages (Cao et al., 2018). Pilot studies of macrophage cell therapy have been performed in a small number of transplant patients (Hutchinson et al., 2011) and autologous monocytes derived from plasmapheresis and expanded *ex vivo* have been administered to patients with liver cirrhosis with limited side effects, though large efficacy studies are yet to be performed (Moroni et al., 2019). The ONE trial in patients undergoing renal transplantation employed six different cell-based therapy approaches, including administration of recipient Treg cells and regulatory macrophages induced by treatment of donor peripheral blood monocytes with macrophage colony-stimulating factor and IFNγ (Sawitzki et al., 2020). Combined analysis of all six approaches suggested that cell-based therapy could facilitate earlier weaning of immunosuppression with fewer infections and no increased risk of acute rejection.

Given the challenges of utilizing macrophages in cell therapy, more reductionist strategies to replicate their beneficial effects have been investigated including the use of microvesicles. Microvesicles are small, membrane-encapsulated structures that are released and taken up by most cell types including macrophages and kidney cells, and which can convey inter-cellular signaling across large distances; hence, they may be harnessed to deliver targeted therapy to the kidney (Tang et al., 2019b). For example, treatment of the macrophage RAW cell line *in vitro* with dexamethasone resulted in release of dexamethasone-containing microvesicles into the supernatant. When these microvesicles were administered to mice, they attenuated renal inflammation and fibrosis in models of LPS- or adriamycin-induced renal injury and the therapeutic effect was greater and side-effects fewer than equimolar doses of systemic dexamethasone (Tang et al., 2019c). Similarly, microvesicles loaded with the anti-inflammatory cytokine IL-10 attenuated renal ischemia-reperfusion injury, in part by promoting a reparative M2 macrophage polarization in the tubulointerstitium (Tang et al., 2020).

Trials of immune modulators would be greatly facilitated if researchers had the ability to serially monitor the efficacy of the drug in dampening kidney inflammation using non-invasive biomarkers, thereby obviating the need for renal biopsies. Serum/plasma biomarkers of inflammation including soluble TNF receptors-1 and -2 or monocyte chemoattractant protein-1 (MCP-1) are associated with poor outcomes in patients with CKD, however, they are likely to reflect inflammation in multiple organs. Analysis of the urine is more likely to be informative of inflammation specific to the kidney, and the urinary concentrations of several inflammatory molecules including complement (Zhao et al., 2019), MCP-1 (Ding et al., 2018) and CD163 (O'Reilly et al., 2016) correlate with the severity of intrinsic renal inflammation as determined in renal biopsies across a range of kidney diseases and may be used to detect flares of vasculitis or lupus nephritis. Clinical trials of anti-inflammatory therapies in kidney disease now routinely employ such non-invasive biomarkers of renal inflammation to monitor drug efficacy. For example, the JAK inhibitor baricitinib reduced plasma soluble TNF receptor and urinary MCP-1 levels in patients with diabetic kidney disease. A particularly interesting feature of the trial of the Syk kinase inhibitor, *fostamatinib,* in patients with IgA nephropathy is the fact that serial biopsies have been conducted NCT02112838. This will enable correlation between any changes observed in intra-renal inflammation with serum and urinary biomarkers.

9. Conclusions

The MNP system plays a key role in renal injury and repair and the increasing understanding of the mechanisms by which these cells are recruited to the kidney and are activated into pro- or anti-inflammatory phenotypes offer promise of new therapies to treat kidney disease; however, most therapies are not yet sufficiently selective for individual phenotypes, and hence may have off target effects. In addition, the timing of such therapies is critical, given that MNP cells are highly plastic and may readily switch phenotypes, taking their cues from the renal environment, which may differ between AKI, through transition to CKD. This is critical in human disease, where, unlike pre-clinical models, there is likely to be a complex interaction of successive waves of injury and concurrent attempts at repair. New technologies, such as scRNA-Seq have identified new MNP subsets in pre-clinical models, however, the MNP heterogeneity in the human kidney during injury and repair is less well characterized. Future studies should focus on human kidney disease and harness the findings of scRNA-Seq studies to identify therapies that can more specifically target pro-inflammatory pathways or augment reparative macrophage subsets to prevent progression of kidney disease and promote renal repair.

References

Abraham, A.P., Ma, F.Y., Mulley, W.R., Ozols, E., Nikolic-Paterson, D.J., 2012. Macrophage infiltration and renal damage are independent of matrix metalloproteinase 12 in the obstructed kidney. Nephrology (Carlton) 17, 322–329.

Abraham, A.P., Ma, F.Y., Mulley, W.R., Nikolic-Paterson, D.J., Tesch, G.H., 2018. Matrix metalloproteinase-12 deficiency attenuates experimental crescentic anti-glomerular basement membrane glomerulonephritis. Nephrology (Carlton) 23, 183–189.

Alikhan, M.A., Jones, C.V., Williams, T.M., Beckhouse, A.G., Fletcher, A.L., Kett, M.M., Sakkal, S., Samuel, C.S., Ramsay, R.G., Deane, J.A., Wells, C.A., Little, M.H., Hume, D.A., Ricardo, S.D., 2011. Colony-stimulating factor-1 promotes kidney growth and repair via alteration of macrophage responses. Am. J. Pathol. 179, 1243–1256.

Amos, L.A., Ma, F.Y., Tesch, G.H., Liles, J.T., Breckenridge, D.G., Nikolic-Paterson, D.J., Han, Y., 2018. ASK1 inhibitor treatment suppresses p38/JNK signalling with reduced kidney inflammation and fibrosis in rat crescentic glomerulonephritis. J. Cell. Mol. Med. 22, 4522–4533.

Arai, S., Kitada, K., Yamazaki, T., Takai, R., Zhang, X., Tsugawa, Y., Sugisawa, R., Matsumoto, A., Mori, M., Yoshihara, Y., Doi, K., Maehara, N., Kusunoki, S., Takahata, A., Noiri, E., Suzuki, Y., Yahagi, N., Nishiyama, A., Gunaratnam, L., Takano, T., Miyazaki, T., 2016. Apoptosis inhibitor of macrophage protein enhances intraluminal debris clearance and ameliorates acute kidney injury in mice. Nat. Med. 22, 183–193.

Arazi, A., Rao, D.A., Berthier, C.C., Davidson, A., Liu, Y., Hoover, P.J., Chicoine, A., Eisenhaure, T.M., Jonsson, A.H., Li, S., Lieb, D.J., Zhang, F., Slowikowski, K., Browne, E.P., Noma, A., Sutherby, D., Steelman, S., Smilek, D.E., Tosta, P., Apruzzese, W., Massarotti, E., Dall'era, M., Park, M., Kamen, D.L., Furie, R.A., Payan-Schober, F., Pendergraft 3rd, W.F., McInnis, E.A., Buyon, J.P., Petri, M.A., Putterman, C., Kalunian, K.C., Woodle, E.S., Lederer, J.A., Hildeman, D.A., Nusbaum, C., Raychaudhuri, S., Kretzler, M., Anolik, J.H., Brenner, M.B., Wofsy, D., Hacohen, N., Diamond, B., In, A.M.P., S. L. E. N., 2019. The immune cell landscape in kidneys of patients with lupus nephritis. Nat. Immunol. 20, 902–914.

Auffray, C., Fogg, D., Garfa, M., Elain, G., Join-Lambert, O., Kayal, S., Sarnacki, S., Cumano, A., Lauvau, G., Geissmann, F., 2007. Monitoring of blood vessels and tissues by a population of monocytes with patrolling behavior. Science 317, 666–670.

Bain, C.C., Bravo-Blas, A., Scott, C.L., Perdiguero, E.G., Geissmann, F., Henri, S., Malissen, B., Osborne, L.C., Artis, D., Mowat, A.M., 2014. Constant replenishment from circulating monocytes maintains the macrophage pool in the intestine of adult mice. Nat. Immunol. 15, 929–937.

Bain, C.C., Gibson, D.A., Steers, N.J., Boufea, K., Louwe, P.A., Doherty, C., Gonzalez-Huici, V., Gentek, R., Magalhaes-Pinto, M., Shaw, T., Bajenoff, M., Benezech, C., Walmsley, S.R., Dockrell, D.H., Saunders, P.T.K., Batada, N.N., Jenkins, S.J., 2020. Rate of replenishment and microenvironment contribute to the sexually dimorphic phenotype and function of peritoneal macrophages. Sci. Immunol. 5 (48) (eabc4466).

Berry, M.R., Mathews, R.J., Ferdinand, J.R., Jing, C., Loudon, K.W., Wlodek, E., Dennison, T.W., Kuper, C., Neuhofer, W., Clatworthy, M.R., 2017. Renal sodium gradient orchestrates a dynamic antibacterial defense zone. Cell 170, 860–874 (e19).

Berthier, C.C., Zhang, H., Schin, M., Henger, A., Nelson, R.G., Yee, B., Boucherot, A., Neusser, M.A., Cohen, C.D., Carter-Su, C., Argetsinger, L.S., Rastaldi, M.P., Brosius, F.C., Kretzler, M., 2009. Enhanced expression of Janus kinase-signal transducer and activator of transcription pathway members in human diabetic nephropathy. Diabetes 58, 469–477.

Braga, T.T., Correa-Costa, M., Silva, R.C., Cruz, M.C., Hiyane, M.I., Da Silva, J.S., Perez, K.R., Cuccovia, I.M., Camara, N.O.S., 2018. CCR2 contributes to the recruitment of monocytes and leads to kidney inflammation and fibrosis development. Inflammopharmacology 26, 403–411.

Brahler, S., Zinselmeyer, B.H., Raju, S., Nitschke, M., Suleiman, H., Saunders, B.T., Johnson, M.W., Bohner, A.M.C., Viehmann, S.F., Theisen, D.J., Kretzer, N.M., Briseno, C.G., Zaitsev, K., Ornatsky, O., Chang, Q., Carrero, J.A., Kopp, J.B.,

Artyomov, M.N., Kurts, C., Murphy, K.M., Miner, J.H., Shaw, A.S., 2018. Opposing roles of dendritic cell subsets in experimental GN. J. Am. Soc. Nephrol. 29, 138–154.

Broekema, M., Harmsen, M.C., Van Luyn, M.J., Koerts, J.A., Petersen, A.H., Van Kooten, T.G., Van Goor, H., Navis, G., Popa, E.R., 2007. Bone marrow-derived myofibroblasts contribute to the renal interstitial myofibroblast population and produce procollagen I after ischemia/reperfusion in rats. J. Am. Soc. Nephrol. 18, 165–175.

Buchtler, S., Grill, A., Hofmarksrichter, S., Stockert, P., Schiechl-Brachner, G., Rodriguez Gomez, M., Neumayer, S., Schmidbauer, K., Talke, Y., Klinkhammer, B.M., Boor, P., Medvinsky, A., Renner, K., Castrop, H., Mack, M., 2018. Cellular origin and functional relevance of collagen I production in the kidney. J. Am. Soc. Nephrol. 29, 1859–1873.

Cao, Q., Wang, C., Zheng, D., Wang, Y., Lee, V.W., Wang, Y.M., Zheng, G., Tan, T.K., Yu, D., Alexander, S.I., Harris, D.C., Wang, Y., 2011. IL-25 induces M2 macrophages and reduces renal injury in proteinuric kidney disease. J. Am. Soc. Nephrol. 22, 1229–1239.

Cao, Q., Wang, Y., Zheng, D., Sun, Y., Wang, C., Wang, X.M., Lee, V.W., Wang, Y., Zheng, G., Tan, T.K., Wang, Y.M., Alexander, S.I., Harris, D.C., 2014. Failed renoprotection by alternatively activated bone marrow macrophages is due to a proliferation-dependent phenotype switch in vivo. Kidney Int. 85, 794–806.

Cao, Q., Wang, Y., Wang, C., Wang, X.M., Lee, V.W.S., Zheng, G., Zhao, Y., Alexander, S.I., Harris, D.C.H., 2018. Therapeutic potential of regulatory macrophages generated from peritoneal dialysate in adriamycin nephropathy. Am. J. Physiol. Renal Physiol. 314, F561–F571.

Carlin, L.M., Stamatiades, E.G., Auffray, C., Hanna, R.N., Glover, L., Vizcay-Barrena, G., Hedrick, C.C., Cook, H.T., Diebold, S., Geissmann, F., 2013. Nr4a1-dependent Ly6C(low) monocytes monitor endothelial cells and orchestrate their disposal. Cell 153, 362–375.

Chen, T., Cao, Q., Wang, Y., Harris, D.C.H., 2019. M2 macrophages in kidney disease: biology, therapies, and perspectives. Kidney Int. 95, 760–773.

Chertow, G.M., Pergola, P.E., Chen, F., Kirby, B.J., Sundy, J.S., Patel, U.D., Investigators, G.-U.-., 2019. Effects of Selonsertib in patients with diabetic kidney disease. J. Am. Soc. Nephrol. 30, 1980–1990.

Chow, F.Y., Nikolic-Paterson, D.J., Ozols, E., Atkins, R.C., Tesch, G.H., 2005. Intercellular adhesion molecule-1 deficiency is protective against nephropathy in type 2 diabetic db/db mice. J. Am. Soc. Nephrol. 16, 1711–1722.

Chow, F.Y., Nikolic-Paterson, D.J., Ozols, E., Atkins, R.C., Rollin, B.J., Tesch, G.H., 2006. Monocyte chemoattractant protein-1 promotes the development of diabetic renal injury in streptozotocin-treated mice. Kidney Int. 69, 73–80.

Chow, F.Y., Nikolic-Paterson, D.J., Ma, F.Y., Ozols, E., Rollins, B.J., Tesch, G.H., 2007. Monocyte chemoattractant protein-1-induced tissue inflammation is critical for the development of renal injury but not type 2 diabetes in obese db/db mice. Diabetologia 50, 471–480.

Coca, S.G., Nadkarni, G.N., Huang, Y., Moledina, D.G., Rao, V., Zhang, J., Ferket, B., Crowley, S.T., Fried, L.F., Parikh, C.R., 2017. Plasma biomarkers and kidney function decline in early and established diabetic kidney disease. J. Am. Soc. Nephrol. 28, 2786–2793.

Conway, B.R., O'Sullivan, E.D., Cairns, C., O'Sullivan, J., Simpson, D.J., Salzano, A., Connor, K., Ding, P., Humphries, D., Stewart, K., Teenan, O., Pius, R., Henderson, N.C., Benezech, C., Ramachandran, P., Ferenbach, D., Hughes, J., Chandra, T., Denby, L., 2020. Kidney single-cell atlas reveals myeloid heterogeneity in progression and regression of kidney disease. J. Am. Soc. Nephrol. 31, 2833–2854.

Cormican, S., Griffin, M.D., 2020. Human monocyte subset distinctions and function: insights from gene expression analysis. Front. Immunol. 11, 1070.

Cros, J., Cagnard, N., Woollard, K., Patey, N., Zhang, S.Y., Senechal, B., Puel, A., Biswas, S.K., Moshous, D., Picard, C., Jais, J.P., D'cruz, D., Casanova, J.L., Trouillet, C., Geissmann, F., 2010. Human Cd14dim monocytes patrol and sense nucleic acids and viruses via TLR7 and TLR8 receptors. Immunity 33, 375–386.

Danobeitia, J.S., Ziemelis, M., Ma, X., Zitur, L.J., Zens, T., Chlebeck, P.J., Van Amersfoort, E.S., Fernandez, L.A., 2017. Complement inhibition attenuates acute kidney injury after ischemia-reperfusion and limits progression to renal fibrosis in mice. PLoS One 12, e0183701.

de Zeeuw, D., Bekker, P., Henkel, E., Hasslacher, C., Gouni-Berthold, I., Mehling, H., Potarca, A., Tesar, V., Heerspink, H.J., Schall, T.J., CCX140-B Diabetic Nephropathy Study Group, 2015. The effect of CCR2 inhibitor CCX140-B on residual albuminuria in patients with type 2 diabetes and nephropathy: a randomised trial. Lancet Diabetes Endocrinol. 3, 687–696.

Dhillon, P., Park, J., Hurtado Del Pozo, C., Li, L., Doke, T., Huang, S., Zhao, J., Kang, H.M., Shrestra, R., Balzer, M.S., Chatterjee, S., Prado, P., Han, S.Y., Liu, H., Sheng, X., Dierickx, P., Batmanov, K., Romero, J.P., Prosper, F., Li, M., Pei, L., Kim, J., Montserrat, N., Susztak, K., 2021. The nuclear receptor ESRRA protects from kidney disease by coupling metabolism and differentiation. Cell Metab. 33, 379–394. e8.

Dick, S.A., Macklin, J.A., Nejat, S., Momen, A., Clemente-Casares, X., Althagafi, M.G., Chen, J., Kantores, C., Hosseinzadeh, S., Aronoff, L., Wong, A., Zaman, R., Barbu, I., Besla, R., Lavine, K.J., Razani, B., Ginhoux, F., Husain, M., Cybulsky, M.I., Robbins, C.S., Epelman, S., 2019. Self-renewing resident cardiac macrophages limit adverse remodeling following myocardial infarction. Nat. Immunol. 20, 29–39.

Ding, Y., Nie, L.M., Pang, Y., Wu, W.J., Tan, Y., Yu, F., Zhao, M.H., 2018. Composite urinary biomarkers to predict pathological tubulointerstitial lesions in lupus nephritis. Lupus 27, 1778–1789.

Dong, X., Swaminathan, S., Bachman, L.A., Croatt, A.J., Nath, K.A., Griffin, M.D., 2007. Resident dendritic cells are the predominant TNF-secreting cell in early renal ischemia-reperfusion injury. Kidney Int. 71, 619–628.

Duffield, J.S., Ware, C.F., Ryffel, B., Savill, J., 2001. Suppression by apoptotic cells defines tumor necrosis factor-mediated induction of glomerular mesangial cell apoptosis by activated macrophages. Am. J. Pathol. 159, 1397–1404.

Duffield, J.S., Tipping, P.G., Kipari, T., Cailhier, J.F., Clay, S., Lang, R., Bonventre, J.V., Hughes, J., 2005. Conditional ablation of macrophages halts progression of crescentic glomerulonephritis. Am. J. Pathol. 167, 1207–1219.

Elmarakby, A.A., Quigley, J.E., Pollock, D.M., Imig, J.D., 2006. Tumor necrosis factor alpha blockade increases renal Cyp2c23 expression and slows the progression of renal damage in salt-sensitive hypertension. Hypertension 47, 557–562.

Epelman, S., Lavine, K.J., Beaudin, A.E., Sojka, D.K., Carrero, J.A., Calderon, B., Brija, T., Gautier, E.L., Ivanov, S., Satpathy, A.T., Schilling, J.D., Schwendener, R., Sergin, I., Razani, B., Forsberg, E.C., Yokoyama, W.M., Unanue, E.R., Colonna, M., Randolph, G.J., Mann, D.L., 2014. Embryonic and adult-derived resident cardiac macrophages are maintained through distinct mechanisms at steady state and during inflammation. Immunity 40, 91–104.

Feng, Y., Ren, J., Gui, Y., Wei, W., Shu, B., Lu, Q., Xue, X., Sun, X., He, W., Yang, J., Dai, C., 2018. Wnt/beta-catenin-promoted macrophage alternative activation contributes to kidney fibrosis. J. Am. Soc. Nephrol. 29, 182–193.

Ferenbach, D.A., Sheldrake, T.A., Dhaliwal, K., Kipari, T.M., Marson, L.P., Kluth, D.C., Hughes, J., 2012. Macrophage/monocyte depletion by clodronate, but not diphtheria toxin, improves renal ischemia/reperfusion injury in mice. Kidney Int. 82, 928–933.

Flanc, R.S., Ma, F.Y., Tesch, G.H., Han, Y., Atkins, R.C., Bennett, B.L., Friedman, G.C., Fan, J.H., Nikolic-Paterson, D.J., 2007. A pathogenic role for JNK signaling in experimental anti-GBM glomerulonephritis. Kidney Int. 72, 698–708.
Flyvbjerg, A., 2017. The role of the complement system in diabetic nephropathy. Nat. Rev. Nephrol. 13, 311–318.
Geissmann, F., Jung, S., Littman, D.R., 2003. Blood monocytes consist of two principal subsets with distinct migratory properties. Immunity 19, 71–82.
Gentek, R., Ghigo, C., Hoeffel, G., Bulle, M.J., Msallam, R., Gautier, G., Launay, P., Chen, J., Ginhoux, F., Bajenoff, M., 2018. Hemogenic endothelial fate mapping reveals dual developmental origin of mast cells. Immunity 48, 1160–1171. e5.
Ginhoux, F., Guilliams, M., 2016. Tissue-resident macrophage ontogeny and homeostasis. Immunity 44, 439–449.
Gordon, S., 2003. Alternative activation of macrophages. Nat. Rev. Immunol. 3, 23–35.
Gottschalk, C., Kurts, C., 2015. The debate about dendritic cells and macrophages in the kidney. Front. Immunol. 6, 435.
Hamann, J., Koning, N., Pouwels, W., Ulfman, L.H., Van Eijk, M., Stacey, M., Lin, H.H., Gordon, S., Kwakkenbos, M.J., 2007. EMR1, the human homolog of F4/80, is an eosinophil-specific receptor. Eur. J. Immunol. 37, 2797–2802.
Han, H.I., Skvarca, L.B., Espiritu, E.B., Davidson, A.J., Hukriede, N.A., 2019. The role of macrophages during acute kidney injury: destruction and repair. Pediatr. Nephrol. 34, 561–569.
Henderson, N.C., MacKinnon, A.C., Farnworth, S.L., Poirier, F., Russo, F.P., Iredale, J.P., Haslett, C., Simpson, K.J., Sethi, T., 2006. Galectin-3 regulates myofibroblast activation and hepatic fibrosis. Proc. Natl. Acad. Sci. U. S. A. 103, 5060–5065.
Henderson, N.C., MacKinnon, A.C., Farnworth, S.L., Kipari, T., Haslett, C., Iredale, J.P., Liu, F.T., Hughes, J., Sethi, T., 2008. Galectin-3 expression and secretion links macrophages to the promotion of renal fibrosis. Am. J. Pathol. 172, 288–298.
Hoeffel, G., Ginhoux, F., 2018. Fetal monocytes and the origins of tissue-resident macrophages. Cell. Immunol. 330, 5–15.
Hoeffel, G., Chen, J., Lavin, Y., Low, D., Almeida, F.F., See, P., Beaudin, A.E., Lum, J., Low, I., Forsberg, E.C., Poidinger, M., Zolezzi, F., Larbi, A., Ng, L.G., Chan, J.K., Greter, M., Becher, B., Samokhvalov, I.M., Merad, M., Ginhoux, F., 2015. C-Myb(+) erythro-myeloid progenitor-derived fetal monocytes give rise to adult tissue-resident macrophages. Immunity 42, 665–678.
Huen, S.C., Huynh, L., Marlier, A., Lee, Y., Moeckel, G.W., Cantley, L.G., 2015. Gm-CSF promotes macrophage alternative activation after renal ischemia/reperfusion injury. J. Am. Soc. Nephrol. 26, 1334–1345.
Hughes, J., Cailhier, J.F., Watson, S., Savill, J.S., 2004. Apoptosis in glomerulonephritis. Rheum. Dis. Clin. North Am. 30 (3), 655–676.
Hutchinson, J.A., Riquelme, P., Sawitzki, B., Tomiuk, S., Miqueu, P., Zuhayra, M., Oberg, H.H., Pascher, A., Lutzen, U., Janssen, U., Broichhausen, C., Renders, L., Thaiss, F., Scheuermann, E., Henze, E., Volk, H.D., Chatenoud, L., Lechler, R.I., Wood, K.J., Kabelitz, D., Schlitt, H.J., Geissler, E.K., Fandrich, F., 2011. Cutting edge: immunological consequences and trafficking of human regulatory macrophages administered to renal transplant recipients. J. Immunol. 187, 2072–2078.
Ikezumi, Y., Atkins, R.C., Nikolic-Paterson, D.J., 2003a. Interferon-gamma augments acute macrophage-mediated renal injury via a glucocorticoid-sensitive mechanism. J. Am. Soc. Nephrol. 14, 888–898.
Ikezumi, Y., Hurst, L.A., Masaki, T., Atkins, R.C., Nikolic-Paterson, D.J., 2003b. Adoptive transfer studies demonstrate that macrophages can induce proteinuria and mesangial cell proliferation. Kidney Int. 63, 83–95.

Ikezumi, Y., Hurst, L., Atkins, R.C., Nikolic-Paterson, D.J., 2004. Macrophage-mediated renal injury is dependent on signaling via the JNK pathway. J. Am. Soc. Nephrol. 15, 1775–1784.

Ikezumi, Y., Suzuki, T., Yamada, T., Hasegawa, H., Kaneko, U., Hara, M., Yanagihara, T., Nikolic-Paterson, D.J., Saitoh, A., 2015. Alternatively activated macrophages in the pathogenesis of chronic kidney allograft injury. Pediatr. Nephrol. 30, 1007–1017.

Jiao, B., An, C., Tran, M., Du, H., Wang, P., Zhou, D., Wang, Y., 2021. Pharmacological inhibition of STAT6 ameliorates myeloid fibroblast activation and alternative macrophage polarization in renal fibrosis. Front. Immunol. 12, 735014.

Jose, M.D., Le Meur, Y., Atkins, R.C., Chadban, S.J., 2003. Blockade of macrophage colony-stimulating factor reduces macrophage proliferation and accumulation in renal allograft rejection. Am. J. Transplant. 3, 294–300.

Kanamori, H, et al., 2007. Inhibition of MCP-1/CCR2 pathway ameliorates the development of diabetic nephropathy. Biochem. Biophys. Res. Commun. 360 (4), 772–777. https://doi.org/10.1016/j.bbrc.2007.06.148.

Kaneko, Y., Sakatsume, M., Xie, Y., Kuroda, T., Igashima, M., Narita, I., Gejyo, F., 2003. Macrophage metalloelastase as a major factor for glomerular injury in anti-glomerular basement membrane nephritis. J. Immunol. 170, 3377–3385.

Kawakami, T., Lichtnekert, J., Thompson, L.J., Karna, P., Bouabe, H., Hohl, T.M., Heinecke, J.W., Ziegler, S.F., Nelson, P.J., Duffield, J.S., 2013. Resident renal mononuclear phagocytes comprise five discrete populations with distinct phenotypes and functions. J. Immunol. 191, 3358–3372.

Kikuchi, Y., Kobayashi, S., Hemmi, N., Ikee, R., Hyodo, N., Saigusa, T., Namikoshi, T., Yamada, M., Suzuki, S., Miura, S., 2004. Galectin-3-positive cell infiltration in human diabetic nephropathy. Nephrol. Dial. Transplant. 19, 602–607.

Kim, M.G., Kim, S.C., Ko, Y.S., Lee, H.Y., Jo, S.K., Cho, W., 2015. The role of M2 macrophages in the progression of chronic kidney disease following acute kidney injury. PLoS One 10, e0143961.

Kipari, T., Cailhier, J.F., Ferenbach, D., Watson, S., Houlberg, K., Walbaum, D., Clay, S., Savill, J., Hughes, J., 2006. Nitric oxide is an important mediator of renal tubular epithelial cell death in vitro and in murine experimental hydronephrosis. Am. J. Pathol. 169, 388–399.

Kirita, Y., Wu, H., Uchimura, K., Wilson, P.C., Humphreys, B.D., 2020. Cell profiling of mouse acute kidney injury reveals conserved cellular responses to injury. Proc. Natl. Acad. Sci. U. S. A. 117, 15874–15883.

Klahr, S., Morrissey, J.J., 1998. The role of growth factors, cytokines, and vasoactive compounds in obstructive nephropathy. Semin. Nephrol. 18, 622–632.

Klessens, C.Q.F., Zandbergen, M., Wolterbeek, R., Bruijn, J.A., Rabelink, T.J., Bajema, I.M., Dht, I.J., 2017. Macrophages in diabetic nephropathy in patients with type 2 diabetes. Nephrol. Dial. Transplant. 32, 1322–1329.

Kramann, R., Machado, F., Wu, H., Kusaba, T., Hoeft, K., Schneider, R.K., Humphreys, B.D., 2018. Parabiosis and single-cell RNA sequencing reveal a limited contribution of monocytes to myofibroblasts in kidney fibrosis. JCI Insight 3 (9), e99561.

Kruger, T., Benke, D., Eitner, F., Lang, A., Wirtz, M., Hamilton-Williams, E.E., Engel, D., Giese, B., Muller-Newen, G., Floege, J., Kurts, C., 2004. Identification and functional characterization of dendritic cells in the healthy murine kidney and in experimental glomerulonephritis. J. Am. Soc. Nephrol. 15, 613–621.

Kurts, C., Ginhoux, F., Panzer, U., 2020. Kidney dendritic cells: fundamental biology and functional roles in health and disease. Nat. Rev. Nephrol. 16, 391–407.

Lang, A., Benke, D., Eitner, F., Engel, D., Ehrlich, S., Breloer, M., Hamilton-Williams, E., Specht, S., Hoerauf, A., Floege, J., Von Bonin, A., Kurts, C., 2005. Heat shock protein

60 is released in immune-mediated glomerulonephritis and aggravates disease: in vivo evidence for an immunologic danger signal. J. Am. Soc. Nephrol. 16, 383–391.
Le Meur, Y., Tesch, G.H., Hill, P.A., Mu, W., Foti, R., Nikolic-Paterson, D.J., Atkins, R.C., 2002. Macrophage accumulation at a site of renal inflammation is dependent on the M-CSF/c-fms pathway. J. Leukoc. Biol. 72, 530–537.
LeBleu, V.S., Taduri, G., O'Connell, J., Teng, Y., Cooke, V.G., Woda, C., Sugimoto, H., Kalluri, R., 2013. Origin and function of myofibroblasts in kidney fibrosis. Nat. Med. 19, 1047–1053.
Lee, S., Huen, S., Nishio, H., Nishio, S., Lee, H.K., Choi, B.S., Ruhrberg, C., Cantley, L.G., 2011. Distinct macrophage phenotypes contribute to kidney injury and repair. J. Am. Soc. Nephrol. 22, 317–326.
Lever, J.M., Hull, T.D., Boddu, R., Pepin, M.E., Black, L.M., Adedoyin, O.O., Yang, Z., Traylor, A.M., Jiang, Y., Li, Z., Peabody, J.E., Eckenrode, H.E., Crossman, D.K., Crowley, M.R., Bolisetty, S., Zimmerman, K.A., Wende, A.R., Mrug, M., Yoder, B.K., Agarwal, A., George, J.F., 2019. Resident macrophages reprogram toward a developmental state after acute kidney injury. JCI Insight 4 (2), e125503.
Li, J., Deane, J.A., Campanale, N.V., Bertram, J.F., Ricardo, S.D., 2007. The contribution of bone marrow-derived cells to the development of renal interstitial fibrosis. Stem Cells 25, 697–706.
Li, L., Huang, L., Sung, S.S., Vergis, A.L., Rosin, D.L., Rose Jr., C.E., Lobo, P.I., Okusa, M.D., 2008. The chemokine receptors CCR2 and CX3CR1 mediate monocyte/macrophage trafficking in kidney ischemia-reperfusion injury. Kidney Int. 74, 1526–1537.
Li, Z.L., Lv, L.L., Tang, T.T., Wang, B., Feng, Y., Zhou, L.T., Cao, J.Y., Tang, R.N., Wu, M., Liu, H., Crowley, S.D., Liu, B.C., 2019. HIF-1alpha inducing exosomal microRNA-23a expression mediates the cross-talk between tubular epithelial cells and macrophages in tubulointerstitial inflammation. Kidney Int. 95, 388–404.
Liang, H., Zhang, Z., Yan, J., Wang, Y., Hu, Z., Mitch, W.E., Wang, Y., 2017. The IL-4 receptor alpha has a critical role in bone marrow-derived fibroblast activation and renal fibrosis. Kidney Int. 92, 1433–1443.
Liles, J.T., Corkey, B.K., Notte, G.T., Budas, G.R., Lansdon, E.B., Hinojosa-Kirschenbaum, F., Badal, S.S., Lee, M., Schultz, B.E., Wise, S., Pendem, S., Graupe, M., Castonguay, L., Koch, K.A., Wong, M.H., Papalia, G.A., French, D.M., Sullivan, T., Huntzicker, E.G., Ma, F.Y., Nikolic-Paterson, D.J., Altuhaifi, T., Yang, H., Fogo, A.B., Breckenridge, D.G., 2018. ASK1 contributes to fibrosis and dysfunction in models of kidney disease. J. Clin. Invest. 128, 4485–4500.
Lim, A.K., Ma, F.Y., Nikolic-Paterson, D.J., Thomas, M.C., Hurst, L.A., Tesch, G.H., 2009. Antibody blockade of c-fms suppresses the progression of inflammation and injury in early diabetic nephropathy in obese db/db mice. Diabetologia 52, 1669–1679.
Lin, S.L., Kisseleva, T., Brenner, D.A., Duffield, J.S., 2008. Pericytes and perivascular fibroblasts are the primary source of collagen-producing cells in obstructive fibrosis of the kidney. Am. J. Pathol. 173, 1617–1627.
Lin, S.L., Castano, A.P., Nowlin, B.T., Lupher Jr., M.L., Duffield, J.S., 2009. Bone marrow Ly6Chigh monocytes are selectively recruited to injured kidney and differentiate into functionally distinct populations. J. Immunol. 183, 6733–6743.
Lin, S.L., Li, B., Rao, S., Yeo, E.J., Hudson, T.E., Nowlin, B.T., Pei, H., Chen, L., Zheng, J.J., Carroll, T.J., Pollard, J.W., McMahon, A.P., Lang, R.A., Duffield, J.S., 2010. Macrophage Wnt7b is critical for kidney repair and regeneration. Proc. Natl. Acad. Sci. U. S. A. 107, 4194–4199.
Liu, C., Wu, C., Yang, Q., Gao, J., Li, L., Yang, D., Luo, L., 2016. Macrophages mediate the repair of brain vascular rupture through direct physical adhesion and mechanical traction. Immunity 44, 1162–1176.

Liu, Z., Gu, Y., Chakarov, S., Bleriot, C., Kwok, I., Chen, X., Shin, A., Huang, W., Dress, R.J., Dutertre, C.A., Schlitzer, A., Chen, J., Ng, L.G., Wang, H., Liu, Z., Su, B., Ginhoux, F., 2019. Fate mapping via Ms4a3-expression history traces monocyte-derived cells. Cell 178, 1509–1525. e19.

Liu, F., Dai, S., Feng, D., Qin, Z., Peng, X., Sakamuri, S., Ren, M., Huang, L., Cheng, M., Mohammad, K.E., Qu, P., Chen, Y., Zhao, C., Zhu, F., Liang, S., Aktas, B.H., Yang, X., Wang, H., Katakam, P.V.G., Busija, D.W., Fischer, T., Datta, P.K., Rappaport, J., Gao, B., Qin, X., 2020. Distinct fate, dynamics and niches of renal macrophages of bone marrow or embryonic origins. Nat. Commun. 11, 2280.

Lloyd, C.M., Minto, A.W., Dorf, M.E., Proudfoot, A., Wells, T.N., Salant, D.J., Gutierrez-Ramos, J.C., 1997. RANTES and monocyte chemoattractant protein-1 (MCP-1) play an important role in the inflammatory phase of crescentic nephritis, but only MCP-1 is involved in crescent formation and interstitial fibrosis. J. Exp. Med. 185, 1371–1380.

Lu, T.C., Wang, Z.H., Feng, X., Chuang, P.Y., Fang, W., Shen, Y., Levy, D.E., Xiong, H., Chen, N., He, J.C., 2009. Knockdown of Stat3 activity in vivo prevents diabetic glomerulopathy. Kidney Int. 76, 63–71.

Lu, J., Cao, Q., Zheng, D., Sun, Y., Wang, C., Yu, X., Wang, Y., Lee, V.W., Zheng, G., Tan, T.K., Wang, X., Alexander, S.I., Harris, D.C., Wang, Y., 2013. Discrete functions of M2a and M2c macrophage subsets determine their relative efficacy in treating chronic kidney disease. Kidney Int. 84, 745–755.

Lv, L.L., Tang, P.M., Li, C.J., You, Y.K., Li, J., Huang, X.R., Ni, J., Feng, M., Liu, B.C., Lan, H.Y., 2017. The pattern recognition receptor, Mincle, is essential for maintaining the M1 macrophage phenotype in acute renal inflammation. Kidney Int. 91, 587–602.

Lv, L.L., Feng, Y., Wu, M., Wang, B., Li, Z.L., Zhong, X., Wu, W.J., Chen, J., Ni, H.F., Tang, T.T., Tang, R.N., Lan, H.Y., Liu, B.C., 2020. Exosomal miRNA-19b-3p of tubular epithelial cells promotes M1 macrophage activation in kidney injury. Cell Death Differ. 27, 210–226.

Ma, F.Y., Flanc, R.S., Tesch, G.H., Han, Y., Atkins, R.C., Bennett, B.L., Friedman, G.C., Fan, J.H., Nikolic-Paterson, D.J., 2007. A pathogenic role for c-Jun amino-terminal kinase signaling in renal fibrosis and tubular cell apoptosis. J. Am. Soc. Nephrol. 18, 472–484.

Ma, F.Y., Liu, J., Nikolic-Paterson, D.J., 2009. The role of stress-activated protein kinase signaling in renal pathophysiology. Braz. J. Med. Biol. Res. 42, 29–37.

Ma, F.Y., Blease, K., Nikolic-Paterson, D.J., 2016. A role for spleen tyrosine kinase in renal fibrosis in the mouse obstructed kidney. Life Sci. 146, 192–200.

Machida, Y., Kitamoto, K., Izumi, Y., Shiota, M., Uchida, J., Kira, Y., Nakatani, T., Miura, K., 2010. Renal fibrosis in murine obstructive nephropathy is attenuated by depletion of monocyte lineage, not dendritic cells. J. Pharmacol. Sci. 114, 464–473.

MacKinnon, A.C., Farnworth, S.L., Hodkinson, P.S., Henderson, N.C., Atkinson, K.M., Leffler, H., Nilsson, U.J., Haslett, C., Forbes, S.J., Sethi, T., 2008. Regulation of alternative macrophage activation by galectin-3. J. Immunol. 180, 2650–2658.

Mass, E., Gentek, R., 2021. Fetal-derived immune cells at the roots of lifelong pathophysiology. Front. Cell Dev. Biol. 9, 648313.

McAdoo, S.P., Reynolds, J., Bhangal, G., Smith, J., McDaid, J.P., Tanna, A., Jackson, W.D., Masuda, E.S., Cook, H.T., Pusey, C.D., Tam, F.W., 2014. Spleen tyrosine kinase inhibition attenuates autoantibody production and reverses experimental autoimmune Gn. J. Am. Soc. Nephrol. 25, 2291–2302.

McAdoo, S.P., Bhangal, G., Page, T., Cook, H.T., Pusey, C.D., Tam, F.W., 2015. Correlation of disease activity in proliferative glomerulonephritis with glomerular spleen tyrosine kinase expression. Kidney Int. 88, 52–60.

McAdoo, S.P., Prendecki, M., Tanna, A., Bhatt, T., Bhangal, G., McDaid, J., Masuda, E.S., Cook, H.T., Tam, F.W.K., Pusey, C.D., 2020. Spleen tyrosine kinase inhibition is an effective treatment for established vasculitis in a pre-clinical model. Kidney Int. 97, 1196–1207.

Montgomery, T.A., Xu, L., Mason, S., Chinnadurai, A., Lee, C.G., Elias, J.A., Cantley, L.G., 2017. Breast regression Protein-39/chitinase 3-like 1 promotes renal fibrosis after kidney injury via activation of myofibroblasts. J. Am. Soc. Nephrol. 28, 3218–3226.

Moreth, K., Frey, H., Hubo, M., Zeng-Brouwers, J., Nastase, M.V., Hsieh, L.T., Haceni, R., Pfeilschifter, J., Iozzo, R.V., Schaefer, L., 2014. Biglycan-triggered TLR-2- and TLR-4-signaling exacerbates the pathophysiology of ischemic acute kidney injury. Matrix Biol. 35, 143–151.

Moroni, F., Dwyer, B.J., Graham, C., Pass, C., Bailey, L., Ritchie, L., Mitchell, D., Glover, A., Laurie, A., Doig, S., Hargreaves, E., Fraser, A.R., Turner, M.L., Campbell, J.D.M., McGowan, N.W.A., Barry, J., Moore, J.K., Hayes, P.C., Leeming, D.J., Nielsen, M.J., Musa, K., Fallowfield, J.A., Forbes, S.J., 2019. Safety profile of autologous macrophage therapy for liver cirrhosis. Nat. Med. 25, 1560–1565.

Mosser, D.M., Edwards, J.P., 2008. Exploring the full spectrum of macrophage activation. Nat. Rev. Immunol. 8, 958–969.

Munro, D.A.D., Hughes, J., 2017. The origins and functions of tissue-resident macrophages in kidney development. Front. Physiol. 8, 837.

Munro, D.A.D., Wineberg, Y., Tarnick, J., Vink, C.S., Li, Z., Pridans, C., Dzierzak, E., Kalisky, T., Hohenstein, P., Davies, J.A., 2019. Macrophages restrict the nephrogenic field and promote endothelial connections during kidney development. Elife 8, e43271.

Murray, P.J., Wynn, T.A., 2011. Protective and pathogenic functions of macrophage subsets. Nat. Rev. Immunol. 11, 723–737.

Musumeci, A., Lutz, K., Winheim, E., Krug, A.B., 2019. What makes a pdc: recent advances in understanding plasmacytoid dc development and heterogeneity. Front. Immunol. 10, 1222.

Muto, Y., Wilson, P.C., Ledru, N., Wu, H., Dimke, H., Waikar, S.S., Humphreys, B.D., 2021. Single cell transcriptional and chromatin accessibility profiling redefine cellular heterogeneity in the adult human kidney. Nat. Commun. 12, 2190.

Nelson, P.J., Rees, A.J., Griffin, M.D., Hughes, J., Kurts, C., Duffield, J., 2012. The renal mononuclear phagocytic system. J. Am. Soc. Nephrol. 23, 194–203.

O'Sullivan, E.D., Mylonas, K.J., Hughes, J., Ferenbach, D.A., 2019. Complementary roles for single-nucleus and single-cell RNA sequencing in kidney disease research. J. Am. Soc. Nephrol. 30, 712–713.

Oh, D.J., Dursun, B., He, Z., Lu, L., Hoke, T.S., Ljubanovic, D., Faubel, S., Edelstein, C.L., 2008. Fractalkine receptor (CX3CR1) inhibition is protective against ischemic acute renal failure in mice. Am. J. Physiol. Renal Physiol. 294, F264–F271.

O'Reilly, V.P., Wong, L., Kennedy, C., Elliot, L.A., O'Meachair, S., Coughlan, A.M., O'Brien, E.C., Ryan, M.M., Sandoval, D., Connolly, E., Dekkema, G.J., Lau, J., Abdulahad, W.H., Sanders, J.S., Heeringa, P., Buckley, C., O'Brien, C., Finn, S., Cohen, C.D., Lindemeyer, M.T., Hickey, F.B., O'Hara, P.V., Feighery, C., Moran, S.M., Mellotte, G., Clarkson, M.R., Dorman, A.J., Murray, P.T., Little, M.A., 2016. Urinary soluble CD163 in active renal Vasculitis. J. Am. Soc. Nephrol. 27, 2906–2916.

Park, J., Shrestha, R., Qiu, C., Kondo, A., Huang, S., Werth, M., Li, M., Barasch, J., Susztak, K., 2018. Single-cell transcriptomics of the mouse kidney reveals potential cellular targets of kidney disease. Science 360, 758–763.

Park, J.G., Lee, C.R., Kim, M.G., Kim, G., Shin, H.M., Jeon, Y.H., Yang, S.H., Kim, D.K., Joo, K.W., Choi, E.Y., Kim, H.R., Kwak, C., Kim, Y.S., Choi, M., Lee, D.S.,

Han, S.S., 2020. Kidney residency of VISTA-positive macrophages accelerates repair from ischemic injury. Kidney Int. 97, 980–994.

Peng, X., Zhang, J., Xiao, Z., Dong, Y., Du, J., 2015. CX3CL1-CX3CR1 interaction increases the population of Ly6C(−)CX3CR1(hi) macrophages contributing to unilateral ureteral obstruction-induced fibrosis. J. Immunol. 195, 2797–2805.

Pillay, J., Tak, T., Kamp, V.M., Koenderman, L., 2013. Immune suppression by neutrophils and granulocytic myeloid-derived suppressor cells: similarities and differences. Cell. Mol. Life Sci. 70, 3813–3827.

Rae, F., Woods, K., Sasmono, T., Campanale, N., Taylor, D., Ovchinnikov, D.A., Grimmond, S.M., Hume, D.A., Ricardo, S.D., Little, M.H., 2007. Characterisation and trophic functions of murine embryonic macrophages based upon the use of a Csf1r-EGFP transgene reporter. Dev. Biol. 308, 232–246.

Ramachandran, P., Pellicoro, A., Vernon, M.A., Boulter, L., Aucott, R.L., Ali, A., Hartland, S.N., Snowdon, V.K., Cappon, A., Gordon-Walker, T.T., Williams, M.J., Dunbar, D.R., Manning, J.R., Van Rooijen, N., Fallowfield, J.A., Forbes, S.J., Iredale, J.P., 2012. Differential Ly-6C expression identifies the recruited macrophage phenotype, which orchestrates the regression of murine liver fibrosis. Proc. Natl. Acad. Sci. U. S. A. 109, E3186–E3195.

Ramesh, G., Reeves, W.B., 2005. p38 MAP kinase inhibition ameliorates cisplatin nephrotoxicity in mice. Am. J. Physiol. Renal Physiol. 289, F166–F174.

Ramessur Chandran, S., Tesch, G.H., Han, Y., Woodman, N., Mulley, W.R., Kanellis, J., Blease, K., Ma, F.Y., Nikolic-Paterson, D.J., 2015. Spleen tyrosine kinase contributes to acute renal allograft rejection in the rat. Int. J. Exp. Pathol. 96, 54–62.

Ricardo, S.D., Van Goor, H., Eddy, A.A., 2008. Macrophage diversity in renal injury and repair. J. Clin. Invest. 118, 3522–3530.

Ryan, J., Kanellis, J., Blease, K., Ma, F.Y., Nikolic-Paterson, D.J., 2016a. Spleen tyrosine kinase signaling promotes myeloid cell recruitment and kidney damage after renal ischemia/reperfusion injury. Am. J. Pathol. 186, 2032–2042.

Ryan, J., Ma, F.Y., Han, Y., Ozols, E., Kanellis, J., Tesch, G.H., Nikolic-Paterson, D.J., 2016b. Myeloid cell-mediated renal injury in rapidly progressive glomerulonephritis depends upon spleen tyrosine kinase. J. Pathol. 238, 10–20.

Ryu, M., Kulkarni, O.P., Radomska, E., Miosge, N., Gross, O., Anders, H.J., 2011. Bacterial CpG-DNA accelerates Alport glomerulosclerosis by inducing an M1 macrophage phenotype and tumor necrosis factor-alpha-mediated podocyte loss. Kidney Int. 79, 189–198.

Salei, N., Rambichler, S., Salvermoser, J., Papaioannou, N.E., Schuchert, R., Pakalniskyte, D., Li, N., Marschner, J.A., Lichtnekert, J., Stremmel, C., Cernilogar, F.M., Salvermoser, M., Walzog, B., Straub, T., Schotta, G., Anders, H.J., Schulz, C., Schraml, B.U., 2020. The kidney contains ontogenetically distinct dendritic cell and macrophage subtypes throughout development that differ in their inflammatory properties. J. Am. Soc. Nephrol. 31, 257–278.

Sasmono, R.T., Oceandy, D., Pollard, J.W., Tong, W., Pavli, P., Wainwright, B.J., Ostrowski, M.C., Himes, S.R., Hume, D.A., 2003. A macrophage colony-stimulating factor receptor-green fluorescent protein transgene is expressed throughout the mononuclear phagocyte system of the mouse. Blood 101, 1155–1163.

Sauter, K.A., Pridans, C., Sehgal, A., Tsai, Y.T., Bradford, B.M., Raza, S., Moffat, L., Gow, D.J., Beard, P.M., Mabbott, N.A., Smith, L.B., Hume, D.A., 2014. Pleiotropic effects of extended blockade of CSF1R signaling in adult mice. J. Leukoc. Biol. 96, 265–274.

Sawitzki, B., Harden, P.N., Reinke, P., Moreau, A., Hutchinson, J.A., Game, D.S., Tang, Q., Guinan, E.C., Battaglia, M., Burlingham, W.J., Roberts, I.S.D., Streitz, M., Josien, R., Boger, C.A., Scotta, C., Markmann, J.F., Hester, J.L.,

Juerchott, K., Braudeau, C., James, B., Contreras-Ruiz, L., Van Der Net, J.B., Bergler, T., Caldara, R., Petchey, W., Edinger, M., Dupas, N., Kapinsky, M., Mutzbauer, I., Otto, N.M., Ollinger, R., Hernandez-Fuentes, M.P., Issa, F., Ahrens, N., Meyenberg, C., Karitzky, S., Kunzendorf, U., Knechtle, S.J., Grinyo, J., Morris, P.J., Brent, L., Bushell, A., Turka, L.A., Bluestone, J.A., Lechler, R.I., Schlitt, H.J., Cuturi, M.C., Schlickeiser, S., Friend, P.J., Miloud, T., Scheffold, A., Secchi, A., Crisalli, K., Kang, S.M., Hilton, R., Banas, B., Blancho, G., Volk, H.D., Lombardi, G., Wood, K.J., Geissler, E.K., 2020. Regulatory cell therapy in kidney transplantation (the one study): a harmonised design and analysis of seven non-randomised, single-arm, phase 1/2A trials. Lancet 395, 1627–1639.

Sayyed, S.G., 2011. An orally active chemokine receptor CCR2 antagonist prevents glomerulosclerosis and renal failure in type 2 diabetes. Kidney Int. 80 (1), 68–78. https://doi.org/10.1038/ki.2011.102.

Schaefer, L., Babelova, A., Kiss, E., Hausser, H.J., Baliova, M., Krzyzankova, M., Marsche, G., Young, M.F., Mihalik, D., Gotte, M., Malle, E., Schaefer, R.M., Grone, H.J., 2005. The matrix component biglycan is proinflammatory and signals through toll-like receptors 4 and 2 in macrophages. J. Clin. Invest. 115, 2223–2233.

Schraml, B.U., Van Blijswijk, J., Zelenay, S., Whitney, P.G., Filby, A., Acton, S.E., Rogers, N.C., Moncaut, N., Carvajal, J.J., Reis, E., Sousa, C., 2013. Genetic tracing via DNGR-1 expression history defines dendritic cells as a hematopoietic lineage. Cell 154, 843–858.

Schrauben, S.J., Shou, H., Zhang, X., Anderson, A.H., Bonventre, J.V., Chen, J., Coca, S., Furth, S.L., Greenberg, J.H., Gutierrez, O.M., Ix, J.H., Lash, J.P., Parikh, C.R., Rebholz, C.M., Sabbisetti, V., Sarnak, M.J., Shlipak, M.G., Waikar, S.S., Kimmel, P.L., Vasan, R.S., Feldman, H.I., Schelling, J.R., CKD BiomarkersConsortium, The Chronic Renal Insufficiency Cohort Study Investigators, 2021. Association of Multiple Plasma Biomarker Concentrations with progression of prevalent diabetic kidney disease: findings from the chronic renal insufficiency cohort (CRIC) study. J. Am. Soc. Nephrol. 32, 115–126.

Shaw, T.N., Houston, S.A., Wemyss, K., Bridgeman, H.M., Barbera, T.A., Zangerle-Murray, T., Strangward, P., Ridley, A.J.L., Wang, P., Tamoutounour, S., Allen, J.E., Konkel, J.E., Grainger, J.R., 2018. Tissue-resident macrophages in the intestine are long lived and defined by Tim-4 and CD4 expression. J. Exp. Med. 215, 1507–1518.

Shen, B., Liu, X., Fan, Y., Qiu, J., 2014. Macrophages regulate renal fibrosis through modulating Tgfbeta superfamily signaling. Inflammation 37, 2076–2084.

Sheng, X., Qiu, C., Liu, H., Gluck, C., Hsu, J.Y., He, J., Hsu, C.Y., Sha, D., Weir, M.R., Isakova, T., Raj, D., Rincon-Choles, H., Feldman, H.I., Townsend, R., Li, H., Susztak, K., 2020. Systematic integrated analysis of genetic and epigenetic variation in diabetic kidney disease. Proc. Natl. Acad. Sci. U. S. A. 117, 29013–29024.

Snelgrove, S.L., Kausman, J.Y., Lo, C., Lo, C., Ooi, J.D., Coates, P.T., Hickey, M.J., Holdsworth, S.R., Kurts, C., Engel, D.R., Kitching, A.R., 2012. Renal dendritic cells adopt a pro-inflammatory phenotype in obstructive uropathy to activate T cells but do not directly contribute to fibrosis. Am. J. Pathol. 180, 91–103.

Sola, A., Weigert, A., Jung, M., Vinuesa, E., Brecht, K., Weis, N., Brune, B., Borregaard, N., Hotter, G., 2011. Sphingosine-1-phosphate signalling induces the production of Lcn-2 by macrophages to promote kidney regeneration. J. Pathol. 225, 597–608.

Stewart, B.J., Ferdinand, J.R., Young, M.D., Mitchell, T.J., Loudon, K.W., Riding, A.M., Richoz, N., Frazer, G.L., Staniforth, J.U.L., Vieira Braga, F.A., Botting, R.A., Popescu, D.M., Vento-Tormo, R., Stephenson, E., Cagan, A., Farndon, S.J., Polanski, K., Efremova, M., Green, K., Del Castillo Velasco-Herrera, M., Guzzo, C., Collord, G., Mamanova, L., Aho, T., Armitage, J.N., Riddick, A.C.P., Mushtaq, I., Farrell, S., Rampling, D., Nicholson, J., Filby, A., Burge, J., Lisgo, S., Lindsay, S.,

Bajenoff, M., Warren, A.Y., Stewart, G.D., Sebire, N., Coleman, N., Haniffa, M., Teichmann, S.A., Behjati, S., Clatworthy, M.R., 2019. Spatiotemporal immune zonation of the human kidney. Science 365, 1461–1466.

Sung, S.J., Ge, Y., Dai, C., Wang, H., Fu, S.M., Sharma, R., Hahn, Y.S., Yu, J., Le, T.H., Okusa, M.D., Bolton, W.K., Lawler, J.R., 2017. Dependence of glomerulonephritis induction on novel intraglomerular alternatively activated bone marrow-derived macrophages and mac-1 and PD-L1 in lupus-prone NZM2328 mice. J. Immunol. 198, 2589–2601.

Tamoutounour, S., Henri, S., Lelouard, H., De Bovis, B., De Haar, C., Van Der Woude, C.J., Woltman, A.M., Reyal, Y., Bonnet, D., Sichien, D., Bain, C.C., Mowat, A.M., Sousa, R.E., C., Poulin, L. F., Malissen, B. & Guilliams, M., 2012. CD64 distinguishes macrophages from dendritic cells in the gut and reveals the Th1-inducing role of mesenteric lymph node macrophages during colitis. Eur. J. Immunol. 42, 3150–3166.

Tan, T.K., Zheng, G., Hsu, T.T., Lee, S.R., Zhang, J., Zhao, Y., Tian, X., Wang, Y., Wang, Y.M., Cao, Q., Wang, Y., Lee, V.W., Wang, C., Zheng, D., Alexander, S.I., Thompson, E., Harris, D.C., 2013. Matrix metalloproteinase-9 of tubular and macrophage origin contributes to the pathogenesis of renal fibrosis via macrophage recruitment through osteopontin cleavage. Lab. Invest. 93, 434–449.

Tan, S.M., Ziemann, M., Thallas-Bonke, V., Snelson, M., Kumar, V., Laskowski, A., Nguyen, T.V., Huynh, K., Clarke, M.V., Libianto, R., Baker, S.T., Skene, A., Power, D.A., MacIsaac, R.J., Henstridge, D.C., Wetsel, R.A., El-Osta, A., Meikle, P.J., Wilson, S.G., Forbes, J.M., Cooper, M.E., Ekinci, E.I., Woodruff, T.M., Coughlan, M.T., 2020. Complement C5a induces renal injury in diabetic kidney disease by disrupting mitochondrial metabolic agility. Diabetes 69, 83–98.

Tanaka, M., Saka-Tanaka, M., Ochi, K., Fujieda, K., Sugiura, Y., Miyamoto, T., Kohda, H., Ito, A., Miyazawa, T., Matsumoto, A., Aoe, S., Miyamoto, Y., Tsuboi, N., Maruyama, S., Suematsu, M., Yamasaki, S., Ogawa, Y., Suganami, T., 2020. C-type lectin Mincle mediates cell death-triggered inflammation in acute kidney injury. J. Exp. Med. 217, e20192230.

Tang, P.M., Zhou, S., Li, C.J., Liao, J., Xiao, J., Wang, Q.M., Lian, G.Y., Li, J., Huang, X.R., To, K.F., Ng, C.F., Chong, C.C., Ma, R.C., Lee, T.L., Lan, H.Y., 2018. The proto-oncogene tyrosine protein kinase Src is essential for macrophage-myofibroblast transition during renal scarring. Kidney Int. 93, 173–187.

Tang, P.M., Nikolic-Paterson, D.J., Lan, H.Y., 2019a. Macrophages: versatile players in renal inflammation and fibrosis. Nat. Rev. Nephrol. 15, 144–158.

Tang, T.T., Lv, L.L., Lan, H.Y., Liu, B.C., 2019b. Extracellular vesicles: opportunities and challenges for the treatment of renal diseases. Front. Physiol. 10, 226.

Tang, T.T., Lv, L.L., Wang, B., Cao, J.Y., Feng, Y., Li, Z.L., Wu, M., Wang, F.M., Wen, Y., Zhou, L.T., Ni, H.F., Chen, P.S., Gu, N., Crowley, S.D., Liu, B.C., 2019c. Employing macrophage-derived microvesicle for kidney-targeted delivery of dexamethasone: An efficient therapeutic strategy against renal inflammation and fibrosis. Theranostics 9, 4740–4755.

Tang, T.T., Wang, B., Wu, M., Li, Z.L., Feng, Y., Cao, J.Y., Yin, D., Liu, H., Tang, R.N., Crowley, S.D., Lv, L.L., Liu, B.C., 2020. Extracellular vesicle-encapsulated IL-10 as novel nanotherapeutics against ischemic Aki. Sci. Adv. 6 (33)eaaz0748.

Tecklenborg, J., Clayton, D., Siebert, S., Coley, S.M., 2018. The role of the immune system in kidney disease. Clin. Exp. Immunol. 192, 142–150.

Tesch, G.H., 2010. Macrophages and diabetic nephropathy. Semin. Nephrol. 30, 290–301.

Timoshanko, J.R., Sedgwick, J.D., Holdsworth, S.R., Tipping, P.G., 2003. Intrinsic renal cells are the major source of tumor necrosis factor contributing to renal injury in murine crescentic glomerulonephritis. J. Am. Soc. Nephrol. 14, 1785–1793.

Tomita, N., Morishita, R., Lan, H.Y., Yamamoto, K., Hashizume, M., Notake, M., Toyosawa, K., Fujitani, B., Mu, W., Nikolic-Paterson, D.J., Atkins, R.C., Kaneda, Y., Higaki, J., Ogihara, T., 2000. In vivo administration of a nuclear transcription factor-kappaB decoy suppresses experimental crescentic glomerulonephritis. J. Am. Soc. Nephrol. 11, 1244–1252.

Tuttle, K.R., Brosius 3rd, F.C., Adler, S.G., Kretzler, M., Mehta, R.L., Tumlin, J.A., Tanaka, Y., Haneda, M., Liu, J., Silk, M.E., Cardillo, T.E., Duffin, K.L., Haas, J.V., Macias, W.L., Nunes, F.P., Janes, J.M., 2018. Jak1/JAK2 inhibition by baricitinib in diabetic kidney disease: results from a phase 2 randomized controlled clinical trial. Nephrol. Dial. Transplant. 33, 1950–1959.

Van Furth, R., Cohn, Z.A., 1968. The origin and kinetics of mononuclear phagocytes. J. Exp. Med. 128, 415–435.

Viehmann, S.F., Bohner, A.M.C., Kurts, C., Brahler, S., 2018. The multifaceted role of the renal mononuclear phagocyte system. Cell. Immunol. 330, 97–104.

Waddell, L.A., Lefevre, L., Bush, S.J., Raper, A., Young, R., Lisowski, Z.M., McCulloch, M.E.B., Muriuki, C., Sauter, K.A., Clark, E.L., Irvine, K.M., Pridans, C., Hope, J.C., Hume, D.A., 2018. ADGRE1 (EMR1, F4/80) is a rapidly-evolving gene expressed in mammalian monocyte-macrophages. Front. Immunol. 9, 2246.

Wang, Y., Wang, Y.P., Zheng, G., Lee, V.W., Ouyang, L., Chang, D.H., Mahajan, D., Coombs, J., Wang, Y.M., Alexander, S.I., Harris, D.C., 2007. Ex vivo programmed macrophages ameliorate experimental chronic inflammatory renal disease. Kidney Int. 72, 290–299.

Wang, S., Meng, X.M., Ng, Y.Y., Ma, F.Y., Zhou, S., Zhang, Y., Yang, C., Huang, X.R., Xiao, J., Wang, Y.Y., Ka, S.M., Tang, Y.J., Chung, A.C., To, K.F., Nikolic-Paterson, D.J., Lan, H.Y., 2016. TGF-beta/Smad3 signalling regulates the transition of bone marrow-derived macrophages into myofibroblasts during tissue fibrosis. Oncotarget 7, 8809–8822.

Wang, Y.Y., Jiang, H., Pan, J., Huang, X.R., Wang, Y.C., Huang, H.F., To, K.F., Nikolic-Paterson, D.J., Lan, H.Y., Chen, J.H., 2017. Macrophage-to-myofibroblast transition contributes to interstitial fibrosis in chronic renal allograft injury. J. Am. Soc. Nephrol. 28, 2053–2067.

Wilson, H.M., Stewart, K.N., Brown, P.A.J., Anegon, I., Chettibi, S., Rees, A.J., Kluth, D.C., 2002. Bone-marrow-derived macrophages genetically modified to produce IL-10 reduce injury in experimental glomerulonephritis. Mol. Ther. 6, 710–717.

Wilson, H.M., Chettibi, S., Jobin, C., Walbaum, D., Rees, A.J., Kluth, D.C., 2005. Inhibition of macrophage nuclear factor-kappaB leads to a dominant anti-inflammatory phenotype that attenuates glomerular inflammation in vivo. Am. J. Pathol. 167, 27–37.

Wilson, P.C., Wu, H., Kirita, Y., Uchimura, K., Ledru, N., Rennke, H.G., Welling, P.A., Waikar, S.S., Humphreys, B.D., 2019. The single-cell transcriptomic landscape of early human diabetic nephropathy. Proc. Natl. Acad. Sci. U. S. A.

Woroniecka, K.I., Park, A.S., Mohtat, D., Thomas, D.B., Pullman, J.M., Susztak, K., 2011. Transcriptome analysis of human diabetic kidney disease. Diabetes 60, 2354–2369.

Wu, H., Ma, J., Wang, P., Corpuz, T.M., Panchapakesan, U., Wyburn, K.R., Chadban, S.J., 2010. Hmgb1 contributes to kidney ischemia reperfusion injury. J. Am. Soc. Nephrol. 21, 1878–1890.

Wu, H., Malone, A.F., Donnelly, E.L., Kirita, Y., Uchimura, K., Ramakrishnan, S.M., Gaut, J.P., Humphreys, B.D., 2018. Single-cell transcriptomics of a human kidney allograft biopsy specimen defines a diverse inflammatory response. J. Am. Soc. Nephrol. 29, 2069–2080.

Wu, H., Kirita, Y., Donnelly, E.L., Humphreys, B.D., 2019. Advantages of single-nucleus over single-cell Rna sequencing of adult kidney: rare cell types and novel cell states revealed in fibrosis. J. Am. Soc. Nephrol. 30, 23–32.

Wuthrich, R.P., 1992. Vascular cell adhesion molecule-1 (Vcam-1) expression in murine lupus nephritis. Kidney Int. 42, 903–914.
Xu, L., Sharkey, D., Cantley, L.G., 2019. Tubular gm-CSF promotes late MCP-1/CCR2-mediated fibrosis and inflammation after ischemia/reperfusion injury. J. Am. Soc. Nephrol. 30, 1825–1840.
Yan, J., Zhang, Z., Yang, J., Mitch, W.E., Wang, Y., 2015. JAK3/STAT6 stimulates bone marrow-derived fibroblast activation in renal fibrosis. J. Am. Soc. Nephrol. 26, 3060–3071.
Yiu, W.H., Li, R.X., Wong, D.W.L., Wu, H.J., Chan, K.W., Chan, L.Y.Y., Leung, J.C.K., Lai, K.N., Sacks, S.H., Zhou, W., Tang, S.C.W., 2018. Complement C5a inhibition moderates lipid metabolism and reduces tubulointerstitial fibrosis in diabetic nephropathy. Nephrol. Dial. Transplant. 33, 1323–1332.
Zhang, M.Z., Yao, B., Yang, S., Jiang, L., Wang, S., Fan, X., Yin, H., Wong, K., Miyazawa, T., Chen, J., Chang, I., Singh, A., Harris, R.C., 2012. CSF-1 signaling mediates recovery from acute kidney injury. J. Clin. Invest. 122, 4519–4532.
Zhang, T., Wu, K.Y., Ma, N., Wei, L.L., Garstka, M., Zhou, W., Li, K., 2020. The C5a/C5aR2 axis promotes renal inflammation and tissue damage. JCI Insight 5 (7), e134081.
Zhao, W.T., Huang, J.W., Sun, P.P., Su, T., Tang, J.W., Wang, S.X., Liu, G., Yang, L., 2019. Diagnostic roles of urinary kidney injury molecule 1 and soluble C5b-9 in acute tubulointerstitial nephritis. Am. J. Physiol. Renal Physiol. 317, F584–F592.
Zimmerman, K.A., Bentley, M.R., Lever, J.M., Li, Z., Crossman, D.K., Song, C.J., Liu, S., Crowley, M.R., George, J.F., Mrug, M., Yoder, B.K., 2019. Single-cell RNA sequencing identifies candidate renal resident macrophage gene expression signatures across species. J. Am. Soc. Nephrol. 30, 767–781.

CHAPTER FIVE

Heterogeneity and function of macrophages in the breast during homeostasis and cancer

Eva Hadadi[a,b], Sofie Deschoemaeker[a,b], Gerard Vicente Venegas[a,b], and Damya Laoui[a,b,*]

[a]Myeloid Cell Immunology Lab, VIB Center for Inflammation Research, Brussels, Belgium
[b]Lab of Cellular and Molecular Immunology, Vrije Universiteit Brussel, Brussels, Belgium
*Corresponding author: e-mail address: dlaoui@vub.be

Contents

1. Introduction 150
2. Mammary gland macrophages 151
   2.1 Development and structure of the mammary gland 151
   2.2 Macrophage heterogeneity in the mammary gland 153
   2.3 Localization and role of macrophages in the mammary gland 159
3. Breast cancer 161
4. Macrophages in breast cancer 162
   4.1 Macrophage heterogeneity in breast cancer 162
   4.2 Impact of hypoxia on the heterogeneity and functions of TAMs in breast cancer 167
5. Concluding remarks and future perspectives 170
Competing interests 172
Funding 172
References 172

Abstract

Macrophages are diverse immune cells populating all tissues and adopting a unique tissue-specific identity. Breast macrophages play an essential role in the development and function of the mammary gland over one's lifetime. In the recent years, with the development of fate-mapping, imaging and scRNA-seq technologies we grew a better understanding of the origin, heterogeneity and function of mammary macrophages in homeostasis but also during breast cancer development. Here, we aim to provide a comprehensive review of the latest improvements in studying the macrophage heterogeneity in healthy mammary tissues and breast cancer.

International Review of Cell and Molecular Biology, Volume 367
ISSN 1937-6448
https://doi.org/10.1016/bs.ircmb.2022.01.002

Abbreviations

BM bone marrow
DM ductal macrophages
ECM extracellular matrix
ER estrogen receptor
HA hyaluron
PR progesterone receptor
SM stromal macrophages
TAM tumor-associated macrophages
TDLU terminal duct lobular unit
TEB terminal end buds
TNBC triple negative breast cancer
TRM tissue resident macrophages

1. Introduction

Macrophages are innate immune cells seeded all over the body with essential roles in tissue homeostasis and immune defense mechanisms. Over the last decade it became clear that most tissue macrophages have an embryonic origin. Throughout adulthood, at steady state, these resident cells either maintain their abundance by proliferating or get partially replenished by bone-marrow derived monocytes in an organ specific manner (Ginhoux and Guilliams, 2016; Gomez Perdiguero et al., 2015; Hoeffel and Ginhoux, 2018; Liu et al., 2019).

Macrophages have been implicated in all stages of mammary gland development and functions (Gouon-Evans et al., 2000; Pollard and Hennighausen, 1994; Van Nguyen and Pollard, 2002). As such, macrophages are essential for morphogenesis of the ductal epithelial tree and modulate normal stem/progenitor cell activity, proliferation, alveolar budding and collagen matrix organization in the developing mammary gland (Gouon-Evans et al., 2002; Gyorki et al., 2009; Ingman et al., 2006). In addition, macrophages contribute to tissue clearance during the hormonal cycle or following damage and tissue remodeling during involution (Chua et al., 2010; Dawson et al., 2020).

Recent studies provided deeper insights into the localization and potential roles of distinct macrophage populations at steady state and in breast tumors (Dawson et al., 2020; Ramos et al., 2021; Wang et al., 2020). Tumor-associated macrophages (TAMs) are important components of the breast cancer microenvironment and their density positively correlates with worst patient prognosis (Zhang et al., 2012; Zhao et al., 2017). TAMs are heterogenous and can originate from both tissue resident and infiltrating monocytes. These highly plastic cells can acquire distinct phenotypic and functional profiles in response to localized stimuli and tissue specific microenvironment (Cassetta et al., 2019; Wu et al., 2020). They can exert both pro-tumor (e.g., promotion of cancer cell proliferation/metastasis, stimulation of angiogenesis, reduced therapy responses or interference with $CD8^{+}$ T cell activation) and anti-tumor (e.g., cancer cell killing, antigen presentation or activation of adaptive immune responses) actions, however, in developing tumors TAMs have a predominantly pro-tumoral behavior (Larionova et al., 2020; Noy and Pollard, 2014). Therefore there is an increasing need to understand the existing TAM phenotypes and their functional characteristics to improve and develop novel macrophage-targeting therapies (Bercovici et al., 2019; Cassetta and Pollard, 2018; Xiang et al., 2021).

Here, we review the latest improvements and current understanding of the heterogeneity and function of mammary gland macrophages in tissue homeostasis and tumor progression.

2. Mammary gland macrophages

2.1 Development and structure of the mammary gland

The mammary gland is an exocrine gland present in mammals that can secrete milk for the feeding of young offspring. It is composed of a dynamic tissue that consists of an epithelial structure which is surrounded by a stromal microenvironment containing adipocytes, endothelial cells, fibroblasts and immune cells, including macrophages. This unique tissue undergoes profound morphological changes throughout lifetime. Mammary organogenesis starts during embryonic development, but its development predominantly occurs in the postnatal period (Cardiff and Wellings, 1999; Javed and Lteif, 2013; Macias and Hinck, 2012). During fetal development, a rudimentary ductal system develops, which after birth remains quiescent until puberty, except for an isometric growth and branching. However, while the embryonic development process of the murine mammary gland is highly conserved across mammals (Oftedal, 2002), there are marked structural differences. In the mouse mammary gland, with the onset of puberty, the hormonal changes induce an increase in the fibrous and fatty tissue of the stroma followed by the formation of proliferative terminal end buds (TEBs) (Fig. 1A). This process is accompanied with an extensive ductal branching

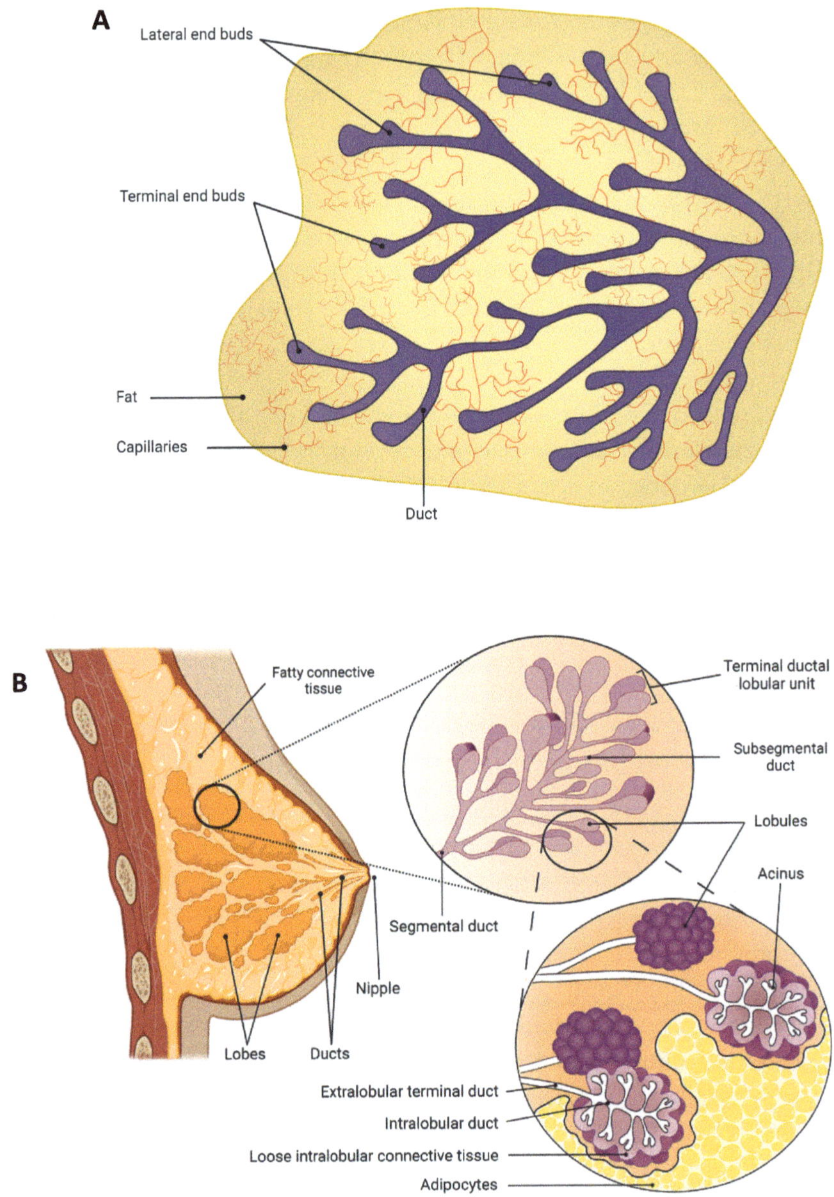

Fig. 1 Organization of the mice (A) and human (B) mammary gland. Diagram of the terminal ductal lobular unit. *This figure was created using BioRender.com.*

and elongation until the developing tree reaches the boundaries of the mammary fat pad and occupies up to 60% of the available stroma. In the human breast, instead of a single epithelial tree, several ductal trees sprout from the mammary bud creating a more complex ductal branching system comparted to the mouse mammary gland (Cardiff and Wellings, 1999; Dontu and Ince, 2015). The primary duct branches into segmental and subsegmental ducts, which further divides into terminal ducts. Each terminal duct ends with numerous blind-ended ductulus, called acini. This group of acini derived from one terminal duct along with the surrounding intralobular stroma is called the terminal duct lobular unit (TDLU), which is the functional unit of the human mammary gland (Fig. 1B). These structural differences might partially be a result of the distinct stromal composition of the human and the mouse mammary tissue, as it has been demonstrated that the epithelial branching patterns are regulated by mesenchymal signals (Kahata et al., 2018; Parmar and Cunha, 2004). In the human mammary gland, the group of acini is embedded in loose intralobular connective tissue surrounded by the interlobular dense, fibrous stroma while the murine ductal tree is situated in an adipose-rich environment. In both species, with every ovarian cycle, the epithelium of the homeostatic female adult mammary gland undergoes additional proliferation, differentiation, and regression leading to the generation of tertiary side branches and alveolar buds (Sternlicht et al., 2006). During pregnancy, an enormous branching and alveolar budding process takes place followed by the differentiation and formation of complex milk-producing alveolar units. Following weaning of the offspring, the mammary gland returns to the homeostatic pre-pregnant resting state through massive cell death and tissue remodeling, called involution (Cardiff and Wellings, 1999; Jindal et al., 2020). The development and the characteristics of mammary gland have been reviewed in more detail elsewhere (Dawson and Visvader, 2021; Fu et al., 2020; Javed and Lteif, 2013; Parmar and Cunha, 2004b; Slepicka et al., 2021; Sternlicht, 2005).

2.2 Macrophage heterogeneity in the mammary gland

The importance of macrophages in mammary gland development and homeostasis has been well recognized for a long time (Gouon-Evans et al., 2000; O'Brien et al., 2012; Pollard and Hennighausen, 1994; Ridker et al., 2017; Van Nguyen and Pollard, 2002). However, the origin and the spatial distribution of distinct macrophage populations during different developmental states were only recently described (Dawson et al., 2020; He et al., 2020; Jäppinen et al., 2019; Stewart et al., 2019; Wang et al., 2020).

A lineage tracing study by Jäppinen and colleagues showed that macrophages colonize the murine mammary gland during embryonic development. Under homeostatic conditions, these fetal-derived tissue-resident macrophages also dominate the mammary tissue throughout adulthood (Jäppinen et al., 2019). Using macrophage depletion methods and transgenic models, this study showed the presence of two distinct tissue resident $CD45^{+}SiglecF^{-}CD11b^{+}$ macrophage populations in the adult mammary tissue being (i) fetal liver derived $CD206^{Hi}$ macrophages and (ii) bone-marrow derived $CD206^{Low/Neg}$ macrophages generated during adulthood. Interestingly, several studies also described a $CD11c^{+}CD11b^{Low}$ macrophage population localized in the mammary epithelium (Dawson et al., 2020; Plaks et al., 2015; Wang et al., 2020). Dawson and colleagues defined the $CD11c^{+}CD11b^{Low}$ macrophage population as ductal macrophages (DM) due to their specific ductal association and intra-epithelial localization in the basal-luminal interface. This DM population, in addition to their CD45, CD64 and F4/80 expression, was further characterized as $Ly6C^{-}Lyve1^{Low/Neg}CD206^{Neg}CX3CR1^{Hi}MHCII^{+}$. In parallel, this study also described $CD11b^{+}CD11c^{Low}CD206^{+}$ stromal macrophages (SM), which could be further subdivided based on their $Lyve1^{Int\text{-}Hi}$ and $MHCII^{Hi\text{-}Low}$ expression levels into $Lyve1^{Int}$ $MHCII^{Hi}$ SM1 and $Lyve1^{Hi}$ $MHCII^{Low}$ SM2 (Dawson et al., 2020). Despite the different gating strategies and limited number of cell surface markers used for the phenotypic identification in both studies, the DM and SM populations potentially correspond respectively to the $CD206^{Low/Neg}$ adult BM-derived and $CD206^{Hi}$ fetal-derived macrophages described by Jäppinen and colleagues (Table 1). However, the overlap between the DM and $CD206^{Low/Neg}$ populations is only partial since Jäppinen and colleagues pre-gated on the $CD11b^{+}$ cells, suggesting that a proportion of the $CD11b^{low}$ DMs might have been excluded from their analysis. Analysis of the Ms4a3-Cre/Rosa-tdTomato fate-mapping model, in which exclusively the monocyte-derived macrophages, but not the embryonic macrophages, are labeled (Liu et al., 2019), showed that majority of DMs in the pre-pubertal mammary glands were of embryonic origin. However, due to the rapid ductal expansion during puberty, the infiltrating BM-derived monocytes rapidly differentiated into DMs to keep up with the functional requirement of epithelial surveillance resulting in a significant drop of the proportion of DM of embryonic origin (from 80% to 46% by 6 weeks of age) (Dawson et al., 2020). Throughout adulthood, DMs similarly to SMs, showed a slow rate of replenishment by BM-derived cells, with the slowest turnover rate being observed in the $CD206^{+}Lyve1^{Hi}MHCII^{Low}$ SM1 population.

Table 1 Macrophage signatures in healthy mammary tissues and breast cancer in mouse and human.

Species	Type of macropage	Health state	Name used by authors	Platform	Defining markers	Additional highlighted markers	Reference
Mouse	Stromal/stromal-like	Naïve	Fetal-derived	CyTOF/FACS	**F4/80^{Hi} CD64^{Hi}Siglec-1^{Hi} CD206^{Hi}** (gated on CD45^{+} SiglecF^{-} CD11b^{+})	CD274	Jäppinen et al. (2019)
Mouse	Stromal/stromal-like	Naïve	SM1	3D confocal imaging/bulk mRNAseq	CD11c^{Lo}CD11b^{+}CD206^{+}**MHCII^{Hi} Lyve1^{Int}** (gated on CD45^{+}CD64^{+}F4/80^{+})		Dawson et al. (2020)
			SM2		CD11c^{Lo}CD11b^{+}CD206^{+}**MHCII^{Lo}Lyve1^{Hi}** (gated on CD45^{+}CD64^{+}F4/80^{+})		
		PyMT, Neu, Wnt1 tumor	SM-like	3D confocal imaging/FACS	**Ly6C^{+} CD11c^{Lo}** (gated on CD45^{+}Ly6G^{-}CD64^{+}CD24^{Lo})	MHCII$^{+/-}$ CX3CR1$^{+/-}$	
Mouse	Stromal/stromal-like	Naïve	F4/80^{+}Lyve-1^{+}	Bulk mRNAseq	F4/80^{+}CD11b^{+}**Lyve1**$^{+}$CD206^{+}	*Mrc1, Sparc, Gas6, Igf1, Cd163, Apoe, Cd209g*	Wang et al. (2020)
			F4/80^{+}Lyve-1^{-}		F4/80^{+}CD11b^{+}**Lyve1**$^{-}$CD206$^{+/-}$	*Il1b, Cd74, H2Ab1*	
Mouse	Stromal/Stromal-like	Naïve (young/aged)	Ma	sc-RNAseq/IHC	***Cd163, Mrc1, Cd209f, Adgre1***	*Lyve1,Cd209g*	Li et al. (2020)
Human	Stromal/stromal-like	Luminal breast cancer	FOLR2^{+}	scRNA-seq/bulk mRNAseq/CyTOF	***FOLR2*, *SEPP1***, *SLC40A1, MRC1*, ***LYVE1*** (scRNA-seq on CD14^{+})	*APOE, TIMD4, MAF, CD163, CXCL12*/CyTOF: LYVE1, MRC1	Ramos et al. (2021)
Mouse		PyMT	*Folr2*$^{+}$ *Mrc1*$^{+}$	scRNA-seq	***Folr2*, *Mrc1***, *Slc40a1, Maf* (scRNA-seq CD45^{+}CD3/19^{-}B220^{-}NKP46^{-})	Timd4, SpiC	
Mouse	Stromal/stromal-like	p53$^{-/-}$ PN1a lesions	MC0	scRNA-seq/IF	*Cd209g, Lyve1, Tim4d, Gas6, Mrc1*/**CD206^{+} Lyve-1^{+}** Csfr1^{+}	*Hmox1, Pf4, Fgfr1, Nrp2*	Ibrahim et al. (2020)
			MC6		*Gas6, Axl, Tgfb1, Ccl7, Ccl2, Mrc1*/ **CD206^{+}Lyve-1^{-}** Gas6^{+}Csfr1^{+}	*Ccl3, Ccl8, Ptp1b, Igf*	

Continued

Table 1 Macrophage signatures in healthy mammary tissues and breast cancer in mouse and human.—cont'd

Species	Type of macropage	Health state	Name used by authors	Platform	Defining markers	Additional highlighted markers	Reference
Human	Stromal/stromal-like	Breast cancer (luminal and TN)	CD206^{+}HLA-DR^{Int}	CyTOF/IF	**CD206^{+}HLA-DR^{Int}**	CD36$^{Int/Hi}$, CD86$^{Int/Hi}$	Wagner et al. (2019)
			PD-L1 + TAMs		**CD64^{Hi} HLA-DR^{Hi} CD38^{+}** CD169^{+} CD204$^{+/-}$	CD163^{Lo}	
Mouse	Ductal/ductal-like	Naïve	BM-derived	CyTOF/FACS	**F4/80^{Int} CD64^{Int} Siglec-1^{Int} CD206$^{Lo/-}$ CX3XR1^{Hi}** (gated on CD45^{+}SiglecF^{-}CD11b^{+})	MHCII^{+}	Jäppinen et al. (2019)
Mouse	Ductal/ductal-like	Naïve	DM	3D confocal imaging/bulk mRNAseq	CD11c^{+}CD11b^{Lo}Ly6C^{-} **CD206-MHCII^{+} Lyve-1$^{Lo/Neg}$**	CX3CR1^{Hi}/*Tmem119*, *Vcam1*, *Mmp12*, *Mmp13*, *Mmp14*	Dawson et al. (2020)
		PyMT, Neu, Wnt1 tumor	DM-like	3D confocal imaging/FACS	**Ly6C^{-} CD11c^{+}** (scRNA-seq on CD45^{+}Ly6G^{-}CD64^{+}CD24^{Lo})	CX3CR1^{Hi}, TMEM119^{+}	
Mouse	Ductal/ductal-like	Naïve (young/aged)	Mb	sc-RNAseq/ IHC	***Mmp12*, *Mmp13***, *Spic*, *MHCII genes (H2-Aa, H2-Ab1, H2-Dma, H2-DMb1, H2-Eb1, H2-Oa)*	*Vcam*, *Cd14*, *Cd52*, *Cd63*, *Cd72*, *Cd74*, *Cd207*, *Il1a*, *Il1b*, *Il10*, *Il12b*, *Cxcl2*, *Cxcl16*,	Li et al. (2020)
Human	Ductal/ductal-like	Luminal breast cancer	TREM2^{+}CADM1^{+}	scRNA-seq/ bulk mRNAseq/ CyTOF	***TREM2*, *SPP1***, *CADM1* (scRNA-seq on CD14^{+})	*APOE*, *C3*, *ISG15*, *CD9*, *ITGAV*, *FN1*, *TLR3*, *MARCKS*, *STAT2*	Ramos et al. (2021)
Mouse		PyMT	*Cadm1*$^{+}$ *Cx3cr1*$^{+}$	scRNA-seq	***Cadm1*, *Cx3cr1***, *Havcr2*, *Ifi44* (gated on CD45^{+}CD3/19^{-}B220^{-}NKP46^{-})	*Tgfb1*, *Itgav*, *Marcks*, *Tlr3*, *Stat2*	
Mouse	Ductal/ductal-like	p53$^{-/-}$ PN1a lesions	MC3	scRNA-seq/IF	***Itgax*, *Cx3cr1*, *Tmem119***/CD206^{-}Csfr11^{+}*Ccr2*$^{+}$	*Axl*, *Hexb*, *Cd86*, *Tnf*, *Il1b*, *Tgfb1*, *Cxcl16*, *Mmp12*, *Mmp13*, *Itgav*, *Pdgf*, *Vcam1*	Ibrahim et al. (2020)
			MC8		***Trem2*, *Fabp5*, *Lgals3***/CD206^{-}Csfr1^{+}Ccr2^{-}	*Cd9*, *Gpnmb*, *Cd36*	

The markers highlighted in bold were used to define or distinguish the key populations in the respective publications.

These observations confirmed that both the DM and SM populations are *bona fide* tissue resident mammary gland macrophages. Similar populations have been described in other studies (Han et al., 2018; Wang et al., 2020): Han and colleagues used unbiased scRNA-seq transcriptome analyses of naïve mammary fat pad where at least two, $Lyz1^{Hi}$ and $C1qc^{Hi}$ macrophage populations were identified (Han et al., 2018). Another scRNA-seq analysis comparing young *versus* aged naïve mammary glands identified two macrophages populations, referred to as Ma (marked by *Mrc1, Cd209f* and *Cd163*) and Mb (marked by *Mmp12, Mmp13* and *Spic*) (Li et al., 2020). Wang and colleagues identified two macrophage populations using scRNA-seq based on Lyve-1 expression within the $CD45^+$ compartment. Furthermore, targeted bulk mRNA-seq analyses was performed on $F4/80^+Lyve\text{-}1^+$ and $F4/80^+Lyve\text{-}1^-$ cells (pre-gated on $CD45^+CD11b^+$) (Wang et al., 2020). Comparing the gene expression data derived from tissue scRNA-seq or bulk mRNA-seq analyses of sorted macrophage populations in mammary tissue suggest that the $C1qc^{Hi}$ (Han et al., 2018), the SM (Dawson et al., 2020) and $F4/80^+Lyve1^+$ (Wang et al., 2020) cells are remarkably similar with highly overlapping gene expression profiles (Table 1). These SMs were all enriched for additional markers like *Cd163*, *Folr2*, *Timd4*, *Cd209b*, *Cd209f*, *Cd209g* or *Pf4*, a signature that was previously associated with perivascular macrophages (Chakarov et al., 2019).

Interestingly, a recent study by Ramos and colleagues described a $FOLR2^+$ *bona fide* tissue resident (TRM) macrophage population in human breast tissue and identified the $CD206^+LYVE1^+$ stromal macrophages as their murine counterparts (Ramos et al., 2021). The comparative analysis of murine and human $FOLR2^+$ macrophages showed a high alignment in their transcriptome, suggesting evolutionary conserved populations (Table 2). These SMs show periductal and perivascular localizations in both species (Dawson et al., 2020; Jäppinen et al., 2019; Ramos et al., 2021).

The recently characterized $CD11c^+CD11b^{Low}Cx3cr1^+Vcam1^{Hi}$ DM population (Dawson et al., 2020) does not completely align with all the similar populations identified by unbiased scRNA-seq studies (Han et al., 2018; Li et al., 2020; Wang et al., 2020), however, there are shared key gene signatures (Table 2). Potentially, this inconsistency roots partially in the depth of the analysis and in the scarcity of this population in virgin mammary fat pads compared to the stromal macrophage populations (Dawson et al., 2020; Ramos et al., 2021). In humans, this population might be identified by their CADM1 and TREM2 expression (Ramos et al., 2021), but at present these cells are poorly investigated in healthy human breast tissue. $TREM2^+$

Table 2 Comparative analysis of the list of differentially expressed genes in stromal and ductal macrophages described in the following publications: Cassetta et al. (2019), Dawson et al. (2020), Han et al. (2018), Ibrahim et al. (2020), Li et al. (2020), Mulder et al. (2021), Ramos et al. (2021), and Wang et al. (2020).

Stromal/Stromal-like macrophages	
Gene	**#**
Folr2	8
Ccl8	7
F13a1	7
Dab2	6
Egr1	6
Fcgrt	6
Igf1	6
Lyve1	6
Maf	6
Pla2g2d	6
Pltp	6
Pmp22	6
C4b	5
Cbr2	5
Cd163	5
Cd209d	5
Cd209f	5
Cd209g	5
Clec10a	5
Ednrb	5
Fcna	5
Fxyd2	5
Grn	5
Hmox1	5
Ier3	5
Igfbp4	5
Jun	5
Ltc4s	5
Ly6e	5
Mrc1	5
Pf4	5
Serpinb6a	5

Stromal/Stromal-like macrophages	
Gene	**#**
Aldh2	4
Calml4	4
Ccl2	4
Ccl24	4
Ccl6	4
Ccl9	4
Cd209b	4
Cltc	4
Ctsl	4
Gas6	4
Hal	4
Hspa1a	4
Ifi27l2a	4
Ifitm2	4
Ifitm3	4
Klf4	4
Lgals1	4
RCN3	4
Retnla	4
Rnase4	4
Smagp	4
Snx2	4
Timd4	4
Timp2	4
Txnip	4
Wfdc17	4

Ductal/Ductal-like macrophages	
Gene	**#**
Il1b	6
Ctss	5
Cd74	5
Hexb	5
Acp5	4
Cadm1	4
Ctsz	4
Gm2a	4
Gngt2	4
H2-EB1	4
Mpeg1	4
Tyrobp	4
Aif1	3
Ccrl2	3
Cd14	3
Cd63	3
Csf2ra	3
Cxcl16	3
Cxcl2	3
Cybb	3
H2-Aa	3
H2-Ab1	3
H2-Dma	3
Il1a	3
Itgb5	3
Mmp12	3
Ppfia4	3
Rgs1	3
Sgk1	3
Syngr2	3

Lists of differentially expressed genes were compared to the stromal and ductal gene signatures from Li et al. (2020). "#" stands for the number of studies in which the respective gene was detected as part of the stromal or ductal macrophage gene signature.

macrophages have been described in various tissues at steady state including lung (Leader et al., 2021), skin (Wang et al., 2019) or adipose tissue (Jaitin et al., 2019). However, $TREM2^+$ macrophages are more investigated in the context of cancer where TREM2 is considered as a pro-tumoral tumor-associated macrophage (TAM) marker in a wide range of tumors, including breast cancer (Chiappinelli et al., 2021; Molgora et al., 2020). The $CD11b^+F4/80^+LYVE1^-$ population, which showed lower CD206 expression compared to the $LYVE1^+$ cell, shares its key gene signature (*Cd74, Il1b, H2-Ab1* and *Cxcl16*) with the $CD11c^+$ DMs (Wang et al., 2020). This population is poorly characterized and a similar population has not yet been described in human breast.

2.3 Localization and role of macrophages in the mammary gland

In earlier studies macrophages have been shown to be localized in proximity to the neck of the TEBs during puberty, to the alveolar buds during pregnancy and they were also found dispersed in the adipose stroma (Gouon-Evans et al., 2000, 2002; Schwertfeger et al., 2006). Recent studies, using 3D imaging of cleared mammary tissue demonstrated frequent direct connections in-between mammary epithelial cells and distinct macrophage populations (Dawson et al., 2020; Jäppinen et al., 2019; Stewart et al., 2019).

With respect to macrophages subpopulations, the characterized Lyve1$^+$ CD206$^+$ macrophage population was localized in hyaluron (HA) enriched stromal regions, including the fibrous areas in the adipose tissue and the fibrous capsule surrounding the mammary fat pad (Wang et al., 2020). Interestingly, the majority of macrophages in the HA-rich region of the stroma surrounding the TEB were Lyve1$^-$, suggesting that additional factors regulate the spatial distribution of distinct macrophages subsets. The Lyve1$^+$CD206$^+$ tissue resident macrophage population has been described to negatively regulate collagen deposition in the lung and in arteries (Chakarov et al., 2019; Lim et al., 2018). As such, the absence of Lyve1$^+$ macrophages resulted in the induction of arterial stiffness (Lim et al., 2018) and exacerbated lung and heart fibrosis through enhanced collagen deposition (Chakarov et al., 2019). Similarly, macrophage depletion resulted in increased HA and collagen deposition in the mammary gland (Wang et al., 2020). In addition, the gene expression profile of the Lyve1$^+$CD206$^+$ stromal macrophages was associated with extracellular matrix (ECM) remodeling and repair functions (Dawson et al., 2020; Wang et al., 2020). Previously, HA was also shown to modulate epithelial branching during mammary development (Tolg et al., 2017). Therefore, it would be interesting to further investigate the role of CD206$^+$Lyve1$^-$ macrophages, predominantly localized in the HA-rich stroma proximity to TEBs (Wang et al., 2020) in this process. Importantly, this negative correlation between collagen deposition and the presence of CD206$^+$ macrophages has also been also described in the human breast (Huo et al., 2015).

The intra-epithelial CD11c$^+$ macrophages have been confirmed by several studies to play a unique role in ductal morphogenesis during development and lactation/involution (Dawson and Visvader, 2021; Plaks et al., 2015; Stewart et al., 2019). These uniquely duct-associated macrophages are evenly distributed throughout the mammary epithelium with higher densities at the nipple area (Dawson and Visvader, 2021). These DMs are important for the clearance of apoptotic cells upon epithelium damage. Interestingly, intravital imaging showed that the DMs surveille the entire epithelium through dendrite movements rather than via migration (Dawson and Visvader, 2021). The phagocytic activity of DMs is also essential for mammary gland involution and these cells were shown to play a non-redundant role in tissue remodeling during the post-lactation period (Dawson and Visvader, 2021; Stewart et al., 2019). During pregnancy, the hormonal cues induce an exponential proliferation of DMs which then become the predominant macrophage population in lactating breasts. Besides regulating structural changes and monitoring the breast microenvironment, DMs showed an anti-inflammatory profile during lactation, suggesting a role in promoting an immunosuppressive/immunotolerant environment (Dawson and Visvader, 2021). Interestingly, DM-like Mb macrophages described by Li and colleagues showed an increased abundance in aged mammary glands, thereby potentially contributing to the development of a more immunosuppressive microenvironment (Li et al., 2020). Previous studies using a colony-stimulating factor 1 (CSF-1) nullizygous (*Csf1* ^{op}/*Csf1* ^{op}) mouse model have shown that total macrophage depletion resulted in reduced epithelial branching and mammary gland development (Gouon-Evans et al., 2000; Van Nguyen and Pollard, 2002). However, a more directed approach using the CD11c:DTR transgenic mouse model, which results in the depletion of ductal macrophages, but spares stromal macrophage populations, suggested that these cells indirectly negatively regulate branching morphogenesis and alveolar bud formation. As such, DMs were involved in modulating the presence and activation of $CD4^+$ T helper 1 cells, which were shown to interfere with luminal differentiation via IFNγ signaling (Plaks et al., 2015). These observations suggest that the distinct macrophage populations play opposite roles in orchestrating a balanced normal mammary gland development. Based on the existing knowledge, stromal macrophages appear to be required for epithelial branching by modulating the extracellular matrix composition while the ductal macrophages ensure normal structural development by phagocytic activity. However, additional studies targeting distinct macrophage phenotypes, including spatial transcriptomics, multiplex tissue imaging, depletion studies and functional assays, will be needed to further decipher the heterogeneity and better understand the functional roles of these tissue macrophages in homeostasis and diseases.

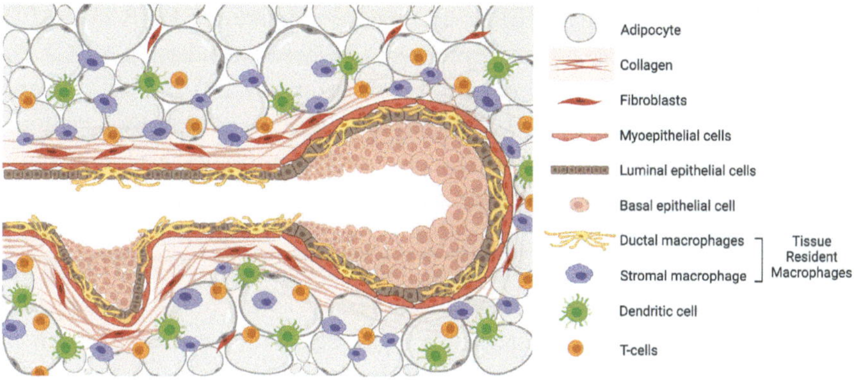

Fig. 2 Macrophages in the mice mammary gland. Representation of a terminal end bud (TEB) and stromal cellular components, surrounded by adipocytes. *This figure was created using BioRender.com.*

Altogether, based on our current knowledge we can clearly identify two spatially, phenotypically and functionally distinct tissue resident mammary gland macrophage populations: LYVE1^{+}CD206^{+}FOLR2^{+} stromal and CD11c^{+}CX3CR1^{+}VCAM1^{Hi}CADM1^{+} ductal F4/80^{+} macrophages in mice (Fig. 2). While the phenotype of the FOLR2^{+} population is highly conserved across mammals, the ductal macrophages are defined as CADM1^{+}TREM2^{+} macrophages in humans. In mouse, a CD206$^{Low/+}$ LYVE1^{-} poorly characterized population has been also described in association with the stroma.

3. Breast cancer

Breast cancer is the most common cancer type worldwide with 2.3 million women receiving a diagnose of breast cancer in 2020 (Sung et al., 2021). Nonetheless, even though these cancers arise in the same organ, breast cancer is a very heterogeneous disease at the molecular level. In the current clinical practice breast cancer is often classified in five molecular subtypes based on the expression of estrogen (ER) and progesterone (PR) hormone receptors, HER2 and the proliferation marker Ki67: (i) Luminal A-like cancers are positive for ER and/or PR, but are HER2 negative and show a low proliferation rate; (ii) Luminal B-like cancers express ER and/or PR, show a high proliferation, but do overexpress HER2; (iii) the HER2 subtype cancers are positive for HER2, but negative for

ER and PR; (iv) luminal HER2 cancers are positive for HER2 and ER and/or PR and (v) basal-like triple negative breast cancers (TNBC) are negative for HER2, ER and PR (Cardoso et al., 2019; Harbeck and Gnant, 2017). Nonetheless, using genomic and transcriptomic analysis, breast cancer can even further be classified in more subtypes, each impacting differently on the survival and treatment responses of patients (Ali et al., 2014; Craven et al., 2021; Curtis et al., 2012). The classification of breast cancer subtypes also defines to some extend the therapies used. Surgery, (neo)adjuvant chemotherapy and radiotherapy alone or in combination are still the mainstay of breast cancer treatment. However, in addition to the classical therapies mentioned, combinations complement approaches, including endocrine therapies, HER2 targeting therapies and immune-checkpoint (PD-L1) inhibitor therapies, are also available respectively for PR^{+}/ER^{+}, $HER2^{+}$ and triple-negative breast cancers. (Cardoso et al., 2019; Harbeck and Gnant, 2017; Mittendorf et al., 2020; Schmid et al., 2020). Irrespective of the type of breast cancer, when patient presents with a germline BRCA1 or BRCA2 mutation, patients can also be treated with PARP inhibitors (Cortesi et al., 2021).

The heterogeneity of breast cancer is suggested to be driven by three main factors: (i) the cell of origin in which the neoplastic mutation occurs, (ii) the multitude of oncogenic driver mutations and (iii) the stromal microenvironment (Ellis and Perou, 2013; Polyak and Kalluri, 2010; Zhang et al., 2017). Each of these factors influences the development, progression and therapy responsiveness of breast cancer and highlights the complexity of cancer treatment. In this section, we more specifically focus on the role macrophages play in these processes.

4. Macrophages in breast cancer

4.1 Macrophage heterogeneity in breast cancer

Besides contributing to the development of mammary glands and their roles in homeostasis, macrophages also play critical roles in breast cancer. Many functions modulated by macrophages, including organogenesis, angiogenesis, immune surveillance or ECM turnover, are altered during mammary tumor progression (Ingman et al., 2006; Lahmar et al., 2016; Leek et al., 1996; Lin et al., 2006; Ruffell and Coussens, 2015). Macrophages are an abundant cell population in breast cancer and their presence in general associated with a poor prognosis and more aggressive tumors (Bense et al., 2017; DeNardo et al., 2011; Mahmoud et al., 2012; Medrek et al., 2012; Yuan et al., 2014; Zhang et al., 2012; Zhao et al., 2017). TAMs are heterogeneous showing relative variability in origin, phenotype and function based on tumor and tissue type. It is widely accepted that in breast cancer the majority of TAMs are *de novo* differentiated monocyte-derived macrophages (Bonapace et al., 2014; Franklin et al., 2014; Movahedi et al., 2010; Qian et al., 2011), however local proliferation has also been shown to contribute to the TAM pool in mouse models (Campbell et al., 2011; Tymoszuk et al., 2014).

TAM heterogeneity has been long time described (Clappaert et al., 2018; Laoui et al., 2011; Movahedi et al., 2010; Pucci et al., 2009), however, recent scRNA-seq studies revealed the existence of novel TAM subpopulations providing insight into their functional role and relation to mammary tissue-resident macrophages across species (Cheng et al., 2021; Dawson et al., 2020; Ibrahim et al., 2020; Kiss et al., 2018; Ramos et al., 2021; Wang et al., 2020; Azizi et al., 2018; Bassez et al., 2021).

Using the PN1a outgrowth line (generated from p53-null transgenic mice; Medina et al., 2002) transplantation model, Ibrahim and colleagues identified six distinct macrophage populations of which four were mature *Adgre1*$^{+}$ (F4/80^{+}) TAM populations both in pre-malignant and well established invasive lesions (Ibrahim et al., 2020). Interestingly, with tumor progression the proportional distribution of these populations evolved, but not the population diversity. The identified populations contained tissue resident macrophage and TAM populations previously described in normal mammary gland and tumor models, respectively. Both *bona fide* TRM, being the SM and DM populations have been identified showing well-defined gene signatures: the SM-like macrophage cluster characterized by *Cd209g, Lyve1, Tim4d, Gas6, Mrc1/Cd206, Folr2* (Ramos et al., 2021; Wang et al., 2020) while the DM-like cluster were described by *Itgax, Cx3cr1, Tmem119, Cxcl16* and *Cadm1* (Dawson et al., 2020; Ramos et al., 2021) expression. Immunostaining analysis using CD206, Lyve1 and CSFR1 marker combinations confirmed that SMs were predominantly peritumoral, while DMs showed intra-tumoral dispersion (Dawson et al., 2020; Ibrahim et al., 2020; Wang et al., 2020). It has been reported that DM-like macrophages dominate tumors (Dawson et al., 2020; Linde et al., 2018; Ramos et al., 2021), however, on the contrary Ibrahim and colleagues showed that their abundance was lower in established tumors. This discrepancy might simply derive from the difference in the experimental models (spontaneous *versus* transplantation) used, either due to the procedure or the removal of the normal residential niche (the epithelial tree and the surrounding stroma/fat) upon mammary fat pad clearance prior to tumor transplantation. It is possible that the lack of the normal epithelial structure or the microenvironment created by the transplanted hyperplastic PN1a impairs the capacity of the newly recruited monocytes to differentiate into DMs. On the other hand, it would be also intriguing to investigate whether the proliferation and migration of DMs from the adjacent tissue into the tumor contribute to the expansion of this subset. In addition, a monocyte-derived $CD206^{+}Lyve1^{-}$ TAM population was identified with an intermediate gene expression profile sharing both SM (*Cd206, Gas6, Folr2, Ccl7, Ccl8, Pf4*) and DM (*Axl, Il1b, Cxcl2, Ccr2, Cxcr1,Cd74, Vcam1, Cxcl16*) signatures (Ibrahim et al., 2020). Immunostaining of tissue sections revealed that $CD206^{+}Lyve1^{-}$ cells are primarily localized in the invasive tumor regions which together with their inflammatory gene expression profile suggests a role in promoting tumor cell invasion (Gomes et al., 2018; Ibrahim et al., 2020). Besides the pro-tumor properties, gene ontology analyses suggest that this population also has anti-tumor characteristics by promoting T cell proliferation and activation (Ibrahim et al., 2020). At last, the previously described $Trem2^{+}$ lipid-associated TAM population (Jaitin et al., 2019) was also identified. This population, which is characterized with the signature of *Trem2, Fabp5, Lgals3, Cd36, Cd9* and *Gpnmb* expression, showed a high abundance in pre-malignant lesions but decreased in abundance in established tumors (Ibrahim et al., 2020). Previously, $CX3CR1^{+}$ and $TREM2^{+}$ tumor-associated myeloid cells were showed to reduce activated $CD8^{+}$ T cell proliferation, suggesting an immunosuppressive role for these subsets (Katzenelenbogen et al., 2020) (Fig. 3).

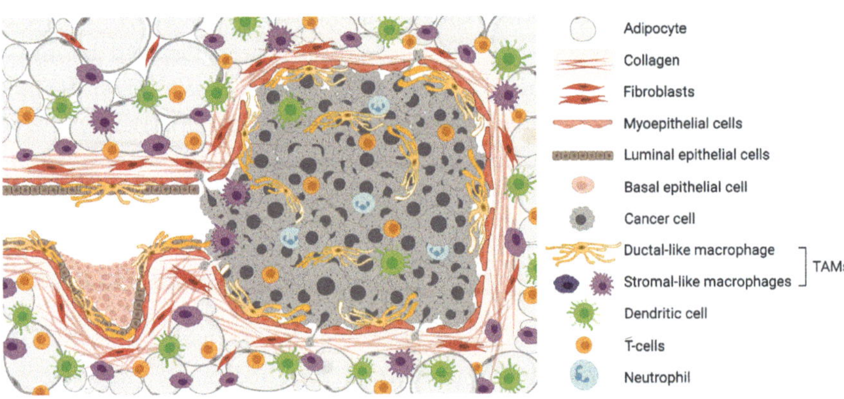

Fig. 3 Macrophages in the mice mammary tumor tissue. Representation of a terminal end bud tumor cellular microenvironment and adipocytes. *This figure was created using BioRender.com.*

TAM heterogeneity is also well documented in human breast cancer. In line with mouse studies, analysis of human breast tumors showed that TAMs are expressing gene signatures associated with both immunostimulatory ("M1-like") and immunosuppressive ("M2-like") activation states, suggesting that different phenotypes might represent distinct states on the activation spectrum (Azizi et al., 2018; Cheng et al., 2021; Wagner et al., 2019). Phenotypical characterization suggested that tissue-resident macrophages are important components of breast tumors and, so far, the described TRM subsets are not restricted to healthy tissues (Azizi et al., 2018; Ramos et al., 2021; Wagner et al., 2019). In line with this, tissue-specific signaling (Cassetta et al., 2019) or tumor cell phenotype (Wagner et al., 2019) were shown to significantly contribute to the diversity of the TAMs landscape. Comparing $CD163^+$ macrophages derived from healthy and tumorous breast by bulk mRNA-seq revealed a TAM-specific signature (including the expression of *FOLR2*), suggesting that TRMs and TRM-like TAMs are transcriptionally distinct, and specifically identified CD169 (Siglec1) and CCL8 as novel marker combination to discriminate TAMs associated with high tumor grade and poor clinical outcomes (Cassetta et al., 2019). Accordingly, $CCL8^{Hi}$ TAM populations have been also described in murine mammary tumors with higher frequencies in invasive tumor regions (Ibrahim et al., 2020). Several "classical" markers (including CD163, CD206, CD204, CD169, CD68, CD36, PD-L1) have been associated with TAMs, however, none of them can label a specific TAM population and even the combination of several of these markers cannot exclusively define distinct TAM subsets (Cassetta et al., 2019; Ramos et al., 2021; Wagner et al., 2019). Recent scRNA-seq analysis by Ramos and colleagues identified APOE as a TAM marker in human breast cancer and described two evolutionary conserved tissue-resident TAM populations, namely $FOLR2^+LYVE1^+MRC1$ $(CD206)^+$ and $TREM2^+SPP1^+CADM1^+$ (Ramos et al., 2021) macrophages. The gene expression profile and spatial distribution of both populations showed high a correlation with their murine counterparts (Dawson et al., 2020; Ibrahim et al., 2020; Wang et al., 2020). The intra-epithelial $TREM2^+CADM1^+$ macrophages are rare in healthy tissue, but they become the prominent subset with tumor progression. $TREM2^+$ macrophages have been described in several cancer types in association with immunosuppression and poor prognosis (Katzenelenbogen et al., 2020; Molgora et al., 2020). On the other hand, the abundance of stromal $FOLR2^+LYVE1^+$ macrophages gradually decreases, with a low abundance in the tumor core and with the highest frequency at the peritumoral regions. The co-localization of $FOLR2^+$ TAMs with lymphoid structures, their positive correlation with $CD8^+$ T cell infiltration and better survival rates suggest an active role in anti-tumor immunity (Ramos et al., 2021). Complementary, a recent cross-tissue scRNA-seq analysis identified three main macrophage populations, including $FOLR2^+$, $TREM2^+$ and $IL4l1^+$ macrophages both in healthy and diseased states, including breast cancer (Mulder et al., 2021). $FOLR2^+$ macrophages appeared to be long term resident macrophages while the $TREM2^+$ and $IL4l1^+$ macrophages were more recently differentiated monocyte-derived cells. This ontological difference was also confirmed in the murine counterparts (Sharma et al., 2020). $IL4l1^+$ macrophages are characterized by $CD38^{Hi}$ and $PD\text{-}L1^{Hi}$ expression showing a similar profile as the $PD\text{-}L1^+$ TAMs described with higher frequencies in high grade ER^+ and ER^- breast tumors (Wagner et al., 2019). Altogether, these three populations represent macrophage subsets that are conserved across tissues and species which could provide the basis for a universal annotation for future studies (Fig. 4). Interestingly, at first glance, the findings by Ramos and colleagues are in contradiction with the observations of Cassetta and

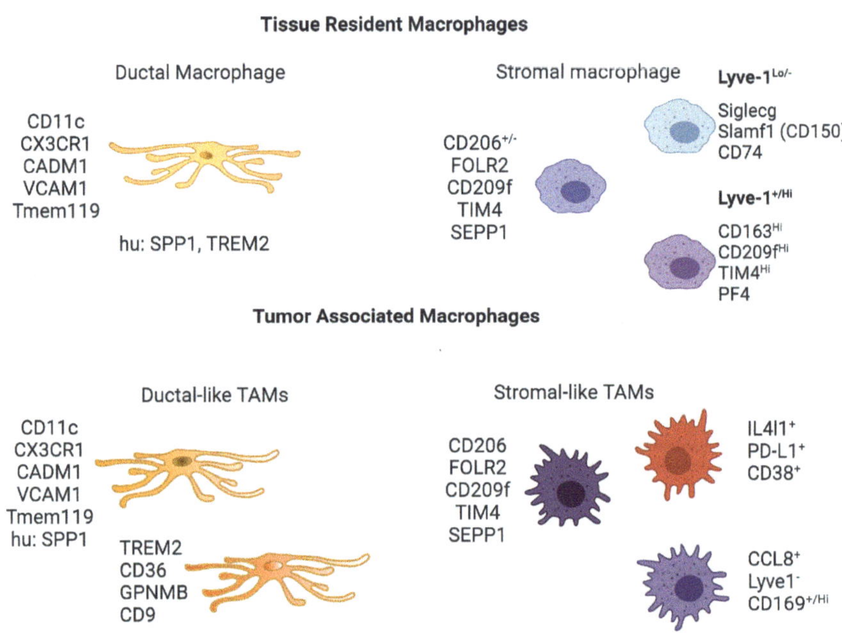

Fig. 4 Ductal and stromal tissue resident macrophages and tumor associated ductal- and stromal-like macrophages in mouse and human breast cancer. *This figure was created using BioRender.com.*

colleagues, the latter showing that the $CD163^+$ TAMs expressing high levels of *CD169, CCL8* and *FOLR2* are associated with poor survival. However, the $CD163^+$ TAMs were also enriched for markers described in distinct TRM and TAM populations including TREM2, CADM1 or CD38 (Cassetta et al., 2019; Mulder et al., 2021; Ramos et al., 2021). Altogether, this suggests that $CD163^+$ TAMs might arise from several distinct TRM or TAM subsets potentially creating a phenotypically heterogenous cluster of TAMs that might exert similar functions.

Clearly, additional studies are required to further dissect the heterogeneity, to validate the spatial distribution of TAM subsets and to clarify their exact origin and functions. An intriguing question that remains to be answered is what are the phenotypic or functional differences between tissue resident macrophage populations derived from healthy or cancerous breast tissues. Spatial transcriptomic and multiplex tissue imaging could be excellent approaches to determine subset specific unique marker combinations and to correlate existing human and murine datasets. Furthermore, systematic functional characterization will be needed to understand how much the phenotypic heterogeneity correlates with functional differences.

4.2 Impact of hypoxia on the heterogeneity and functions of TAMs in breast cancer

A tumor is not one homogenous structure, but rather presents with a high level of intratumor heterogeneity. One of the parameters causing this phenomenon is tumor hypoxia (Ramón y Cajal et al., 2020). When a tumor develops, it soon outgrows its supply of oxygen. As a consequence, a signaling cascade is activated leading to the stabilization of hypoxia-inducible factor (HIF) 1α and HIF2α transcription factors and transcription of angiogenic factors such as VEGF (Harris, 2002). Nonetheless, due to excessive production of pro-angiogenic factors, the vasculature remains immature, resulting in an abnormal and inadequate vascular network that cannot resolve the lack of oxygen (Jain, 2005). The resulting presence of hypoxia is in many cancers negatively associated with patient survival and results in therapy resistance (Walsh et al., 2014). In breast cancer specifically, a median pO_2 of 30 mmHg was observed, while normal breast tissues had a median pO_2 of 65 mmHg (Vaupel et al., 1991). In addition, the pO_2 values and anoxic fraction correlate with the degree of differentiation of breast cancer tumors with grade 1 tumors having a mean pO_2 of 59 mmHg and less than 1% anoxic fraction, while grade 4 tumors presented with a mean pO_2 of 10 mmHg and a 21.5% anoxic fraction (Hohenberger et al., 1998). Furthermore, the surrogate hypoxia markers HIF1α, HIF2α and their target gene CAIX, have been correlated with therapy resistance, worse relapse free survival and worse overall survival (Betof et al., 2012; Bos et al., 2003; Brennan et al., 2006; Chia et al., 2001; Dales et al., 2005; Goonewardene et al., 2002; Helczynska et al., 2008; Hussain et al., 2007). TAMs play a role in mediating the detrimental effects hypoxia has on the survival of cancer patients. They can reside in both normoxic and hypoxic regions of both human and mouse breast tumors. In murine tumor models, this has been shown using immunohistochemistry and flow cytometry in spontaneous and syngeneic orthotopic breast cancer models where hypoxic TAMs are often identified as $MHCII^{Low}$ or $GLUT1^{Hi}$ TAMs (Casazza et al., 2013; Fang et al., 2009; Movahedi et al., 2010; Wenes et al., 2016). In human breast cancer, the first evidence of the presence of TAMs in hypoxic areas of breast cancer tumors was provided using immunohistochemical analysis (Murdoch and Lewis, 2005). In addition, a TAM population expressing a hypoxic gene signature has recently been identified in distinct scRNA-seq datasets of human breast cancer, suggesting their presence in hypoxic regions of the tumor (Azizi et al., 2018; Bassez et al., 2021).

TAMs are recruited into hypoxic regions of the tumor by factors such as Sema3A, Oncostatin M and Eotaxin expressed by cancer cells exposed to the hypoxic microenvironment (Casazza et al., 2013; Tripathi et al., 2014). Sema3A signaling via Nrp1-VEGFR1 complex has been shown to be responsible for recruitment of TAMs in hypoxic regions of the tumor and silencing of Sema3A, similarly to genetic deletion of its receptor Nrp1 in macrophages, resulted in an increased overall presence of TAMs, but reduced presence of pro-tumoral TAMs (Casazza et al., 2013). These results are difficult to reconcile with the recent findings that Sema3A has also been shown—in contrast to hypoxia—to be inversely correlated with the breast cancer grade (Wallerius et al., 2016). In addition, Sema3A overexpression in the cancer cells has been shown to reduce tumor growth in the 4T1 syngeneic breast cancer model by shifting the balance from the more pro-tumoral $MHCII^{Hi}$ TAMs to the anti-tumoral $MHCII^{Low}$ TAMs (Wallerius et al., 2016). The discrepancy between these results requires further investigation, but could potentially be explained by the dual role Sema3A signaling can play as in the absence of Nrp1, Sema3A via its receptors PlexinA1 and PlexinA4 has an opposite function to trap TAMs in hypoxia (Casazza et al., 2013). Therefore, experiments with overexpression of Sema3A should be taken with caution as these could disrupt the balance between ligand and receptors, though further studies are required to investigate these results in depth and understand the role of Sema3A in TAMs in hypoxic regions of breast cancer tumors. The other two factors described to attract macrophages to hypoxic breast cancer cells are Oncostatin M and Eotaxin. Indeed, also in breast cancer patient samples, these two molecules co-localize with HIF1α expression. Blocking of these factors in the syngeneic 4T1 breast cancer model consequently reduces tumor growth, tumor vasculature and CD206 expression, though more extensive analysis on the TAM compartment in hypoxia would be required to confirm *in vitro* experiments that Oncostatin M and Eotaxin are required for *in vivo* recruitment to and polarization of TAMs in hypoxic regions of the tumor (Shrivastava et al., 2018; Tripathi et al., 2014). Other factors suggested to play a role in the attraction of TAMs to hypoxic regions of the tumor are for example VEGF, CCL2, CCL5, EMAP-II, endothelin and SDF1α, though their role in TAM recruitment to hypoxic regions in breast cancer specifically has not been addressed yet (Henze and Mazzone, 2016).

Once TAMs arrive in hypoxic regions, they are trapped there due to the downregulation of several receptors such as Nrp1, CCR1, CCR2 and CCR5. Indeed, Nrp1 is significantly downregulated in hypoxic TAMs in a HIF2a-mediated activation of the Nf-κB pathway, thereby specifically abrogating chemoattraction to SEMA3A (Casazza et al., 2013). In addition to Nrp1, *in vitro* experiments would also suggest a role for CCR1, CCR2 and CCR5 in entrapping TAMs in hypoxic regions of the tumor as the expression of these proteins is downregulated in macrophages in hypoxia (Bosco et al., 2006; Tausendschön et al., 2011). Nonetheless, more *in vivo* studies would be required to understand the role these proteins play in entrapping TAMs in hypoxic regions of breast cancer tumors. Indeed, in contrast to the *in vitro* data, CCR1 has also been shown to be increased upon hypoxia in the monocytic THP-1 cell line suggesting a cell type/context dependent regulation of CCR1 (Scotton et al., 2020). In addition, while in the TS/A breast cancer model CCR2 levels were higher in MHCII^{low} "hypoxic" TAMs compared to MHCII^{Hi} "normoxic" TAMs in line with the suggested role of CCR2 in trapping TAMs in hypoxia, this was the opposite in the 3LL-R lung cancer model, again suggesting a potential context dependent regulation (Laoui et al., 2014; Movahedi et al., 2010).

Once trapped in hypoxic regions of the tumor, the phenotype of TAMs is altered significantly. More specifically, hypoxic TAMs present with a higher angiogenic potential and an increased anti-inflammatory capacity (Casazza et al., 2013; Movahedi et al., 2010). Indeed, exclusion of macrophages from hypoxic regions of the tumor leads to a reduction in vasculature and increased anti-tumor immune response and as a consequence decreased tumor growth (Casazza et al., 2013). The factors mediating this response are plentiful and are likely the result of a combination of factors. Gene expression analysis in $MHCII^{Low}$ and $MHCII^{Hi}$ TAMs derived from the TS/A breast cancer model indicated differential regulation of a multitude of genes with $MHCII^{Hi}$ TAMs expressing more angiostatic factors and lymphocyte attracting chemokines, while $MHCII^{Lo}$ TAMs rather presented with increased expression of monocyte/macrophage chemoattractants (Movahedi et al., 2010). The pathways regulating these changes are also diverse. Deletion of HIF1α, one of the most obvious candidates to regulate TAM changes in hypoxia has been shown to delay progression of the spontaneous MMTV-PyMT breast cancer model and macrophages deficient for HIF1α caused less T-cell suppression under hypoxic conditions, potentially explaining the delayed tumor progression (Doedens et al., 2010). Nonetheless, HIF1α deletion in the myeloid cell compartment does not affect VEGF or vascularization in the MMTV-PyMT breast cancer model, confirming that multiple pathways are at play (Doedens et al., 2010). As such, hypoxia not only leads to HIF1α and HIF2α stabilization, but can also act in a HIF-independent manner on for example the Nf-κB pathway (D'Ignazio et al., 2017). In addition, hypoxia also leads to a shift in metabolism in cancer cells towards a more glycolytic metabolism, resulting in an increased lactate production. Exposure of TAMs to lactate has been shown to enhance their pro-tumoral function to increase breast cancer metastasis (Chen et al., 2017). In addition, metabolic stress caused by hypoxia can also activate the unfolded protein response, leading to mTOR inhibition, activation of ATF6, eIF2a, XBP1 and ATF4 transcription factors thereby influencing gene transcription in TAMs extensively (Díaz-Bulnes et al., 2019). In conclusion, hypoxia has a major impact on TAM activity and function, though the regulation of this phenomenon is a highly complex interplay between factors/pathways influencing each other that has not been completely unraveled yet. In addition, as likely tissue specific phenomena are taking place as well, extrapolation of data from one cancer type to another always has to be taken with caution.

5. Concluding remarks and future perspectives

Macrophages in breast and breast cancer have already been studied relatively intensively and novel technologies are further increasing our understanding about the macrophage populations present in breast and breast cancer. Nonetheless, there are still many shortcomings to the current data that require further investigation.

Firstly, scRNA-seq has provided us with novel data on the macrophage populations present in breast and breast cancer tissues (Azizi et al., 2018; Bassez et al., 2021). Nonetheless, as the heterogeneity within breast cancer is high, it would be anticipated that also macrophage populations within these cancer types differ fundamentally. Currently however, partly due to the cost and large amount of data processing required, there has not yet been any analysis focusing on the differences in TAM populations in different breast cancer subtypes. Nevertheless, major efforts are being made to make the technology more accessible (De Rop et al., 2021) and to improve data analysis (Luecken and Theis, 2019; Saelens et al., 2019), increasing the feasibility to perform larger studies. In addition, complementing scRNA-seq with spatial transcriptomic analysis (Bassiouni et al., 2021; Chen et al., 2015; Nagasawa et al., 2021; Park et al., 2021), will allow to determine the localization of the identified TAM populations within the normal or breast cancer tissue.

This will, however, only be a starting point due to the descriptive nature of these studies without addressing the biological function of differentially expressed genes in different TAM populations. Research identifying and studying the role of novel factors determining TAM function will be required. Unfortunately, however, current studies investigating TAM ontogeny or function have been performed in a limited number of breast cancer models and therefore do not represent the diversity of breast cancer in patients. Recently however, it was shown that upon p53 deletion in breast cancer, Wnt ligand secretion was increased and induced the production of IL-1b in TAMs, thereby driving systemic inflammation and breast cancer metastasis (Wellenstein et al., 2019). This research was possible by using a panel of 16 genetically engineered breast cancer models each having their own characteristics (Wellenstein et al., 2019). Indeed, by making use of these models, it has already become clear that the origin of basal breast cancers could also be luminal (Skibinski and Kuperwasser, 2015). A further characterization of TAM subsets in these models using scRNA-seq and spatial transcriptomics could then further allow comparison to the human TAM subsets. Subsequently allowing also the functional study of TAM subsets in more clinically relevant breast cancer models.

Overall, although we are only at the start of unravelling the complexity of TAM regulation in breast and breast cancer, the recent advances in technologies will allow us to study TAMs more in depth than ever before. This will open a new era of research to discover novel potential targets to tackle immunosuppressive TAMs in breast cancer and to develop novel therapies that can increase breast cancer survival.

Competing interests

The authors declare that they have no conflict of interest.

Funding

E.H. is supported by an FWO postdoctoral fellowship, S.D. is supported by a postdoctoral fellowship from Stichting tegen kanker, D.L. is supported by grants from FWO, Kom op Tegen Kanker, Stichting tegen kanker and Vrije Universiteit Brussel.

References

Ali, H.R., Rueda, O.M., Chin, S.F., Curtis, C., Dunning, M.J., Aparicio, S.A.J.R., Caldas, C., 2014. Genome-driven integrated classification of breast cancer validated in over 7,500 samples. Genome Biol. 15, 1–14. https://doi.org/10.1186/s13059-014-0431-1.

Azizi, E., Carr, A.J., Plitas, G., Cornish, A.E., Konopacki, C., Prabhakaran, S., Nainys, J., Wu, K., Kiseliovas, V., Setty, M., Choi, K., Fromme, R.M., Dao, P., McKenney, P.T., Wasti, R.C., Kadaveru, K., Mazutis, L., Rudensky, A.Y., Pe'er, D., 2018. Single-cell map of diverse immune phenotypes in the breast tumor microenvironment. Cell 174, 1293–1308, e36. https://doi.org/10.1016/j.cell.2018.05.060.

Bassez, A., Vos, H., Van Dyck, L., Floris, G., Arijs, I., Desmedt, C., Boeckx, B., Vanden Bempt, M., Nevelsteen, I., Lambein, K., Punie, K., Neven, P., Garg, A.D., Wildiers, H., Qian, J., Smeets, A., Lambrechts, D., 2021. Anti-PD1 treatment of patients with breast cancer. Nat. Med. https://doi.org/10.1038/s41591-021-01323-8. Springer US.

Bassiouni, R., Gibbs, L.D., Craig, D.W., Carpten, J.D., McEachron, T.A., 2021. Applicability of spatial transcriptional profiling to cancer research. Mol. Cell 81, 1631–1639. https://doi.org/10.1016/j.molcel.2021.03.016.

Bense, R.D., Sotiriou, C., Piccart-Gebhart, M.J., Haanen, J.B.A.G., van Vugt, M.A.T.M., de Vries, E.G.E., Schröder, C.P., Fehrmann, R.S.N., 2017. Relevance of tumor-infiltrating immune cell composition and functionality for disease outcome in breast cancer. J. Natl. Cancer Inst. 109. https://doi.org/10.1093/jnci/djw192.

Bercovici, N., Guérin, M.V., Trautmann, A., Donnadieu, E., 2019. The remarkable plasticity of macrophages: a chance to fight cancer. Front. Immunol. 10, 1–9. https://doi.org/10.3389/fimmu.2019.01563.

Betof, A.S., Rabbani, Z.N., Hardee, M.E., Kim, S.J., Broadwater, G., Bentley, R.C., Snyder, S.A., Vujaskovic, Z., Oosterwijk, E., Harris, L.N., Horton, J.K., Dewhirst, M.W., Blackwell, K.L., 2012. Carbonic anhydrase IX is a predictive marker of doxorubicin resistance in early-stage breast cancer independent of HER2 and TOP2A amplification. Br. J. Cancer 106, 916–922. https://doi.org/10.1038/bjc.2012.32.

Bonapace, L., Coissieux, M.-M., Wyckoff, J., Mertz, K.D., Varga, Z., Junt, T., Bentires-Alj, M., 2014. Cessation of CCL2 inhibition accelerates breast cancer metastasis by promoting angiogenesis. Nature 515, 130–133. https://doi.org/10.1038/nature13862.

Bos, R., Van der Groep, P., Greijer, A.E., Shvarts, A., Meijer, S., Pinedo, H.M., Semenza, G.L., Van Diest, P.J., Van der Wall, E., 2003. Levels of hypoxia-inducible factor-1α independently predict prognosis in patients with lymph node negative breast carcinoma. Cancer 97, 1573–1581. https://doi.org/10.1002/cncr.11246.

Bosco, M.C., Puppo, M., Santangelo, C., Anfosso, L., Pfeffer, U., Fardin, P., Battaglia, F., Varesio, L., 2006. Hypoxia modifies the transcriptome of primary human monocytes:

modulation of novel immune-related genes and identification of CC-chemokine ligand 20 as a new hypoxia-inducible gene. J. Immonol. 177 (3), 1941–1955. https://doi.org/10.4049/jimmunol.177.3.1941.

Brennan, D.J., Jirstrom, K., Kronblad, Å., Millikan, R.C., Landberg, G., Duffy, M.J., Rydén, L., Gallagher, W.M., O'Brien, S.L., 2006. CA IX is an independent prognostic marker in premenopausal breast cancer patients with one to three positive lymph nodes and a putative marker of radiation resistance. Clin. Cancer Res. 12, 6421–6431. https://doi.org/10.1158/1078-0432.CCR-06-0480.

Campbell, M.J., Tonlaar, N.Y., Garwood, E.R., Huo, D., Moore, D.H., Khramtsov, A.I., Au, A., Baehner, F., Chen, Y., Malaka, D.O., Lin, A., Adeyanju, O.O., Li, S., Gong, C., McGrath, M., Olopade, O.I., Esserman, L.J., 2011. Proliferating macrophages associated with high grade, hormone receptor negative breast cancer and poor clinical outcome. Breast Cancer Res. Treat. 128, 703–711. https://doi.org/10.1007/s10549-010-1154-y.

Cardiff, R.D., Wellings, S.R., 1999. The comparative pathology of human and mouse mammary glands. J. Mammary Gland Biol. Neoplasia 4 (1), 105–122.

Cardoso, F., Kyriakides, S., Ohno, S., Penault-Llorca, F., Poortmans, P., Rubio, I.T., Zackrisson, S., Senkus, E., 2019. Early breast cancer: ESMO clinical practice guidelines for diagnosis, treatment and follow-up. Ann. Oncol. 30, 1194–1220. https://doi.org/10.1093/annonc/mdz173.

Casazza, A., Laoui, D., Wenes, M., Rizzolio, S., Bassani, N., Mambretti, M., Deschoemaeker, S., Van Ginderachter, J.A., Tamagnone, L., Mazzone, M., 2013. Impeding macrophage entry into hypoxic tumor areas by Sema3A/Nrp1 signaling blockade inhibits angiogenesis and restores antitumor immunity. Cancer Cell 24, 695–709. https://doi.org/10.1016/j.ccr.2013.11.007.

Cassetta, L., Pollard, J.W., 2018. Targeting Macrophages. Nature Publishing Group, https://doi.org/10.1038/nrd.2018.169.

Cassetta, L., Fragkogianni, S., Sims, A.H., Swierczak, A., Forrester, L.M., Zhang, H., Soong, D.Y.H., Cotechini, T., Anur, P., Lin, E.Y., Fidanza, A., Lopez-Yrigoyen, M., Millar, M.R., Urman, A., Ai, Z., Spellman, P.T., Hwang, E.S., Dixon, J.M., Wiechmann, L., Coussens, L.M., Smith, H.O., Pollard, J.W., 2019. Human tumor-associated macrophage and monocyte transcriptional landscapes reveal cancer-specific reprogramming, biomarkers, and therapeutic targets. Cancer Cell 35, 588–602.e10. https://doi.org/10.1016/j.ccell.2019.02.009.

Chakarov, S., Lim, H.Y., Tan, L., Lim, S.Y., See, P., Lum, J., Zhang, X.M., Foo, S., Nakamizo, S., Duan, K., Kong, W.T., Gentek, R., Balachander, A., Carbajo, D., Bleriot, C., Malleret, B., Tam, J.K.C., Baig, S., Shabeer, M., Toh, S.A.E.S., Schlitzer, A., Larbi, A., Marichal, T., Malissen, B., Chen, J., Poidinger, M., Kabashima, K., Bajenoff, M., Ng, L.G., Angeli, V., Ginhoux, F., 2019. Two distinct interstitial macrophage populations coexist across tissues in specific subtissular niches. Science 363. https://doi.org/10.1126/science.aau0964.

Chen, K.H., Boettiger, A.N., Moffitt, J.R., Wang, S., Zhuang, X., 2015. RNA imaging. Spatially resolved, highly multiplexed RNA profiling in single cells. Science (New York, N.Y.) 348, aaa6090. https://doi.org/10.1126/science.aaa6090. .

Chen, P., Zuo, H., Xiong, H., Kolar, M.J., Chu, Q., Saghatelian, A., Siegwart, D.J., Wan, Y., 2017. Gpr132 sensing of lactate mediates tumor—macrophage interplay to promote breast cancer metastasis. Proc. Natl. Acad. Sci. U. S. A. 114 (3), 580–585. https://doi.org/10.1073/pnas.1614035114.

Cheng, S., Li, Z., Gao, R., Xing, B., Gao, Y., Yang, Y., Qin, S., Zhang, L., Ouyang, H., Du, P., Jiang, L., Zhang, B., Yang, Y., Wang, X., Ren, X., Bei, J.X., Hu, X., Bu, Z., Ji, J., Zhang, Z., 2021. A pan-cancer single-cell transcriptional atlas of tumor infiltrating myeloid cells. Cell 184, 792–809, e23. https://doi.org/10.1016/j.cell.2021.01.010.

Chia, S.K., Wykoff, C.C., Watson, P.H., Han, C., Leek, R.D., Pastorek, J., Gatter, K.C., Ratcliffe, P., Harris, A.L., 2001. Prognostic significance of a novel hypoxia-regulated marker, carbonic anhydrase IX, in invasive breast carcinoma. J. Clin. Oncol. 19, 3660–3668. https://doi.org/10.1200/JCO.2001.19.16.3660.

Chiappinelli, K., Ali, S.M., Peng, W., Cheng, X., Wang, X., Nie, K., Cheng, L., Zhang, Z., Hu, Y., 2021. Systematic pan-cancer analysis identifies TREM2 as an immunological and prognostic biomarker. Front. Immunol. 12, 646523. https://doi.org/10.3389/fimmu.2021.646523. www.frontiersin.org.

Chua, A.C.L., Hodson, L.J., Moldenhauer, L.M., Robertson, S.A., Ingman, W.V., 2010. Dual roles for macrophages in ovarian cycle-associated development and remodelling of the mammary gland epithelium. Development 137, 4229–4238. https://doi.org/10.1242/DEV.059261.

Clappaert, E.J., Murgaski, A., Damme, H.V., Kiss, M., Laoui, D., 2018. Diamonds in the rough: harnessing tumor-associated myeloid cells for cancer therapy. Front. Immunol. 9. https://doi.org/10.3389/fimmu.2018.02250.

Cortesi, L., Rugo, H.S., Jackisch, C., 2021. An overview of PARP inhibitors for the treatment of breast cancer. Target. Oncol. 16, 255–282. https://doi.org/10.1007/s11523-021-00796-4.

Craven, K.E., Polar, Y.G., Badve, S.S., 2021. OPEN CIBERSORT analysis of TCGA and METABRIC identifies subgroups with better outcomes in triple negative breast cancer. Sci. Rep. 1–19. https://doi.org/10.1038/s41598-021-83913-7.

Curtis, C., Shah, S.P., Chin, S., Turashvili, G., Rueda, O.M., Dunning, M.J., Speed, D., Lynch, A.G., Samarajiwa, S., Yuan, Y., Gräf, S., Ha, G., Haffari, G., Bashashati, A., Russell, R., McKinney, S., Langerød, A., Green, A., Provenzano, E., Wishart, G., Pinder, S., Watson, P., Markowetz, F., Murphy, L., Ellis, I., Purushotham, A., Børresen-Dale, A.-L., Brenton, J.D., Tavaré, S., Caldas, C., Aparicio, S., 2012. The genomic and transcriptomic architecture of 2,000 breast tumours reveals novel subgroups. Nature 486, 346–352. https://doi.org/10.1038/nature10983.

D'Ignazio, L., Batie, M., Rocha, S., 2017. Hypoxia and inflammation in cancer, focus on HIF and NF-κB. Biomedicine 5. https://doi.org/10.3390/biomedicines5020021.

Dales, J.P., Garcia, S., Meunier-Carpentier, S., Andrac-Meyer, L., Haddad, O., Lavaut, M.N., Allasia, C., Bonnier, P., Charpin, C., 2005. Overexpression of hypoxia-inducible factor HIF-1α predicts early relapse in breast cancer: retrospective study in a series of 745 patients. Int. J. Cancer 116, 734–739. https://doi.org/10.1002/ijc.20984.

Dawson, C.A., Visvader, J.E., 2021. The cellular organization of the mammary gland: insights from microscopy the role of diverse cell-types in mammary gland development. J. Mammary Gland Biol. Neoplasia 1, 3. https://doi.org/10.1007/s10911-021-09483-6.

Dawson, C.A., Pal, B., Vaillant, F., Gandolfo, L.C., Liu, Z., Bleriot, C., Ginhoux, F., Smyth, G.K., Lindeman, G.J., Mueller, S.N., Rios, A.C., Visvader, J.E., 2020. Tissue-resident ductal macrophages survey the mammary epithelium and facilitate tissue remodelling. Nat. Cell Biol. 22, 546–558. https://doi.org/10.1038/s41556-020-0505-0.

De Rop, F.V., Ismail, J.N., Bravo González-Blas, C., Hulselmans, G.J., Flerin, C.C., Janssens, J., Theunis, K., Christiaens, V.M., Marcassa, G., de Wit, J., Poovathingal, S., Aerts, S., 2021. HyDrop: droplet-based sc ATAC-seq and scRNA-seq using dissolvable hydrogel beads. bioRxiv. https://doi.org/10.1101/2021.06.04.447104.

DeNardo, D.G., Brennan, D.J., Rexhepaj, E., Ruffell, B., Shiao, S.L., Madden, S.F., Gallagher, W.M., Wadhwani, N., Keil, S.D., Junaid, S.A., Rugo, H.S., Hwang, E.S., Jirström, K., West, B.L., Coussens, L.M., 2011. Leukocyte complexity predicts breast cancer survival and functionally regulates response to chemotherapy. Cancer Discov. 1, 54–67. https://doi.org/10.1158/2159-8274.CD-10-0028.

Díaz-Bulnes, P., Saiz, M.L., López-Larrea, C., Rodríguez, R.M., 2019. Crosstalk between hypoxia and ER stress response: a key regulator of macrophage polarization. Front. Immunol. 10, 2951. https://doi.org/10.3389/fimmu.2019.02951.

Doedens, A.L., Stockmann, C., Rubinstein, M.P., Liao, D., Zhang, N., DeNardo, D.G., Coussens, L.M., Karin, M., Goldrath, A.W., Johnson, R.S., 2010. Macrophage expression of hypoxia-inducible factor-1α suppresses T-cell function and promotes tumor progression. Cancer Res. 70, 7465–7475. https://doi.org/10.1158/0008-5472.CAN-10-1439.

Dontu, G., Ince, T.A., 2015. Of mice and women: a comparative tissue biology perspective of breast stem cells and differentiation. J. Mammary Gland Biol. Neoplasia 20, 51. https://doi.org/10.1007/S10911-015-9341-4.

Ellis, M.J., Perou, C.M., 2013. The genomic landscape of breast cancer as a therapeutic roadmap. Cancer Discov. 3, 27–34. https://doi.org/10.1158/2159-8290.CD-12-0462.

Fang, H.Y., Hughes, R., Murdoch, C., Coffelt, S.B., Biswas, S.K., Harris, A.L., Johnson, R.S., Imityaz, H.Z., Simon, M.C., Fredlund, E., Greten, F.R., Rius, J., Lewis, C.E., 2009. Hypoxia-inducible factors 1 and 2 are important transcriptional effectors in primary macrophages experiencing hypoxia. Blood 114, 844–859. https://doi.org/10.1182/blood-2008-12-195941.

Franklin, R.A., Liao, W., Sarkar, A., Kim, M.V., Bivona, M.R., Liu, K., Pamer, E.G., Li, M.O., 2014. The cellular and molecular origin of tumor-associated macrophages. Science 344, 921–925. https://doi.org/10.1126/SCIENCE.1252510.

Fu, N.Y., Nolan, E., Lindeman, G.J., Visvader, J.E., 2020. Stem cells and the differentiation hierarchy in mammary gland development. Physiol. Rev. 100 (2), 489–523. https://doi.org/10.1152/PHYSREV.00040.2018.

Ginhoux, F., Guilliams, M., 2016. Tissue-resident macrophage ontogeny and homeostasis. Immunity 44, 439–449. https://doi.org/10.1016/j.immuni.2016.02.024.

Gomes, A.M., Carron, E.C., Mills, K.L., Dow, A.M., Gray, Z., Fecca, C.R., Lakey, M.A., Carmeliet, P., Kittrell, F., Medina, D., Machado, H.L., 2018. Stromal Gas6 promotes the progression of premalignant mammary cells. Oncogene 38 (14), 2437–2450. https://doi.org/10.1038/s41388-018-0593-5.

Gomez Perdiguero, E., Klapproth, K., Schulz, C., Busch, K., Azzoni, E., Crozet, L., Garner, H., Trouillet, C., de Bruijn, M.F., Geissmann, F., Rodewald, H.-R., 2015. Tissue-resident macrophages originate from yolk-sac-derived erythro-myeloid progenitors. Nature 518, 547–551. https://doi.org/10.1038/nature13989.

Goonewardene, T.I., Sowter, H.M., Harris, A.L., 2002. Hypoxia-induced pathways in breast cancer. Microsc. Res. Tech. 59, 41–48. https://doi.org/10.1002/jemt.10175.

Gouon-Evans, V., Rothenberg, M.E., Pollard, J.W., 2000. Postnatal mammary gland development requires macrophages and eosinophils. Development 2282, 2269–2282.

Gouon-Evans, V., Lin, E.Y., Pollard, J.W., 2002. Requirement of macrophages and eosinophils and their cytokines/chemokines for mammary gland development. Breast Cancer Res. 4 (4), 1–10. https://doi.org/10.1186/BCR441.

Gyorki, D.E., Asselin-Labat, M.-L., van Rooijen, N., Lindeman, G.J., Visvader, J.E., 2009. Resident macrophages influence stem cell activity in the mammary gland. Breast Cancer Res. 11 (4), 1–6. https://doi.org/10.1186/BCR2353.

Han, X., Wang, R., Zhou, Y., Fei, L., Sun, H., Lai, S., Saadatpour, A., Zhou, Z., Chen, H., Ye, F., Huang, D., Xu, Y., Huang, W., Jiang, M., Jiang, X., Mao, J., Chen, Y., Lu, C., Xie, J., Fang, Q., Wang, Y., Yue, R., Li, T., Huang, H., Orkin, S.H., Yuan, G.C., Chen, M., Guo, G., 2018. Mapping the mouse cell atlas by microwell-seq. Cell 172, 1091–1107.e17. https://doi.org/10.1016/j.cell.2018.02.001.

Harbeck, N., Gnant, M., 2017. Breast cancer. Lancet 389, 1134–1150. https://doi.org/10.1016/S0140-6736(16)31891-8.

Harris, A.L., 2002. Hypoxia—a key regulatory factor in tumour growth. Nat. Rev. Cancer 2, 38–47. https://doi.org/10.1038/nrc704.

He, D., Wang, D., Lu, P., Yang, N., Xue, Z., Zhu, X., Zhang, P., Fan, G., 2020. Single-cell RNA sequencing reveals heterogeneous tumor and immune cell populations in early-stage lung adenocarcinomas harboring EGFR mutations. Oncogene 40, 355–368. https://doi.org/10.1038/s41388-020-01528-0.

Helczynska, K., Larsson, A.M., Mengelbier, L.H., Bridges, E., Fredlund, E., Borgquist, S., Landberg, G., Påhlman, S., Jirström, K., 2008. Hypoxia-inducible factor-2α correlates to distant recurrence and poor outcome in invasive breast cancer. Cancer Res. 68, 9212–9220. https://doi.org/10.1158/0008-5472.CAN-08-1135.

Henze, A., Mazzone, M., 2016. The impact of hypoxia on tumor-associated macrophages. J. Clin. Invest. 126, 3672–3679. https://doi.org/10.1172/JCI84427.

Hoeffel, G., Ginhoux, F., 2018. Fetal monocytes and the origins of tissue-resident macrophages. Cell. Immunol. 330, 5–15. https://doi.org/10.1016/j.cellimm.2018.01.001.

Hohenberger, P., Felgner, C., Haensch, W., Schlag, P.M., 1998. Tumor oxygenation correlates with molecular growth determinants in breast cancer. Breast Cancer Res. Treat. 48, 97–106. https://doi.org/10.1023/A:1005921513083.

Huo, C.W., Chew, G., Hill, P., Huang, D., Ingman, W., Hodson, L., Brown, K.A., Magenau, A., Allam, A.H., McGhee, E., Timpson, P., Henderson, M.A., Thompson, E.W., Britt, K., 2015. High mammographic density is associated with an increase in stromal collagen and immune cells within the mammary epithelium. Breast Cancer Res. 17 (1), 1–20. https://doi.org/10.1186/S13058-015-0592-1.

Hussain, S.A., Ganesan, R., Reynolds, G., Gross, L., Stevens, A., Pastorek, J., Murray, P.G., Perunovic, B., Anwar, M.S., Billingham, L., James, N.D., Spooner, D., Poole, C.J., Rea, D.W., Palmer, D.H., 2007. Hypoxia-regulated carbonic anhydrase IX expression is associated with poor survival in patients with invasive breast cancer. Br. J. Cancer 96, 104–109. https://doi.org/10.1038/sj.bjc.6603530.

Ibrahim, A.M., Moss, M.A., Gray, Z., Rojo, M.D., Burke, C.M., Schwertfeger, K.L., dos Santos, C.O., Machado, H.L., 2020. Diverse macrophage populations contribute to the inflammatory microenvironment in premalignant lesions during localized invasion. Front. Oncol. 10, 1–16. https://doi.org/10.3389/fonc.2020.569985.

Ingman, W.V., Wyckoff, J., Gouon-Evans, V., Condeelis, J., Pollard, J.W., 2006. Macrophages promote collagen fibrillogenesis around terminal end buds of the developing mammary gland. Dev. Dyn. 235, 3222–3229. https://doi.org/10.1002/DVDY.20972.

Jain, R.K., 2005. Normalization of tumor vasculature: an emerging concept in anti-angiogenic theraphy. Science 307, 58–62.

Jaitin, D.A., Adlung, L., Thaiss, C.A., Weiner, A., Li, B., Descamps, H., Lundgren, P., Bleriot, C., Liu, Z., Deczkowska, A., Keren-Shaul, H., David, E., Zmora, N., Eldar, S.M., Lubezky, N., Shibolet, O., Hill, D.A., Lazar, M.A., Colonna, M., Ginhoux, F., Shapiro, H., Elinav, E., Amit, I., 2019. Lipid-associated macrophages control metabolic homeostasis in a Trem2-dependent manner. Cell 178, 686–698, e14. https://doi.org/10.1016/J.CELL.2019.05.054.

Jäppinen, N., Félix, I., Lokka, E., Tyystjärvi, S., Pynttäri, A., Lahtela, T., Gerke, H., Elima, K., Rantakari, P., Salmi, M., 2019. Fetal-derived macrophages dominate in adult mammary glands. Nat. Commun. 10, 1–12. https://doi.org/10.1038/s41467-018-08065-1.

Javed, A., Lteif, A., 2013. Development of the human. Breast. https://doi.org/10.1055/s-0033-1343989.

Jindal, S., Narasimhan, J., Borges, V.F., Schedin, P., 2020. Characterization of weaning-induced breast involution in women: implications for young women's breast cancer. NPJ Breast Cancer 6. https://doi.org/10.1038/S41523-020-00196-3.

Kahata, K., Maturi, V., Moustakas, A., 2018. TGF-β family signaling in ductal differentiation and branching morphogenesis. Cold Spring Harb. Perspect. Biol. 10. https://doi.org/10.1101/CSHPERSPECT.A031997.

Katzenelenbogen, Y., Sheban, F., Katzenelenbogen, Y., Sheban, F., Yalin, A., Yofe, I., Svetlichnyy, D., Jaitin, D.A., 2020. Coupled scRNA-Seq and intracellular protein activity reveal an immunosuppressive role of TREM2 in cancer article coupled scRNA-Seq and intracellular protein activity reveal an immunosuppressive role of TREM2 in cancer. Cell 1–14. https://doi.org/10.1016/j.cell.2020.06.032.

Kiss, M., Van Gassen, S., Movahedi, K., Saeys, Y., Laoui, D., 2018. Myeloid cell heterogeneity in cancer: not a single cell alike. Cell. Immunol. 330, 188–201. https://doi.org/10.1016/j.cellimm.2018.02.008.

Lahmar, Q., Keirsse, J., Laoui, D., Movahedi, K., Van Overmeire, E., Van Ginderachter, J.A., 2016. Tissue-resident versus monocyte-derived macrophages in the tumor microenvironment. Biochim. Biophys. Acta Rev. Cancer 1865, 23–34. https://doi.org/10.1016/j.bbcan.2015.06.009.

Laoui, D., Movahedi, K., van Overmeire, E., van den Bossche, J., Schouppe, E., Mommer, C., Nikolaou, A., Morias, Y., de Baetselier, P., van Ginderachter, J.A., 2011. Tumor-associated macrophages in breast cancer: distinct subsets, distinct functions. Int. J. Dev. Biol. 55. https://doi.org/10.1387/ijdb.113371dl.

Laoui, D., Van Overmeire, E., Di Conza, G., Aldeni, C., Keirsse, J., Morias, Y., Movahedi, K., Houbracken, I., Schouppe, E., Elkrim, Y., Karroum, O., Jordan, B., Carmeliet, P., Gysemans, C., DeBaetselier, P., Mazzone, M., Van Ginderachter, J.A., 2014. Tumor hypoxia does not drive differentiation of tumor-associated macrophages but rather fine-tunes the M2-like macrophage population. Cancer Res. 74, 24–30. https://doi.org/10.1158/0008-5472.CAN-13-1196.

Larionova, I., Tuguzbaeva, G., Ponomaryova, A., Stakheyeva, M., Cherdyntseva, N., Pavlov, V., Choinzonov, E., Kzhyshkowska, J., 2020. Tumor-associated macrophages in human breast, colorectal, lung, ovarian and prostate cancers. Front. Oncol. 0, 2232. https://doi.org/10.3389/FONC.2020.566511.

Leader, A.M., Grout, J.A., Chang, C., Maier, B., Walker, L., Lansky, A., Leberichel, J., Malissen, N., Davila, M., Martin, J., Magri, G., Tuballes, K., Zhao, Z., Samstein, R., Amore, N.R.D., Thurston, G., Kamphorst, A., Flores, R., Wang, P., Beasley, M.B., Salmon, H., Rahman, A.H., 2021. Single-cell analysis of human non-small cell lung cancer lesions refines tumor classification and patient stratification. Cancer Cell 39 (12), 1594–1609. https://doi.org/10.1016/j.ccell.2021.10.009.

Leek, R.D., Lewis, C.E., Whitehouse, R., Greenall, M., Clarke, J., Harris, A.L., 1996. Association of macrophage infiltration with angiogenesis and prognosis in invasive breast carcinoma. Cancer Res. 56, 4625–4629.

Li, C.M.-C.A., Shapiro, H., Tsiobikas, C., Pinello, L., Regev, A., Brugge Correspondence, J.S., 2020. Aging-associated alterations in mammary epithelia and stroma revealed by single-cell RNA sequencing. Cell Rep. https://doi.org/10.1016/j.celrep.2020.108566.

Lim, H.Y., Lim, S.Y., Tan, C.K., Thiam, C.H., Goh, C.C., Carbajo, D., Chew, S.H.S., See, P., Chakarov, S., Wang, X.N., Lim, L.H., Johnson, L.A., Lum, J., Fong, C.Y., Bongso, A., Biswas, A., Goh, C., Evrard, M., Yeo, K.P., Basu, R., Wang, J.K., Tan, Y., Jain, R., Tikoo, S., Choong, C., Weninger, W., Poidinger, M., Stanley, R.E., Collin, M., Tan, N.S., Ng, L.G., Jackson, D.G., Ginhoux, F., Angeli, V., 2018. Hyaluronan receptor LYVE-1-expressing macrophages maintain arterial tone through hyaluronan-mediated regulation of smooth muscle cell collagen. Immunity 49, 326–341, e7. https://doi.org/10.1016/J.IMMUNI.2018.06.008.

Lin, E.Y., Li, J.-F., Gnatovskiy, L., Deng, Y., Zhu, L., Grzesik, D.A., Qian, H., Xue, X., Pollard, J.W., 2006. Macrophages regulate the angiogenic switch in a mouse model of breast cancer. Cancer Res. 66, 11238–11246. https://doi.org/10.1158/0008-5472.CAN-06-1278.

Linde, N., Casanova-Acebes, M., Sosa, M.S., Mortha, A., Rahman, A., Farias, E., Harper, K., Tardio, E., Reyes Torres, I., Jones, J., Condeelis, J., Merad, M., Aguirre-Ghiso, J.A., 2018. Macrophages orchestrate breast cancer early dissemination and metastasis. Nat. Commun. 9 (1), 1–14. https://doi.org/10.1038/s41467-017-02481-5.

Liu, Z., Gu, Y., Chakarov, S., Bleriot, C., Kwok, I., Chen, X., Shin, A., Huang, W., Dress, R.J., Dutertre, C.A., Schlitzer, A., Chen, J., Ng, L.G., Wang, H., Liu, Z., Su, B., Ginhoux, F., 2019. Fate mapping via Ms4a3-expression history traces monocyte-derived cells. Cell 178, 1509–1525.e19. https://doi.org/10.1016/j.cell.2019.08.009.

Luecken, M.D., Theis, F.J., 2019. Current best practices in single-cell RNA-Seq analysis: a tutorial. Mol. Syst. Biol. 15, e8746. https://doi.org/10.15252/msb.20188746.

Macias, H., Hinck, L., 2012. Mammary gland development. WIREs Dev. Biol. 1, 533–557. https://doi.org/10.1002/wdev.35.

Mahmoud, S.M.A., Lee, A.H.S., Paish, E.C., Macmillan, R.D., Ellis, I.O., Green, A.R., 2012. Tumour-infiltrating macrophages and clinical outcome in breast cancer. J. Clin. Pathol. 65, 159–163. https://doi.org/10.1136/jclinpath-2011-200355.

Medina, D., Kittrell, F.S., Shepard, A., Stephens, L.C., Jiang, C., Lu, J., Allred, D.C., McCarthy, M., Ullrich, R.L., 2002. Biological and genetic properties of the p53 null preneoplastic mammary epithelium. FASEB J. 16, 881–883. https://doi.org/10.1096/fj.01-0885fje.

Medrek, C., Pontén, F., Jirström, K., Leandersson, K., 2012. The presence of tumor associated macrophages in tumor stroma as a prognostic marker for breast cancer patients. BMC Cancer 12, 306. https://doi.org/10.1186/1471-2407-12-306.

Mittendorf, E.A., Zhang, H., Barrios, C.H., Saji, S., Jung, K.H., Hegg, R., Koehler, A., Sohn, J., Iwata, H., Telli, M.L., Ferrario, C., Punie, K., Penault-Llorca, F., Patel, S., Duc, A.N., Liste-Hermoso, M., Maiya, V., Molinero, L., Chui, S.Y., Harbeck, N., 2020. Neoadjuvant atezolizumab in combination with sequential nab-paclitaxel and anthracycline-based chemotherapy versus placebo and chemotherapy in patients with early-stage triple-negative breast cancer (IMpassion031): a randomised, double-blind, phase 3 tria. Lancet 396, 1090–1100. https://doi.org/10.1016/S0140-6736(20)31953-X.

Molgora, M., Esaulova, E., Vermi, W., Artyomov, M.N., Schreiber, R.D., Colonna, M., Molgora, M., Esaulova, E., Vermi, W., Hou, J., Chen, Y., Luo, J., Brioschi, S., 2020. TREM2 modulation remodels the tumor myeloid landscape enhancing anti-PD-1 immunotherapy article TREM2 modulation remodels the tumor myeloid landscape enhancing anti-PD-1 immunotherapy. Cell 1–15. https://doi.org/10.1016/j.cell.2020.07.013.

Movahedi, K., Laoui, D., Gysemans, C., Baeten, M., Stangé, G., Den Van Bossche, J., Mack, M., Pipeleers, D., In't Veld, P., De Baetselier, P., Van Ginderachter, J.A., 2010. Different tumor microenvironments contain functionally distinct subsets of macrophages derived from Ly6C (high) monocytes. Cancer Res. 70, 5728–5739. https://doi.org/10.1158/0008-5472.CAN-09-4672.

Mulder, K., Patel, A.A., Kong, W.T., Cile Piot, C., Halitzki, E., Dunsmore, G., Khalilnezhad, S., Irac, S.E., Dubuisson, A., Chevrier, M., Zhang, X.M., Kit, J., Tam, C., Kiat, T., Lim, H., Men, R., Wong, M., Pai, R., Ibrahim, A., Khalil, S., 2021. Cross-tissue single-cell landscape of human monocytes and macrophages in health and disease. Immunity 10, 11. https://doi.org/10.1016/j.immuni.2021.07.007.

Murdoch, C., Lewis, C.E., 2005. Macrophage migration and gene expression in response to tumor hypoxia. Int. J. Cancer 117, 701–708. https://doi.org/10.1002/ijc.21422.

Nagasawa, S., Kuze, Y., Maeda, I., Kojima, Y., Motoyoshi, A., Onishi, T., Iwatani, T., Yokoe, T., Koike, J., Chosokabe, M., Kubota, M., Seino, H., Suzuki, A., Seki, M., Tsuchihara, K., Inoue, E., Tsugawa, K., Ohta, T., Suzuki, Y., 2021. Genomic profiling reveals heterogeneous populations of ductal carcinoma in situ of the breast. Commun. Biol. 4, 438. https://doi.org/10.1038/s42003-021-01959-9.

Noy, R., Pollard, J.W., 2014. Review tumor-associated macrophages: from mechanisms to therapy. Immunity 41, 49–61. https://doi.org/10.1016/j.immuni.2014.06.010.

O'Brien, J., Martinson, H., Durand-Rougely, C., Schedin, P., 2012. Macrophages are crucial for epithelial cell death and adipocyte repopulation during mammary gland involution. Development. https://doi.org/10.1242/dev.071696.

Oftedal, O.T., 2002. The mammary gland and its origin during synapsid evolution. J. Mammary Gland Biol. Neoplasia 7, 225–252. https://doi.org/10.1023/a:1022896515287.

Park, J., Choi, W., Tiesmeyer, S., Long, B., Borm, L.E., Garren, E., Nguyen, T.N., Tasic, B., Codeluppi, S., Graf, T., Schlesner, M., Stegle, O., Eils, R., Ishaque, N., 2021. Cell segmentation-free inference of cell types from in situ transcriptomics data. Nat. Commun. 12, 3545. https://doi.org/10.1038/s41467-021-23807-4.

Parmar, H., Cunha, G.R., 2004. Epithelial–stromal interactions in the mouse and human mammary gland in vivo. Endocr. Relat. Cancer 11, 437–458. https://doi.org/10.1677/ERC.1.00659.

Plaks, V., Boldajipour, B., Linnemann, J.R., Melton, A.C., Krummel, M.F., Correspondence, Z.W., 2015. Adaptive immune regulation of mammary postnatal organogenesis. Dev. Cell 34, 493–504. https://doi.org/10.1016/j.devcel.2015.07.015.

Pollard, J.W., Hennighausen, L., 1994. Colony stimulating factor 1 is required for mammary gland development during pregnancy. Proc. Natl. Acad. Sci. 91, 9312–9316. https://doi.org/10.1073/PNAS.91.20.9312.

Polyak, K., Kalluri, R., 2010. The role of the microenvironment in mammary gland development and cancer. Cold Spring Harb. Perspect. Biol. 2. https://doi.org/10.1101/cshperspect.a003244.

Pucci, F., Venneri, M.A., Biziato, D., Nonis, A., Moi, D., Sica, A., Di Serio, C., Naldini, L., De Palma, M., 2009. A distinguishing gene signature shared by tumor-infiltrating Tie 2-expressing monocytes, blood "resident" monocytes, and embryonic macrophages suggests common functions and developmental relationships. Blood 114, 901–914. https://doi.org/10.1182/blood-2009-01-200931.

Qian, B.Z., Li, J., Zhang, H., Kitamura, T., Zhang, J., Campion, L.R., Kaiser, E.A., Snyder, L.A., Pollard, J.W., 2011. CCL2 recruits inflammatory monocytes to facilitate breast-tumour metastasis. Nature 475, 222–225. https://doi.org/10.1038/nature10138.

Ramón y Cajal, S., Sesé, M., Capdevila, C., Aasen, T., De Mattos-Arruda, L., Diaz-Cano, S.J., Hernandez-Losa, J., Castellvi, J., 2020. Clinical implications of intratumor heterogeneity: challenges and opportunities. J. Mol. Med. 98, 161–177.

Ramos, R.N., Missolo-Koussou, Y., Gerber-Ferder, Y., Bromley, C., Bugatti, M., Gonzalo Núñez, N., Tosello, J.B., Richer, W., Denizeau, J., Sedlik, C., Caudana, P., Kotsias, F., Niborski, L.L., Viel, S., Bohec, M., Lameiras, S., Baulande, S., Lesage, L., Nicolas, A., Meseure, D., Vincent-Salomon, A., Reyal, F., Dutertre, C.-A., Ginhoux, F., Vimeux, L., Donnadieu, E., Buttard, B., Galon, J., Zelenay, S., Vermi, W., Guermonprez, P., Piaggio, E., Helft, J., 2021. Tissue-Resident FOLR2 + Macrophages Associate with Tumor-Infiltrating CD8 + T Cells and with Increased Survival of Breast Cancer Patients. bioRxiv.

Ridker, P.M., MacFadyen, J.G., Thuren, T., Everett, B., Libby, P., Glynn, R., Ridker, P., Lorenzatti, A., Krum, H., Varigos, J., Siostrzonek, P., Sinnaeve, P., Fonseca, F., Nicolau, J., Gotcheva, N., Genest, J., Yong, H., Urina-Triana, M., Milicic, D., Cifkova, R., Vettus, R., Koenig, W., Anker, S.D., Manolis, A.J., Wyss, F.,

Forster, T., Sigurdsson, A., Pais, P., Fucili, A., Ogawa, H., Shimokawa, H., Veze, I., Petrauskiene, B., Salvador, L., Kastelein, J., Cornel, J.H., Klemsdal, T.O., Medina, F., Budaj, A., Vida-Simiti, L., Kobalava, Z., Otasevic, P., Pella, D., Lainscak, M., Seung, K.B., Commerford, P., Dellborg, M., Donath, M., Hwang, J.J., Kultursay, H., Flather, M., Ballantyne, C., Bilazarian, S., Chang, W., East, C., Forgosh, L., Harris, B., Ligueros, M., Bohula, E., Charmarthi, B., Cheng, S., Chou, S., Danik, J., McMahon, G., Maron, B., Ning, M.M., Olenchock, B., Pande, R., Perlstein, T., Pradhan, A., Rost, N., Singhal, A., Taqueti, V., Wei, N., Burris, H., Cioffi, A., Dalseg, A.M., Ghosh, N., Gralow, J., Mayer, T., Rugo, H., Fowler, V., Limaye, A.P., Cosgrove, S., Levine, D., Lopes, R., Scott, J., Hilkert, R., Tamesby, G., Mickel, C., Manning, B., Woelcke, J., Tan, M., Manfreda, S., Ponce, T., Kam, J., Saini, R., Banker, K., Salko, T., Nandy, P., Tawfik, R., O'Neil, G., Manne, S., Jirvankar, P., Lal, S., Nema, D., Jose, J., Collins, R., Bailey, K., Blumenthal, R., Colhoun, H., Gersh, B., 2017. Effect of interleukin-1β inhibition with canakinumab on incident lung cancer in patients with atherosclerosis: exploratory results from a randomised, double-blind, placebo-controlled trial. Lancet 390, 1833–1842. https://doi.org/10.1016/S0140-6736(17)32247-X.

Ruffell, B., Coussens, L.M., 2015. Macrophages and therapeutic resistance in cancer. Cancer Cell 27, 462–472. https://doi.org/10.1016/j.ccell.2015.02.015.

Saelens, W., Cannoodt, R., Todorov, H., Saeys, Y., 2019. A comparison of single-cell trajectory inference methods. Nat. Biotechnol. 37, 547–554. https://doi.org/10.1038/s41587-019-0071-9.

Schmid, P., Cortes, J., Pusztai, L., McArthur, H., Kümmel, S., Bergh, J., Denkert, C., Park, Y.H., Hui, R., Harbeck, N., Takahashi, M., Foukakis, T., Fasching, P.A., Cardoso, F., Untch, M., Jia, L., Karantza, V., Zhao, J., Aktan, G., Dent, R., O'Shaughnessy, J., 2020. Pembrolizumab for early triple-negative breast cancer. N. Engl. J. Med. 382, 810–821. https://doi.org/10.1056/nejmoa1910549.

Schwertfeger, K.L., Rosen, J.M., Cohen, D.A., 2006. Mammary gland macrophages: pleiotropic functions in mammary development. J. Mammary Gland Biol. Neoplasia 11, 229–238. https://doi.org/10.1007/S10911-006-9028-Y.

Scotton, C., Milliken, D., Wilson, J., Raju, S., Balkwill, F., 2020. Analysis of CC chemokine and chemokine receptor expression in solid ovarian tumours. Br. J. Cancer 85, 891–897.

Sharma, A., Seow, J.J.W., Dutertre, C.A., Pai, R., Blériot, C., Mishra, A., Wong, R.M.M., Singh, G.S.N., Sudhagar, S., Khalilnezhad, S., Erdal, S., Teo, H.M., Khalilnezhad, A., Chakarov, S., Lim, T.K.H., Fui, A.C.Y., Chieh, A.K.W., Chung, C.P., Bonney, G.K., Goh, B.K.P., Chan, J.K.Y., Chow, P.K.H., Ginhoux, F., Das Gupta, R., 2020. Onco-fetal reprogramming of endothelial cells drives immunosuppressive macrophages in hepatocellular carcinoma. Cell 183, 377–394. e21 https://doi.org/10.1016/j.cell.2020.08.040.

Shrivastava, R., Singh, V., Asif, M., Negi, M.P.S., Bhadauria, S., 2018. Oncostatin M upregulates HIF-1α in breast tumor associated macrophages independent of intracellular oxygen concentration. Life Sci. 194, 59–66. https://doi.org/10.1016/j.lfs.2017.12.017.

Skibinski, A., Kuperwasser, C., 2015. The origin of breast tumor heterogeneity. Oncogene 34, 5309–5316. https://doi.org/10.1038/onc.2014.475.

Slepicka, P.F., Somasundara, A.V.H., dos Santos, C.O., 2021. The molecular basis of mammary gland development and epithelial differentiation. Semin. Cell Dev. Biol. 114, 93–112. https://doi.org/10.1016/J.SEMCDB.2020.09.014.

Sternlicht, M.D., 2005. Key stages in mammary gland development: the cues that regulate ductal branching morphogenesis. Breast Cancer Res. 8, 1–11. https://doi.org/10.1186/BCR1368.

Sternlicht, M.D., Kouros-Mehr, H., Lu, P., Werb, Z., 2006. Hormonal and local control of mammary branching morphogenesis. Differentiation: Res. Biol. Divers. 74, 365. https://doi.org/10.1111/J.1432-0436.2006.00105.X.

Stewart, T.A., Hughes, K., Hume, D.A., Davis, F.M., 2019. Developmental stage-specific distribution of macrophages in mouse mammary gland. Front. Cell Dev. Biol. 0, 250. https://doi.org/10.3389/FCELL.2019.00250.

Sung, H., Ferlay, J., Siegel, R.L., Laversanne, M., Soerjomataram, I., Jemal, A., Bray, F., 2021. Global cancer statistics 2020: GLOBOCAN estimates of incidence and mortality worldwide for 36 cancers in 185 countries. CA Cancer J. Clin. 0, 1–41. https://doi.org/10.3322/caac.21660.

Tausendschön, M., Dehne, N., Brüne, B., 2011. Hypoxia causes epigenetic gene regulation in macrophages by attenuating Jumonji histone demethylase activity. Cytokine 53, 256–262. https://doi.org/10.1016/j.cyto.2010.11.002.

Tolg, C., Yuan, H., Flynn, S.M., Basu, K., Ma, J., Tse, K.C.K., Kowalska, B., Vulkanesku, D., Cowman, M.K., McCarthy, J.B., Turley, E.A., 2017. Hyaluronan modulates growth factor induced mammary gland branching in a size dependent manner. Matrix Biol. 63, 117–132. https://doi.org/10.1016/J.MATBIO.2017.02.003.

Tripathi, C., Tewari, B.N., Kanchan, R.K., Baghel, K.S., Nautiyal, N., Shrivastava, R., Kaur, H., Bramha Bhatt, M.L., Bhadauria, S., 2014. Macrophages are recruited to hypoxic tumor areas and acquire a pro-angiogenic M2-polarized phenotype via hypoxic cancer cell derived cytokines oncostatin M and eotaxin. Oncotarget 5, 5350–5368. https://doi.org/10.18632/oncotarget.2110.

Tymoszuk, P., Evens, H., Marzola, V., Wachowicz, K., Wasmer, M.-H., Datta, S., Müller-Holzner, E., Fiegl, H., Böck, G., van Rooijen, N., Theurl, I., Doppler, W., 2014. In situ proliferation contributes to accumulation of tumor-associated macrophages in spontaneous mammary tumors. Eur. J. Immunol. 44, 2247–2262. https://doi.org/10.1002/eji.201344304.

Van Nguyen, A., Pollard, J.W., 2002. Colony stimulating factor-1 is required to recruit macrophages into the mammary gland to facilitate mammary ductal outgrowth. Dev. Biol. 247, 11–25. https://doi.org/10.1006/DBIO.2002.0669.

Vaupel, P., Schlenger, K., Knoop, C., Höckel, M., 1991. Oxygenation of human tumors: evaluation of tissue oxygen distribution in breast cancers by computerized O2 tension measurements. Cancer Res. 51, 3316–3322.

Wagner, J., Rapsomaniki, M.A., Chevrier, S., Anzeneder, T., Langwieder, C., Dykgers, A., Rees, M., Ramaswamy, A., Muenst, S., Soysal, S.D., Jacobs, A., Windhager, J., Silina, K., van den Broek, M., Dedes, K.J., Martinez, M.R., Weber, W.P., Bodenmiller, B., 2019. A single-cell atlas of the tumor and immune ecosystem of human breast cancer. Cell 177 (5), 1330–1345.

Wallerius, M., Wallmann, T., Bartish, M., Östling, J., Mezheyeuski, A., Tobin, N.P., Nygren, E., Pangigadde, P., Pellegrini, P., Squadrito, M.L., Ponten, F., Hartman, J., Bergh, J., De Milito, A., De Palma, M., Östman, A., Andersson, J., Rolny, C., 2016. Guidance molecule SEMA3A restricts tumor growth by differentially regulating the proliferation of tumor-associated macrophages. Cancer Res. 76, 3166–3178. https://doi.org/10.1158/0008-5472.CAN-15-2596.

Walsh, J.C., Lebedev, A., Aten, E., Madsen, K., Marciano, L., Kolb, H.C., 2014. The clinical importance of assessing tumor hypoxia: relationship of tumor hypoxia to prognosis and therapeutic opportunities. Antioxid. Redox Signal. 21, 1516–1554. https://doi.org/10.1089/ars.2013.5378.

Wang, E.C.E., Dai, Z., Ferrante, A.W., Drake, C.G., Christiano, A.M., 2019. A subset of TREM2+ dermal macrophages secretes oncostatin M to maintain hair follicle stem cell quiescence and inhibit hair growth. Cell Stem Cell 24, 654–669.e6. https://doi.org/10.1016/J.STEM.2019.01.011.

Wang, Y., Chaffee, T.S., LaRue, R.S., Huggins, D.N., Witschen, P.M., Ibrahim, A.M., Nelson, A.C., Machado, H.L., Schwertfeger, K.L., 2020. Tissue-resident macrophages promote extracellular matrix homeostasis in the mammary gland stroma of nulliparous mice. elife 9, 1–27. https://doi.org/10.7554/eLife.57438.

Wellenstein, M.D., Coffelt, S.B., Duits, D.E.M., van Miltenburg, M.H., Slagter, M., de Rink, I., Henneman, L., Kas, S.M., Prekovic, S., Hau, C.S., Vrijland, K., Drenth, A.P., de Korte-Grimmerink, R., Schut, E., van der Heijden, I., Zwart, W., Wessels, L.F.A., Schumacher, T.N., Jonkers, J., de Visser, K.E., 2019. Loss of p53 triggers WNT-dependent systemic inflammation to drive breast cancer metastasis. Nature 572, 538–542. https://doi.org/10.1038/s41586-019-1450-6.

Wenes, M., Shang, M., Di Matteo, M., Goveia, J., Martín-Pérez, R., Serneels, J., Prenen, H., Ghesquière, B., Carmeliet, P., Mazzone, M., 2016. Macrophage metabolism controls tumor blood vessel morphogenesis and metastasis. Cell Metab. 24, 701–715. https://doi.org/10.1016/j.cmet.2016.09.008.

Wu, K., Lin, K., Li, X., Yuan, X., Xu, P., Ni, P., Xu, D., 2020. Redefining tumor-associated macrophage subpopulations and functions in the tumor microenvironment. Front. Immunol. 11, 1731. https://doi.org/10.3389/FIMMU.2020.01731.

Xiang, X., Wang, J., Lu, D., Xu, X., 2021. Targeting tumor-associated macrophages to synergize tumor immunotherapy. Signal Transduction Targeted Ther. 6 (1), 1–12. https://doi.org/10.1038/s41392-021-00484-9.

Yuan, Z.-Y., Luo, R.-Z., Peng, R.-J., Wang, S.-S., Xue, C., 2014. High infiltration of tumor-associated macrophages in triple-negative breast cancer is associated with a higher risk of distant metastasis. OncoTargets Ther. 7, 1475–1480. https://doi.org/10.2147/OTT.S61838.

Zhang, Q.W., Liu, L., Gong, C.Y., Shi, H.S., Zeng, Y.H., Wang, X.Z., Zhao, Y.W., Wei, Y.Q., 2012. Prognostic significance of tumor-associated macrophages in solid tumor: a meta-analysis of the literature. PLoS One. https://doi.org/10.1371/journal.pone.0050946.

Zhang, M., Lee, A.V., Rosen, J.M., 2017. The cellular origin and evolution of breast cancer. Cold Spring Harb. Perspect. Med. 7. https://doi.org/10.1101/cshperspect.a027128.

Zhao, X., Qu, J., Sun, Y., Wang, J., Liu, X., Wang, F., Zhang, H., Wang, W., Ma, X., Gao, X., Zhang, S., 2017. Prognostic significance of tumor-associated macrophages in breast cancer: a meta-analysis of the literature. Oncotarget 8, 30576–30586. https://doi.org/10.18632/oncotarget.15736.

CHAPTER SIX

Endometrial macrophages in health and disease

Erin Brown[a], Rocío Martínez-Aguilar[b], Jacqueline A. Maybin[b], and Douglas A. Gibson[a,b,*]
[a]Centre for Inflammation Research, University of Edinburgh, Edinburgh, United Kingdom
[b]MRC Centre for Reproductive Health, University of Edinburgh, Edinburgh, United Kingdom
*Corresponding author: e-mail address: d.a.gibson@ed.ac.uk

Contents

1. Introduction 184
2. Endometrial physiology 184
3. Endometrial macrophages—Origin, location and abundance 186
4. Identification and characterization of endometrial macrophages 187
5. Regulation of macrophage function during the menstrual cycle 189
   5.1 Decidualization 190
   5.2 Menstruation 191
6. Endometrial macrophages in women's reproductive health disorders 194
   6.1 Macrophages in endometriosis 194
   6.2 Macrophages in abnormal uterine bleeding 196
7. Future directions 199
8. Conclusions 200
Acknowledgments 202
Conflict of interest 202
References 202

Abstract

Macrophages are present in the endometrium throughout the menstrual cycle and are most abundant during menstruation. Endometrial macrophages contribute to tissue remodeling during establishment of pregnancy and are thought to play key roles in mediating tissue breakdown and repair during menstruation. Despite these important roles, the phenotype and function of endometrial macrophages remains poorly understood. In this review, we summarize approaches used to characterize endometrial macrophage phenotype, current understanding of the functional role of macrophages in normal endometrial physiology as well as the putative contribution of macrophage dysfunction to women's reproductive health disorders.

International Review of Cell and Molecular Biology, Volume 367
ISSN 1937-6448
https://doi.org/10.1016/bs.ircmb.2022.03.011

1. Introduction

Macrophages are abundant in the uterus and have been detected throughout the tissue (Starkey et al., 1991) where they play key roles in regulating tissue homeostasis (reviewed in Gordon and Plüddemann, 2017). It is widely accepted that macrophages are important for endometrial remodeling associated with decidualization and establishment of pregnancy as well as for mediating tissue breakdown and repair during menstruation (Berbic et al., 2009; Critchley et al., 2020; Thiruchelvam et al., 2013). However, the precise phenotype and function of macrophages in the uterus remains poorly understood. Although macrophages are present throughout the female reproductive tract (Chambers et al., 2020; Payne et al., 2012), the primary focus of this review will be on macrophages in the endometrial layer of the uterus in reproductive health and disease. The contribution of macrophages to pregnancy and parturition has been appraised elsewhere (see Chambers et al., 2020; Gomez-Lopez et al., 2014; Ning et al., 2016). In this review, we summarize current understanding of the role of macrophages in normal endometrial physiology as well as the putative contribution of macrophage dysfunction to reproductive disorders.

2. Endometrial physiology

The endometrium is a dynamic sex steroid-dependent organ, undergoing cyclical phases of proliferation, differentiation, breakdown and repair, approximately 400 times in a woman's reproductive lifetime (Cousins et al., 2014). The uterus is composed of an inner lining, known as the endometrium (Dallenbach-Hellweg, 1971), which is surrounded by a muscular layer known as the myometrium. The human endometrium comprises two layers; the *stratum basalis*, and the *stratum functionalis,* and is composed of glandular and luminal epithelial cell layers, endometrial stromal cells, vasculature and a diverse population of immune cells (Fig. 1).

During a non-pregnant menstrual cycle, the endometrium is subject to phases of proliferation, differentiation (decidualization), shedding and repair (Jabbour et al., 2006) driven predominantly by the ovarian-derived sex steroids estradiol and progesterone. During the proliferative phase, estradiol promotes cellular proliferation and revascularization that drives growth of the tissue. Following ovulation, progesterone produced by the corpus

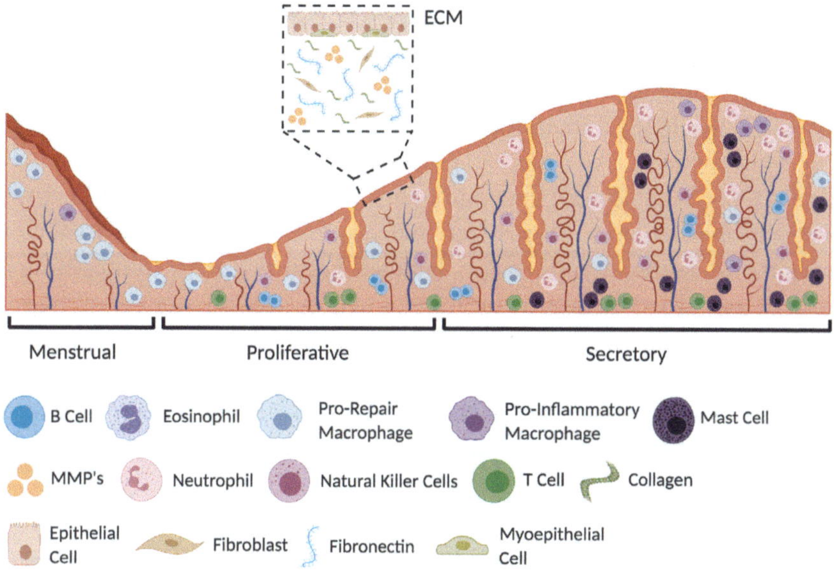

Fig. 1 Spatial distribution of leukocytes across the endometrial cycle. During the menstrual phase, leukocytes are spatially and temporally regulated. Leukocyte abundance substantially increases during the secretory phase with an increase in macrophages detected pre-menstraully in the late secertory phase and during menstruation. *Created with BioRender.com.*

luteum in the ovary stimulates endometrial tissue remodeling and differentiation of endometrial stromal cells – a process termed decidualization. During decidualization, endometrial stromal cells undergo morphologic change from elongated fibroblasts into rounded epithelial-like cells that secrete growth factors and cytokines that support tissue remodeling. Decidualization is important for preparing the endometrium for implantation, and is essential for the establishment of pregnancy in women (Gellersen and Brosens, 2014). Importantly, as decidualization is a terminal differentiation of the endometrial stromal cell compartment, the decidualized tissue is shed in a non-pregnant cycle.

In the absence of pregnancy, the corpus luteum degenerates resulting in a rapid decline in progesterone which triggers menstruation (Critchley et al., 2001). Menstruation results in the rapid breakdown and sloughing of the luminal portion of the endometrium which is shed and subsequently regenerated in the next cycle (reviewed in Critchley et al., 2020). The basalis layer of the endometrium remains intact during the menstrual cycle (Dallenbach-Hellweg, 1971). Menstruation involves a complex cascade of events including a coordinated inflammatory response, degradation of the extracellular matrix (ECM), vascular constriction and myometrial contraction (Lockwood et al., 1998). The spiral arteries undergo intense vasoconstriction resulting in localized hypoxia (Cousins et al., 2016a; Markee, 1978; Maybin et al., 2018). Inflammatory cells including monocytes, macrophages and neutrophils are key mediators of endometrial tissue breakdown and repair (Kelly et al., 2001). To ensure efficient restoration of tissue integrity, tissue breakdown and repair occur synchronously, characterized by rapid re-epithelialization to restore the denuded luminal surface, remodeling of endometrial blood vessels to promote vascular stability, and resolution of inflammatory responses (Ludwig and Spornitz, 1991; Salamonsen, 2003; Serhan and Savill, 2005). Given their importance to tissue homeostasis and repair, macrophages are thought to play a critical role in regulating the events underpinning remodeling and repair in the endometrium.

3. Endometrial macrophages—Origin, location and abundance

Most of our understanding of endometrial macrophages has come from investigating tissue biopsies using immunohistochemistry. These studies demonstrate that mature macrophages are found distributed throughout the tissue (Thiruchelvam et al., 2013). Macrophages are reported to concentrate around endometrial glands during the secretory phase (Garry et al., 2010; Song and Fraser, 1995) and accumulate in perivascular regions and close to the luminal epithelium prior to the onset of menses. Macrophages are present in the endometrium throughout the menstrual cycle, but they are most abundant prior to menses where they account for 6-15% of endometrial cells (Salamonsen and Woolley, 1999). This sharp rise in macrophage number suggests that either *in situ* proliferation of tissue-resident endometrial macrophages and/or recruitment of peripheral blood monocytes could account for changes in macrophage number in the endometrium.

Tissue resident macrophages are uniquely adapted within tissues by local environmental cues (Gordon and Taylor, 2005) and their phenotype is also dictated by their ontogeny. Generally, macrophages are derived from distinct lineages that are either seeded in embryonic life and maintained by self-renewal or maintained by replenishment from bone-marrow precursors throughout adult life. Recent work has shown that developmentally distinct macrophages in the same environment can play discrete roles in tissues (Mowat et al., 2017) but whether this affects the endometrium is not known. Macrophage ontogeny has been well characterized in other organs (Bain and Schridde, 2018; Evren et al., 2020), but lineage tracing studies to define endometrial macrophage subpopulations are lacking.

In the endometrium, macrophages proliferate in response to hormone stimulation and estradiol promotes proliferation of endometrial macrophages during regeneration of the endometrium (Pepe et al., 2017). Human endometrial xenograft studies in mice exposed to an exogenous hormone-induced menstrual cycle showed that donor human leukocytes (human $CD45^+$) within xenografts increased in response to estradiol and progesterone (secretory phase) but that host leukocytes (mouse $CD45^+$) infiltrate into endometrial tissues following withdrawal of hormones and induced menses (Guo et al., 2011). Although macrophages were not specifically assessed in this study, it suggests both local proliferation and peripheral replenishment of immune cells contribute to the regulation of immune cell numbers in the endometrium, and that this may vary depending on stage of the menstrual cycle.

4. Identification and characterization of endometrial macrophages

Classically, macrophages in human tissues have been defined by the presence of single markers including CD68, CD14, and CD11c, whereas mouse studies have commonly used F4/80 and/or CSF1R (Table 1). Although these studies are informative, use of single markers to define macrophage populations occludes potential heterogeneity and the expression of the most commonly used markers is not restricted to single immune cell subsets. For example, CD68 is frequently used as a macrophage marker in human (Betjes et al., 1991; Holness and Simmons, 1993), however, it can also be used to identify monocytes (Holness and Simmons, 1993), precursors to tissue-resident macrophages (Fogg et al., 2006) as well as non-myeloid cells (Travaglione et al., 2002). Similarly, expression of CSF1R is not restricted to macrophages and may identify monocytes, macrophages, neutrophils and dendritic cells depending on the tissue context (Cousins et al., 2016b).

Some studies have used multiple markers to better characterize the phenotype of macrophages in the endometrium. For example, a study by Jensen et al. found in the human uterus that the majority of $CD68^+$ macrophages (>70%) expressed the scavenger receptor CD163 and a small subset expressed CD14 (~10%) (Jensen et al., 2012). The authors also found that

Table 1 Markers used to identify uterine macrophages within the endometrium.

Marker	Identified in	Species	Phenotype (M1/M2)	References
CD11c	Endometrium (normal) First trimester decidua	Human Human	U M2	Tabibzadeh (1990) Houser et al. (2011)
CCR2-CD11c	First trimester decidua	Human	M2	Jiang and Wang (2020)
CD14[a]	Endometrium First trimester decidua Decidua and blood	Human Murine Human	U M2 M2	Biswas Shivhare et al. (2014) Laskarin et al. (2005) Gustafsson et al. (2008)
CD16[a]	Blood derived monocytes Endometrium (normal)	Human Human	M2 M2	Verreck et al. (2006) Jensen et al. (2012)
CD68	Endometrium (normal) Endometrium (normal) Endometrium (normal and endometriotic)	Human Human Human	U M2 M2	Klentzeris et al. (1992) Jensen et al. (2012) Cominelli et al. (2014)
CD136	Blood derived monocytes Endometrium (normal)	Human Human	M2 M2	Verreck et al. (2006) Jensen et al. (2012)
CD206	Endometrium (normal and endometriotic)	Human	M2	Cominelli et al. (2014)
CD209	Decidua and blood	Human	M2	Gustafsson et al. (2008)
F4/80	Yolk sac Endometrium	Murine Murine	U U	Takahashi et al. (1989) Kaitu'u-Lino et al. (2007)
CSF1R	Endometrium	Murine	U	Cousins et al. (2016b)

[a]No longer used as an independent macrophage marker.
Abbreviations: Unknown/unspecified (U), M1/M2 assignation based on reported phenotype.

isolated $CD68^{+}CD163^{+}$ endometrial macrophages secreted pro- and anti-inflammatory cytokines, such as tumor necrosis factor (TNF) and IL-10, as well as pro-angiogenic factors such as vascular endothelial growth factor (VEGF), Fibroblast growth factor (FGF)2 and Platelet-derived growth factor (PDGF) (Jensen et al., 2012). These findings highlight the potential for heterogeneity within endometrial macrophage populations as they exhibit properties of both pro-inflammatory and pro-repair macrophages.

Tissue macrophages rely on environmental cues for regulation of their function (Guilliams et al., 2020) where their local niche induces a specific transcriptomic and proteomic profile. This gives rise to a dynamic spectrum of macrophage phenotypes that can be activated or suppressed by the local niche (Martinez and Gordon, 2014). Macrophages have commonly been grouped into either pro-inflammatory (or M1) and anti-inflammatory/ pro-repair (or M2) phenotypes based on expression of key surface markers and production of certain cytokines. This dichotomous paradigm has historically been used to describe tissue macrophage phenotypes (Martinez et al., 2008) but it is now widely accepted that the M1/M2 paradigm is an over-simplification (Nahrendorf and Swirski, 2016) that does not accurately account for macrophage phenotypes that occur in different contexts (Murray et al., 2014). There is yet to be a standardized nomenclature system that can fully encapsulate the diversity of macrophage origins, stimuli, and microenvironment. More recently, macrophage phenotypes have been reported based upon the stimulus they are exposed to (e.g. IL-4) in order to provide better context (Murray et al., 2014; Orecchioni et al., 2019) or have been described based on the tissue/location they have been isolated from. Despite this, M1 and M2 are common descriptors in the literature relating to endometrial macrophages. For clarity, in this review we have described the marker(s) used in respective studies rather than solely an M1 or M2 assignation to allow a broader interpretation of the findings. In general, endometrial macrophages have been described has having an M2-like profile based on expression of CD163 (Jensen et al., 2012). Studies using human endometrial biopsies show evidence of macrophages expressing molecules involved in repair, such as connective tissue growth factor (CTGF) (Maybin et al., 2012). Endometrial macrophages also express chemokines such as CXCL2, CXCL8 and CXCL4 that may influence leukocyte trafficking within the endometrium (Maybin et al., 2017a; Thiruchelvam et al., 2016). This pro-repair phenotype fits with an anti-inflammatory macrophage polarization, which is in line with previous characterizations of endometrial macrophages as M2 ($CD206^{+}$, $CD163^{+}$) (Cominelli et al., 2014; Quillay et al., 2015).

5. Regulation of macrophage function during the menstrual cycle

Endometrial macrophages encounter physiological fluctuations in sex steroid hormones and local tissue mediators that affect their function

(Simitsidellis et al., 2018). Macrophages have to adapt to the requirements of the tissue to support tissue remodeling and repair during two key processes in the non-pregnant menstrual cycle—decidualization and menstruation. Principally this involves secretion of factors that control tissue remodeling; cytokines, growth factors and matrix metalloproteinases (MMPs), as well as phagocytic clearance of tissue debris and efferocytosis of apoptotic cells.

5.1 Decidualization

Decidualization is essential for the establishment and maintenance of pregnancy. In women, decidualization occurs spontaneously and this process is therefore a key part of non-pregnant cyclical endometrial function. Decidualization precedes the establishment of pregnancy and in the absence of fertilization, menstruation. The tissue remodeling events associated with decidualization are therefore critical for healthy endometrial function.

Macrophages accumulate in the endometrium during the late secretory phase in women, coincident with decidualization. This has also been reported in mouse uterine tissues undergoing induced decidualization (Cousins et al., 2016b). Progesterone has been proposed to promote endometrial leukocyte trafficking/proliferation *in situ* during the secretory phase (Guo et al., 2011) but the lack of progesterone receptor expression on endometrial macrophages precludes direct regulation by progesterone (Stewart et al., 1998). However, progesterone does promote increased expression of chemoattractants during decidualization predominantly secreted by endometrial stromal cells (Dimitriadis et al., 2005). Genomic array analysis of endometrial tissue biopsies demonstrated expression of macrophage inflammatory protein (MIP-1β/CCL4), monocyte-chemotactic protein 3 (MCP-3/CCL7), macrophage-derived chemokine (MDC/CCL22), 6Ckine, and fractalkine (FKN/CX3CL1) were increased in secretory phase biopsies which could account for increases in macrophage abundance at this stage of the menstrual cycle (Jones et al., 2004). Changes in chemokine expression may reflect transient fluctuations in macrophage numbers across the menstrual cycle. For example, expression of the chemoattractant MIP-1β increases during decidualization and declines thereafter (Jones et al., 2004; Kats et al., 2005) which may account for changes in macrophage abundance observed in endometrial tissues during decidualization.

Our understanding of the phenotype and function of macrophages present during decidualization in non-pregnant endometrium has largely been extrapolated based on profiling of macrophages present in maternal decidual tissues in pregnancy. Most studies have focused on assessing decidual tissue from 1st trimester pregnancy and these studies support a role for macrophages in modulating spiral artery remodeling and regulating interaction with invading trophoblast (Lash et al., 2016). Decidual macrophages have been classified into at least two sub-types; $CD14^{+}CD11c^{high}$ and $CD14^{+}CD11c^{low}$ which may reflect their distinct roles in early pregnancy. $CD14^{+}CD11c^{low}$ express M2-like markers, are distributed throughout the decidua and are reported to play roles in ECM regeneration, tissue remodeling and growth (Houser et al., 2011; Jiang and Wang, 2020). $CD11c^{high}$ macrophages, which may be further divided based on presence or absence of the chemokine receptor CCR2, associate with extravillous trophoblasts. $CCR2^{+}$ macrophages have a pro-inflammatory transcriptional profile which may be consistent with their recruitment from peripheral blood and these cells play distinct roles in modulating trophoblast invasion (Jiang and Wang, 2020). To the best of our knowledge, comparative studies between macrophages present in non-pregnant decidualized endometrium and those present in decidual tissues in pregnancy have not been undertaken. However, studies from other cell types such as uterine natural killer cells suggest that macrophage phenotype will be determined by the distinct and changing tissue microenvironment that is present in non-pregnant and pregnant uterine tissues (Strunz et al., 2021).

Macrophages may directly contribute to vascular remodeling in the endometrium via secretion of angiogenic factors including VEGFA, as well as tissue remodeling factors such as MMPs. Macrophages are known to produce MMP-9 (Jeziorska et al., 1996; Salamonsen and Lathbury, 2000), which may facilitate further leukocyte infiltration by altering vascular integrity. Interestingly, in explant studies decidual tissue samples treated with an MMP-2/9 inhibitor showed no evidence of spiral artery remodeling or leukocyte infiltration (Hazan et al., 2010). This study suggested that macrophages mediate early stages of vascular remodeling by promoting vascular smooth muscle de-differentiation and phagocytosis. These functional roles may be common to decidualization-related vascular remodeling, and to changes in vasculature function associated with menstruation.

5.2 Menstruation

Menstruation is the monthly shedding of the luminal portion of the endometrium (functionalis), triggered by the withdrawal of progesterone. Endometrial stromal cells sense this change and orchestrate an inflammatory signaling cascade that initiates endometrial breakdown and repair (Evans and Salamonsen, 2014). This is characterized by transient expression of inflammatory mediators within endometrial tissues including chemokines that promote the recruitment of monocytes, macrophages and neutrophils, as well as up-regulating inflammatory mediators including IL-6 and TNF during the menstrual phase (Armstrong et al., 2017). Macrophages are most abundant during the menstrual phase of the cycle accounting for ~15% of stromal cells (Cominelli et al., 2014) where they play roles in tissue breakdown, remodeling and repair.

Beyond endocrine cues, macrophage function can also be altered by local signals within the tissue during endometrial breakdown. Menstruation is associated with transient and physiological hypoxia that has been reported both in mouse models of simulated menses (Cousins et al., 2016a; Maybin et al., 2018) and *in vivo* in humans (Reavey et al., 2021). *In vitro* studies using human endometrial biopsies have also proven that hypoxia triggers the expression of the VEGF (Maybin et al., 2011a) and other repair factors (CTGF, CXCL8) (Maybin et al., 2011b, 2012). The presence of menstrual endometrial hypoxia, alongside evidence of delay in endometrial repair in the absence of hypoxia (Maybin et al., 2018), suggests a potential role for endometrial macrophages. Although pharmacological inhibition of HIF-1 in a mouse model of menstruation did not alter macrophage abundance (Maybin et al., 2018), it remains to be determined if macrophage phenotype is altered, as is the case in the tumor microenvironment (Laoui et al., 2014).

Macrophages contribute to endometrial shedding via production of MMPs that degrade extracellular matrix (ECM) and by secreting inflammatory mediators (Salamonsen and Lathbury, 2000). It has been shown that MMPs mediate tissue degradation of endometrial explants *in vitro* (Marbaix et al., 1996; Vassilev et al., 2005). Endometrial macrophages are capable of releasing latent and converting enzymes (Thiruchelvam et al., 2013) which, in turn, can activate immature MMPs secreted by other endometrial populations contributing to the breakdown of the endometrial layer (Chambers et al., 2020). Neutrophils also have the capacity to perform these functions (Lathbury and Salamonsen, 2000; Salamonsen et al., 2000), and they too have been detected in areas of active endometrial shedding in several human and mouse studies (Armstrong et al., 2017; Kaitu'u-Lino et al., 2007). This has traditionally positioned neutrophils as crucial mediators of endometrial breakdown. However, the identification of monocytic populations clustering in areas undergoing active shedding has revised this interpretation (Cousins et al., 2016b). Pre-menstrual accumulation of macrophages may help to initiate tissue breakdown via localized production and action of MMPs in response to progesterone withdrawal (Salamonsen et al., 1997).

Macrophage recruitment prior to menstruation is coincident with the up-regulation of MMP-9 (Jeziorska et al., 1996), MMP-12 (Goffin et al., 2003), MMP-14 (Goffin et al., 2003), and MMP-27 (Cominelli et al., 2014). MMP-9 (also referred to as gelatinase B, type IV collagenase) is expressed by endometrial macrophages, neutrophils, stromal, and epithelial cells (Jeziorska et al., 1996; Salamonsen et al., 2000). MMP-9 degrades denatured collagen, a major constituent of the ECM (Gaide Chevronnay et al., 2012). Comparatively, MMP-12 (metalloelastase) peaks during the menstrual phase in endometrial tissues, whereby it mediates elastin digestion (Goffin et al., 2003). Moreover, in mice, MMP-12 has been shown to induce macrophage migration to various tissue sites (Li et al., 2012; Shipley et al., 1996). Therefore, it may be plausible that MMP-12 also promotes macrophage trafficking during the menstrual phase. MMP-14 (MT-MMP-1) expression remains relatively high across the menstrual cycle, but substantially increases during the menstrual phase (Goffin et al., 2003). MMP-27 has received little attention in human studies, but has recently been reported to be significantly up-regulated during the menstrual phase (Cominelli et al., 2014). In addition, expression of MMP-27 in a subset of $CD163^{+}$ $CD206^{+}$ macrophages has been reported which supports a direct role for macrophages in endometrial ECM degradation but this yet to be fully elucidated (Cominelli et al., 2014).

Macrophages may also contribute to endometrial repair. During menstruation, they are located around and within endometrial glands, suggesting a role in clearance and removal of apoptotic epithelial cells (Garry et al., 2010). Mice with myeloid (neutrophil and monocyte/macrophage) depletion using an anti-Gr-1 antibody show delays in endometrial repair (Kaitu'u-Lino et al., 2007). Using a mouse model where GFP expression was controlled by the colony stimulating factor 1 receptor (CSF1R) promoter (Cousins et al., 2016b), we previously defined distinct myeloid populations present during endometrial breakdown and repair, comprising of classical monocytes, monocyte-derived macrophages and tissue-resident macrophages (Cousins et al., 2016b). Interestingly, monocyte-derived macrophages were found specifically located in areas of active repair, while tissue-resident macrophages seem to be restricted to areas that have already undergone repair (Cousins et al., 2016b). As endometrial breakdown and repair occur concurrently (Garry et al., 2009), these studies highlight the important spatial and temporal contribution of macrophages to endometrial function during menstruation.

6. Endometrial macrophages in women's reproductive health disorders

Altered macrophage function in disease states may disrupt normal functioning of the endometrium. Changes to the abundance and/or phenotype of endometrial macrophages have been reported in women's reproductive health disorders such as endometriosis and abnormal uterine bleeding. Elucidating the role of endometrial macrophages in these conditions may provide insights into disease pathogenesis and aid in the identification of novel therapeutic targets.

6.1 Macrophages in endometriosis

Endometriosis is a chronic condition characterized by symptoms of infertility and pelvic pain (Capobianco and Rovere-Querini, 2013), affecting around 1 in 10 women in the UK (Saunders and Horne, 2021). Endometriosis is defined by the presence of endometrial-like tissue outside of the endometrium. This ectopic endometrial-like tissue forms "lesions" (Saunders and Horne, 2021), commonly within the peritoneal cavity, at sites including the ovaries, fallopian tubes, and peritoneal wall (Saunders and Horne, 2021). The etiology of endometriosis is commonly attributed to the process of retrograde menstruation (Yovich et al., 2020) where shed endometrial fragments flow backward into the fallopian tubes and peritoneal cavity, where they can adhere and establish vascular connections, and develop their own nerve supply (Burney and Giudice, 2012). Despite appearing to be an abnormal physiological phenomenon, retrograde menstruation is found to occur in approximately 90% of women (Halme et al., 1984). Therefore, the establishment and persistence of the endometrial fragments is likely to be influenced by alterations in shed endometrial tissue and the microenvironment where lesions are established, underpinned by abnormal immune cell function (Burney and Giudice, 2012).

Endometriosis has been described as a "disease of the macrophage" (Capobianco and Rovere-Querini, 2013). Macrophages are reported to contribute to lesion growth and vascularization (Bacci et al., 2009) in addition to enhancing pain symptoms through nerve innervation

(Forster et al., 2019). Women with endometriosis exhibit increased macrophage numbers in peritoneal fluid, lesions, and in the (eutopic) endometrium compared to women without endometriosis (Berbic et al., 2009; Hill et al., 1988; Khan et al., 2004). Endometrial macrophages are also found to have altered functional properties, appearing to exhibit a pro-inflammatory phenotype in endometriosis patients (Takebayashi et al., 2015; Vallvé-Juanico et al., 2019) characterized by increased secretions of pro-inflammatory cytokines, including TNF (Richter et al., 2005). As well as changes in abundance, immune cell function is altered in endometriosis (reviewed in (Herington et al., 2011)). As mentioned previously, macrophages scavenge and phagocytose sloughed endometrial fragments during menstruation (Garry et al., 2010). Recognition of these fragments is achieved by scavenger molecules including CD36 (Chuang et al., 2009) and CD163 (Etzerodt and Moestrup, 2013), which are found to be decreased in women with endometriosis. Defective scavenging capability may in part explain the persistence of endometrial fragments that form lesions in endometriosis. Macrophage function is compromised by excessive iron (Fe^{2+}) uptake (Lousse et al., 2009). Upon phagocytosis, unbound hemoglobin is catalyzed and Fe^{2+} is released (Soares and Hamza, 2016). Subsequently, Fe^{2+} can either be processed by ferritin or chaperoned out of the cell (reviewed in Soares and Hamza, 2016). In the absence of these regulatory pathways, Fe^{2+} accumulation is extremely toxic and results in apoptosis (Soares and Hamza, 2016). Fe^{2+} overload in macrophages was first reported by Gaulier et al., who histologically observed their abundance in peritoneal lesions (Gaulier et al., 1983). Similarly, women suffering from endometriosis have an increased abundance of Fe^{2+} overloaded macrophages in the peritoneal fluid, and evidence of increased Fe^{2+} storage in peritoneal macrophages (Lousse et al., 2009; Stowell et al., 1997; Van Langendonckt et al., 2002). Furthermore, Fe^{2+} overloaded macrophages are found to have reduced adhesion potential (Sharpe-Timms et al., 2002), potentially occurring as a trade-off to cope with the surplus of toxic iron. This may diminish the capacity of macrophages to target and efficiently phagocytose sloughed fragments during menstruation. Taken together, these changes in macrophage function may contribute to shed endometrial fragments having the opportunity to disperse and adhere at sites outside the endometrial cavity.

Macrophages may also influence pain symptoms in endometriosis. Using a mouse model of endometriosis, Greaves et al. (2015) observed $CD68^{+}$ macrophages concentrate at nerve bundle sites within peritoneal lesions.

Further, they reported that mice treated with estradiol have significantly higher macrophage numbers in their lesion sites (Greaves et al., 2015). Macrophage abundance and localization at lesion sites may contribute to the heightened pain responses observed in women with endometriosis. Recently, two sub-populations of macrophages (CD14^{high} and CD14^{low}, respectively) were described in peritoneal fluid from women undergoing diagnostic laparoscopy for endometriosis (Gibson et al., 2021). The abundance of these peritoneal macrophage populations correlated with pain scores independent of diagnostic outcome at laparoscopy, which suggests changes to macrophages within the peritoneal microenvironment could independently affect pelvic pain. Although further studies are needed, identifying the signaling pathways responsible for macrophage-associated pelvic pain may help to develop therapeutic approaches for targeting these symptoms in women with endometriosis.

6.2 Macrophages in abnormal uterine bleeding

Abnormal uterine bleeding (AUB) is a condition that affects up to one in three women of reproductive age (Shapley et al., 2004). Increased duration or flow during menses leads to greater menstrual blood volume/loss. This may lead to anemia and can necessitate blood transfusion, reflecting the significant physical effects of AUB. AUB also has a significant negative impact on quality of life and is a huge socio-economic burden for women, their families, health services and society (Frick et al., 2009; Prentice, 1999). A global system for AUB, endorsed by The International Federation of Gynaecology and Obstetrics (FIGO) has provided standardization of terminology and a classification system to aid the diagnosis of underlying causes of AUB (Munro et al., 2011, 2018). These can be divided into structural and non-structural causes. Structural causes can be detected during routine examination or imaging and include polyps, adenomyosis, fibroids or malignancy. Non-structural causes include coagulation disorders, ovulatory dysfunction, endometrial or iatrogenic causes. Macrophages may contribute to both structural and non-structural causes of AUB.

Abnormal uterine bleeding of endometrial origin (AUB-E) is a diagnosis of exclusion and the underlying mechanisms remain undefined. As described above, physiological menstruation and endometrial repair rely on appropriate self-limiting inflammatory responses (Maybin et al., 2011). Hence, imbalance of the inflammatory response during menstruation may result in AUB. Whilst performing their classical immune surveillance role, macrophages are major producers of TNF (Parameswaran and Patial, 2010). In the healthy endometrium, TNF concentrations are increased during menstruation (Armstrong et al., 2017) and is thought to promote apoptosis and cell-cell dissociation (Tabibzadeh, 1996), aiding vascular destruction and onset of menstrual bleeding (Tabibzadeh, 1996; Tabibzadeh et al., 1995). However, in patients experiencing heavy menstrual bleeding (HMB), objectively defined as a blood loss greater than 80 mL/cycle, TNF concentrations in menstrual fluid are greater than those from women with normal menstrual bleeding (NMB) (Malik et al., 2006). Moreover, in a mouse model, Shalaby et al. (1989) demonstrated that TNF is capable of inducing vascular necrosis and subsequently, endometrial hemorrhage. Thus, aberrant production and action of TNF secreted from endometrial macrophages may alter endometrial bleeding and blood loss. Macrophage subsets have been shown to secrete VEGF, MMP-9 and TGFβ in repair/wound-healing contexts (Krzyszczyk et al., 2018). VEGF and MMP-9 are reduced in the menstrual effluent from women with HMB compared to NMB (Malik et al., 2006). Furthermore, stromal levels of TGFβ protein during menses have also been shown to be significantly lower in endometrial biopsies of patients with HMB when compared to those with normal menstrual blood loss (Lu et al., 2021; Maybin et al., 2017b). Together, these findings suggest macrophage phenotype and function may be altered in women that experience HMB. Future studies would benefit from profiling the secretome of endometrial macrophages between AUB and healthy controls to better characterize the local inflammatory dysregulation that may contribute to endometrial dysregulation that could underlie HMB.

Studies investigating endometrial vasculature have indicated women suffering from HMB show impaired vascular differentiation (Biswas Shivhare et al., 2014) and have more pronounced endothelial gaps (Mints et al., 2007). VEGF is a potent regulator of vascular angiogenesis, proliferation, and permeability (Meduri et al., 2000). Previously, macrophages have been reported to produce VEGF in the endometrium (McLaren et al., 1996). However, its influence on endometrial vasculature remains controversial. Malik et al. (2006) reported that mRNA levels of VEGFA were significantly reduced in HMB women, whereas Mints and colleagues showed the endometrial capillaries of HMB women had greater intensity of VEGF staining than tissues from NMB women (Mints et al., 2007). Analogous to this, VEGFA expression has been found to positively correlate with the number of endothelial gaps, whilst negatively correlating with the number of pericytes present in the endometrium (Andersson et al., 2015) - an indicator of vessel maturity (Zetterberg et al., 2007). Thus, deficient VEGFA production from endometrial macrophages may contribute to HMB by compromising vascular integrity and maturity. Based on the collective evidence from the endometrium, other tissue sites and a murine model it is plausible to suggest dysregulation of macrophage-associated cytokines including TNF and VEGFA could compromise vasculature function. Abnormal endometrial vasculature may promote excessive bleeding patterns experienced by women suffering from HMB. However, this requires further investigation.

Macrophage phenotype and secretome may also be influenced by crosstalk with other endometrial populations, such as epithelial and stromal cells. For example, *in vitro* studies suggest that co-culture of menstrual endometrial stromal cells with a monocytic cell line results in macrophage polarization toward a regulatory phenotype (Martínez-Aguilar et al., 2020). This regulatory phenotype is characterized by the presence of both pro-inflammatory and pro-repair surface markers and exhibit poor bactericidal activity (Bystrom et al., 2008; Kim and Hematti, 2009). It would be very informative to explore whether endometrial stromal cells from HMB and those with normal menstrual blood loss differ in their macrophage polarization capacity as this could affect the local inflammatory milieu and its regulation.

Patients with HMB have also shown a decreased prostaglandin F2α (PGF2α)/prostaglandin E2 (PGE_2) ratio (Smith et al., 2007), making the vasodilator PGE_2 more bioavailable. This, in addition to the reduced expression of the vasoconstrictor endothelin-1 observed in HMB compared to controls (Marsh et al., 1997), which together may lead to a defective vasoconstriction during the perimenstrual phase. In fact, Non-Steroidal Anti-Inflammatory Drugs (NSAIDs), which target prostaglandin biosynthesis, are a current therapy for AUB (Maybin and Critchley, 2016). As stated previously, hypoxia transiently occurs in the menstrual endometrium and appears to be required for physiological endometrial repair (Cousins et al., 2016a; Maybin et al., 2018; Reavey et al., 2021). Therefore, adequate endometrial vasoconstriction may be required to limit menstrual blood flow during menses and trigger subsequent repair to prevent prolonged menstruation. As macrophages are classical source of PGE_2 (Humes et al., 1977) and endothelins (Ehrenreich et al., 1990), they may also contribute to regulation of vasoconstriction and vasodilatation. It remains to be determined if certain macrophage subsets have specific roles in appropriate vasoconstriction and if others are restricted to endometrial repair. Defining the contribution of macrophages to endometrial physiology and pathology may identify specific therapeutic targets for the management of AUB-E.

Adenomyosis is a disorder framed within the structural causes of AUB (AUB-A) (Munro et al., 2011, 2018). It is defined as the benign invasion of the endometrium into the myometrium (Guo et al., 2021) and there is evidence of macrophage dysregulation both in the eutopic endometrium and in adenomyotic deposits (see systematic review; Bourdon et al., 2021). An increase in macrophage infiltration was detected in the endometrium of women with adenomyosis versus controls and within the adenomyotic deposits. Macrophage number was correlated with the severity and symptoms of the disease (Tremellen and Russell, 2012). Studies on macrophage phenotype in adenomyosis are scarce, but *in vitro* studies suggest that co-culture of adenomyotic epithelial cells from the eutopic endometrium with a monocytic cell line impairs M2 polarization when compared to co-cultures from patients without adenomyosis (An et al., 2017a). Poorly polarized macrophages may also exert effects on adenomyosis epithelial cells from the eutopic endometrium by enhancing epithelial to mesenchymal transition (EMT)-like processes (An et al., 2017b). EMT events likely contribute to the pathogenesis of adenomyosis (Chen et al., 2010), targeting macrophages may be an effective strategy to control adenomyosis progression and symptoms.

7. Future directions

A lack of consensus markers that are restricted to macrophages within the endometrium has likely masked heterogeneity within endometrial macrophage subpopulations (summarized in Table 1). This makes integrating and interpreting results from different studies challenging, particularly those examining human cells/tissues. This issue is less pronounced in mouse models, where there is a greater capacity to use experimental tools to label and trace macrophages in the uterus. Nevertheless, the full potential of these approaches has yet to be realized in uterine tissues, in contrast to other organ systems. Use of mouse models may help identify suitable markers for characterization of macrophage phenotype and function in human tissues. Recent advances in cell labeling/reporter systems (Hawley et al., 2018) and genetic fate-mapping approaches (Liu et al., 2019) will help to address key questions relating to macrophage ontogeny and fate in the uterus.

Using comparative biology approaches should aid integration of this new understanding to identify conserved markers and functional pathways between mouse and human.

Single cell technologies including flow cytometry and single cell RNA sequencing analysis that allow unbiased profiling of endometrial macrophage populations will help to build consensus canonical markers suitable for identifying and characterizing endometrial macrophages. Emerging techniques, such as spatially-resolved transcriptomics and proteomics, should allow characterization of specific areas of active shedding or early repair. This will potentially increase our understanding of the specific function of different subpopulations of myeloid cells in localized areas of the endometrium during menses and repair.

Macrophages have been investigated as a therapeutic strategy in pathological conditions including rheumatoid arthritis (Van Roon et al., 2003), and cancer (Beatty Gregory et al., 2011). Given the pleiotropic nature of endometrial macrophages and their noticeable dysfunction in reproductive health, they have great potential for future therapeutic targeting. The latest technologies and research tools will be needed in order to improve fundamental understanding of macrophage function in the uterus and maximize future opportunities for developing new therapeutic approaches.

8. Conclusions

Within the endometrium, macrophages exist across a spectrum of pro- and anti-inflammatory states and their phenotype and function is highly specific to different tissue niches and to different phases of endometrial function (Fig. 2). This dynamic milieu is fundamental for endometrial physiology, contributing to successful cyclic remodeling, breakdown and repair in the non-pregnant endometrium. Endometrial macrophages are dysregulated and display abnormal phenotype and function in women's reproductive health disorders. Dysfunction of endometrial macrophages is characterized by unbalanced pro-inflammatory signaling which has been reported in endometriosis (Vallvé-Juanico et al., 2019), adenomyosis (An et al., 2017b) and HMB (Berbic et al., 2009). Endometrial macrophages contribute to endometrial disorders by disrupting the inflammatory microenvironment, and may exhibit deficits in scavenging, phagocytosis, and cytokine production that affect key functional processes including vascular and tissue remodeling in the endometrium. The phenotype and function of

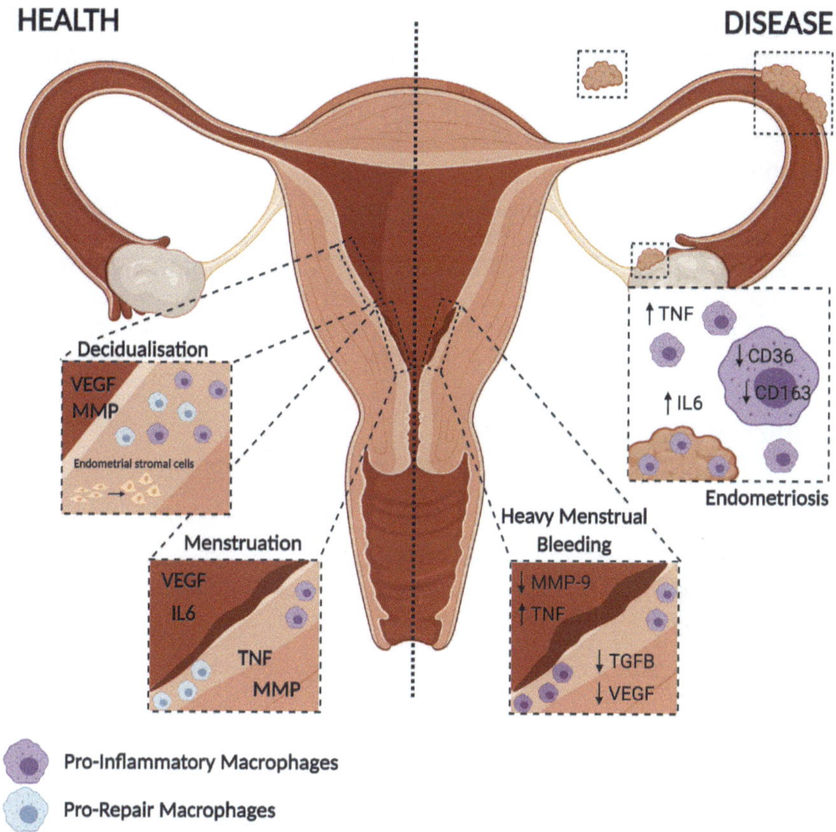

Fig. 2 Phenotype and function of uterine macrophages in health and disease. Macrophages play important roles in the menstrual cycle during decidualization and menstruation where they secrete factors that promote vascular remodelling and tissue repair. Endometrial macrophages exhibit features of pro-inflammatory (purple) and repair (blue) macrophages. In women's reproductive disorders, including endometriosis and heavy menstrual bleeding (HMB) macrophage function is perturbed and their phenotype is skewed toward a pro-inflammatory state which is associated with macrophage dysfunction. In endometriosis this is characterized by increased production of inflammatory mediators such as IL-6 and TNF and decreased expression of scavenger receptors CD36 and CD163 which may promote an environment that allows establishment and persistence of endometriosis lesions. In HMB, endometrial production of MMP-9, TGFB and VEGF is decreased whereas secretion of TNF is increased which is associated with increased blood loss during menstruation. *Created with BioRender.com.*

endometrial macrophages in health and disease is yet to be fully defined but studies from mouse models suggest endometrial macrophages are heterogeneous and are subject to spatial and temporal regulation within the tissue (Cousins et al., 2016b). Understanding the functional processes associated with these cells and the contribution of macrophage subsets to distinct roles within the tissue will help to identify disease-associated pathways and future therapeutic targets for women's reproductive health disorders.

Acknowledgments

Research costs and salaries were supported by Wellcome Trust Fellowship 220656/Z/20/Z to D.A.G. (E.B. and D.A.G.) and Wellcome Trust 100646/Z/12/Z, 209589/Z/17/Z, Wellbeing of Women RG1820 to J.A.M. (R.M.A. and J.A.M.). MRC Centre grants G1002033 and MR/N022556/1. Figures generated using BioRender.com. For the purpose of open access, D.A.G. has applied a Creative Commons Attribution (CC BY) licence to any Author Accepted Manuscript version arising from this submission.

Conflict of interest

The authors declare no competing interests related to the published work.

References

An, M., et al., 2017a. Different macrophages equally induce EMT in endometria of adenomyosis and normal. Reproduction 154 (1), 79–92.

An, M., et al., 2017b. Interaction of macrophages and endometrial cells induces epithelial–mesenchymal transition-like processes in adenomyosis†. Biol. Reprod. 96 (1), 46–57.

Andersson, E., et al., 2015. Low pericyte coverage of endometrial microvessels in heavy menstrual bleeding correlates with the microvessel expression of VEGF-A. Int. J. Mol. Med. 35 (2), 433–438.

Armstrong, G.M., et al., 2017. Endometrial apoptosis and neutrophil infiltration during menstruation exhibits spatial and temporal dynamics that are recapitulated in a mouse model. Sci. Rep. 7 (1), 17416.

Bacci, M., et al., 2009. Macrophages are alternatively activated in patients with endometriosis and required for growth and vascularization of lesions in a mouse model of disease. Am. J. Pathol. 175 (2), 547–556.

Bain, C.C., Schridde, A., 2018. Origin, differentiation, and function of intestinal macrophages. Front. Immunol. 9, 2733.

Beatty Gregory, L., et al., 2011. CD40 agonists alter tumor stroma and show efficacy against pancreatic carcinoma in mice and humans. Science 331 (6024), 1612–1616.

Berbic, M., et al., 2009. Macrophage expression in endometrium of women with and without endometriosis. Hum. Reprod. 24 (2), 325–332.

Betjes, M.G., et al., 1991. Monoclonal antibody EBM11 (anti-CD68) discriminates between dendritic cells and macrophages after short-term culture. Immunobiology 183 (1-2), 79–87.

Biswas Shivhare, S., et al., 2014. Altered vascular smooth muscle cell differentiation in the endometrial vasculature in menorrhagia. Hum. Reprod. 29 (9), 1884–1894.

Bourdon, M., et al., 2021. Immunological changes associated with adenomyosis: a systematic review. Hum. Reprod. Update 27 (1), 108–129.

Burney, R.O., Giudice, L.C., 2012. Pathogenesis and pathophysiology of endometriosis. Fertil. Steril. 98 (3), 511–519.

Bystrom, J., et al., 2008. Resolution-phase macrophages possess a unique inflammatory phenotype that is controlled by cAMP. Blood 112 (10), 4117–4127.

Capobianco, A., Rovere-Querini, P., 2013. Endometriosis, a disease of the macrophage. Front. Immunol. 4, 9.
Chambers, M., et al., 2020. Macrophage plasticity in reproduction and environmental influences on their function. Front. Immunol. 11, 607328.
Chen, Y.-J., et al., 2010. Oestrogen-induced epithelial–mesenchymal transition of endometrial epithelial cells contributes to the development of adenomyosis. J. Pathol. 222 (3), 261–270.
Chuang, P.C., et al., 2009. Downregulation of CD36 results in reduced phagocytic ability of peritoneal macrophages of women with endometriosis. J. Pathol. 219 (2), 232–241.
Cominelli, A., et al., 2014. Matrix metalloproteinase-27 is expressed in CD163+/CD206+ M2 macrophages in the cycling human endometrium and in superficial endometriotic lesions. Mol. Hum. Reprod. 20 (8), 767–775.
Cousins, F.L., et al., 2014. Evidence from a mouse model that epithelial cell migration and mesenchymal-epithelial transition contribute to rapid restoration of uterine tissue integrity during menstruation. PLoS One 9 (1), e86378.
Cousins, F.L., et al., 2016a. Hypoxyprobe™ reveals dynamic spatial and temporal changes in hypoxia in a mouse model of endometrial breakdown and repair. BMC Res Notes 9, 30.
Cousins, F.L., et al., 2016b. Evidence for a dynamic role for mononuclear phagocytes during endometrial repair and remodelling. Sci. Rep. 6 (1), 36748.
Critchley, H.O.D., et al., 2001. The endocrinology of menstruation—a role for the immune system. Clin. Endocrinol. 55 (6), 701–710.
Critchley, H.O.D., et al., 2020. Physiology of the endometrium and regulation of menstruation. Physiol. Rev. 100 (3), 1149–1179.
Dallenbach-Hellweg, G., 1971. The normal histology of the endometrium. In: Dallenbach-Hellweg, G. (Ed.), Histopathology of the Endometrium. Springer Berlin Heidelberg, Berlin, Heidelberg, pp. 20–76.
Dimitriadis, E., et al., 2005. Cytokines, chemokines and growth factors in endometrium related to implantation. Hum. Reprod. Update 11 (6), 613–630.
Ehrenreich, H., et al., 1990. Endothelins, peptides with potent vasoactive properties, are produced by human macrophages. J. Exp. Med. 172 (6), 1741–1748.
Etzerodt, A., Moestrup, S.K., 2013. CD163 and inflammation: biological, diagnostic, and therapeutic aspects. Antioxid. Redox Signal. 18 (17), 2352–2363.
Evans, J., Salamonsen, L.A., 2014. Decidualized human endometrial stromal cells are sensors of hormone withdrawal in the menstrual inflammatory cascade1. Biol. Reprod. 90 (1).
Evren, E., Ringqvist, E., Willinger, T., 2020. Origin and ontogeny of lung macrophages: from mice to humans. Immunology 160 (2), 126–138.
Fogg, D.K., et al., 2006. A clonogenic bone marrow progenitor specific for macrophages and dendritic cells. Science 311 (5757), 83–87.
Forster, R., et al., 2019. Macrophage-derived insulin-like growth factor-1 is a key neurotrophic and nerve-sensitizing factor in pain associated with endometriosis. FASEB J. 33 (10), 11210–11222.
Frick, K.D., et al., 2009. Financial and quality-of-life burden of dysfunctional uterine bleeding among women agreeing to obtain surgical treatment. Womens Health Issues 19 (1), 70–78.
Gaide Chevronnay, H.P., et al., 2012. Regulation of matrix metalloproteinases activity studied in human endometrium as a paradigm of cyclic tissue breakdown and regeneration. Biochim. Biophys. Acta 1824 (1), 146–156.
Garry, R., et al., 2009. A re-appraisal of the morphological changes within the endometrium during menstruation: a hysteroscopic, histological and scanning electron microscopic study. Hum. Reprod. 24 (6), 1393–1401.
Garry, R., et al., 2010. Structural changes in endometrial basal glands during menstruation. BJOG 117 (10), 1175–1185.

Gaulier, A., Jouret-Mourin, A., Marsan, C., 1983. Peritoneal endometriosis. Report of a case with cytologic, cytochemical and histopathologic study. Acta Cytol. 27 (4), 446–449.
Gellersen, B., Brosens, J.J., 2014. Cyclic decidualization of the human endometrium in reproductive health and failure. Endocr. Rev. 35 (6), 851–905.
Gibson, D.A., et al., 2021. Pelvic pain correlates with peritoneal macrophage abundance not endometriosis. Reprod. Fertil. 2 (1), 47–57.
Goffin, F., et al., 2003. Expression pattern of metalloproteinases and tissue inhibitors of matrix-metalloproteinases in cycling human endometrium. Biol. Reprod. 69 (3), 976–984.
Gomez-Lopez, N., et al., 2014. Immune cells in term and preterm labor. Cell. Mol. Immunol. 11 (6), 571–581.
Gordon, S., Plüddemann, A., 2017. Tissue macrophages: heterogeneity and functions. BMC Biol. 15 (1), 53.
Gordon, S., Taylor, P.R., 2005. Monocyte and macrophage heterogeneity. Nat. Rev. Immunol. 5 (12), 953–964.
Greaves, E., et al., 2015. Estradiol is a critical mediator of macrophage-nerve cross talk in peritoneal endometriosis. Am. J. Pathol. 185 (8), 2286–2297.
Guilliams, M., et al., 2020. Establishment and maintenance of the macrophage niche. Immunity 52 (3), 434–451.
Guo, Y., et al., 2011. Comprehensive analysis of leukocytes, vascularization and matrix metalloproteinases in human menstrual xenograft model. PLoS One 6 (2), e16840.
Guo, S., et al., 2021. Hypoxia and its possible relationship with endometrial receptivity in adenomyosis: a preliminary study. Reprod. Biol. Endocrinol. 19 (1), 7.
Gustafsson, C., et al., 2008. Gene expression profiling of human decidual macrophages: evidence for immunosuppressive phenotype. PLoS One 3 (4), e2078.
Halme, J., et al., 1984. Retrograde menstruation in healthy women and in patients with endometriosis. Obstet. Gynecol. 64 (2), 151–154.
Hawley, C.A., et al., 2018. Csflr-mApple transgene expression and ligand binding in vivo reveal dynamics of CSF1R expression within the mononuclear phagocyte system. J. Immunol. 200 (6), 2209.
Hazan, A.D., et al., 2010. Vascular-leukocyte interactions: mechanisms of human decidual spiral artery remodeling in vitro. Am. J. Pathol. 177 (2), 1017–1030.
Herington, J.L., et al., 2011. Immune interactions in endometriosis. Expert. Rev. Clin. Immunol. 7 (5), 611–626.
Hill, J.A., et al., 1988. Characterization of leukocyte subpopulations in the peritoneal fluid of women with endometriosis. Fertil. Steril. 50 (2), 216–222.
Holness, C.L., Simmons, D.L., 1993. Molecular cloning of CD68, a human macrophage marker related to lysosomal glycoproteins. Blood 81 (6), 1607–1613.
Houser, B.L., et al., 2011. Two unique human decidual macrophage populations. J. Immunol. 186 (4), 2633–2642.
Humes, J.L., et al., 1977. Macrophages synthesis and release prostaglandins in response to inflammatory stimuli. Nature 269 (5624), 149–151.
Jabbour, H.N., et al., 2006. Endocrine regulation of menstruation. Endocr. Rev. 27 (1), 17–46.
Jensen, A.L., et al., 2012. A subset of human uterine endometrial macrophages is alternatively activated. Am. J. Reprod. Immunol. 68 (5), 374–386.
Jeziorska, M., et al., 1996. Immunolocalization of the matrix metalloproteinases gelatinase B and stromelysin 1 in human endometrium throughout the menstrual cycle. J. Reprod. Fertil. 107 (1), 43–51.
Jiang, X., Wang, H., 2020. Macrophage subsets at the maternal-fetal interface. Cell. Mol. Immunol. 17 (8), 889–891.

Jones, R.L., et al., 2004. Identification of chemokines important for leukocyte recruitment to the human endometrium at the times of embryo implantation and menstruation. J. Clin. Endocrinol. Metab. 89 (12), 6155–6167.
Kaitu'u-Lino, T.J., Morison, N.B., Salamonsen, L.A., 2007. Neutrophil depletion retards endometrial repair in a mouse model. Cell Tissue Res. 328 (1), 197–206.
Kats, R., et al., 2005. Cycle-dependent expression of macrophage migration inhibitory factor in the human endometrium. Hum. Reprod. 20 (12), 3518–3525.
Kelly, R.W., King, A.E., Critchley, H.O., 2001. Cytokine control in human endometrium. Reproduction 121 (1), 3–19.
Khan, K.N., et al., 2004. Differential macrophage infiltration in early and advanced endometriosis and adjacent peritoneum. Fertil. Steril. 81 (3), 652–661.
Kim, J., Hematti, P., 2009. Mesenchymal stem cell-educated macrophages: a novel type of alternatively activated macrophages. Exp. Hematol. 37 (12), 1445–1453.
Klentzeris, L.D., et al., 1992. Endometrial lymphoid tissue in the timed endometrial biopsy: morphometric and immunohistochemical aspects. Am. J. Obstet. Gynecol. 167 (3), 667–674.
Krzyszczyk, P., et al., 2018. The role of macrophages in acute and chronic wound healing and interventions to promote pro-wound healing phenotypes. Front. Physiol. 9, 419.
Laoui, D., et al., 2014. Tumor hypoxia does not drive differentiation of tumor-associated macrophages but rather fine-tunes the M2-like macrophage population. Cancer Res. 74 (1), 24–30.
Lash, G.E., et al., 2016. Decidual macrophages: key regulators of vascular remodeling in human pregnancy. J. Leukoc. Biol. 100 (2), 315–325.
Laskarin, G., et al., 2005. The presence of functional mannose receptor on macrophages at the maternal-fetal interface. Hum. Reprod. 20 (4), 1057–1066.
Lathbury, L.J., Salamonsen, L.A., 2000. In-vitro studies of the potential role of neutrophils in the process of menstruation. Mol. Hum. Reprod. 6 (10), 899–906.
Li, J., et al., 2012. Macrophage metalloelastase (MMP-12) deficiency mitigates retinal inflammation and pathological angiogenesis in ischemic retinopathy. PLoS One 7 (12), e52699.
Liu, Z., et al., 2019. Fate mapping via Ms4a3-expression history traces monocyte-derived cells. Cell 178 (6), 1509–1525.e19.
Lockwood, C.J., et al., 1998. Matrix metalloproteinase and matrix metalloproteinase inhibitor expression in endometrial stromal cells during progestin-initiated decidualization and menstruation-related progestin withdrawal. Endocrinology 139 (11), 4607–4613.
Lousse, J.C., et al., 2009. Iron storage is significantly increased in peritoneal macrophages of endometriosis patients and correlates with iron overload in peritoneal fluid. Fertil. Steril. 91 (5), 1668–1675.
Lu, Q., et al., 2021. Transforming growth factor (TGF) β and endometrial vascular maturation. Front Cell Dev Biol 9, 640065.
Ludwig, H., Spornitz, U.M., 1991. Microarchitecture of the human endometrium by scanning electron microscopy: menstrual desquamation and remodeling. Ann. N. Y. Acad. Sci. 622, 28–46.
Malik, S., et al., 2006. Reduced levels of VEGF-A and MMP-2 and MMP-9 activity and increased TNF-alpha in menstrual endometrium and effluent in women with menorrhagia. Hum. Reprod. 21 (8), 2158–2166.
Marbaix, E., et al., 1996. Menstrual breakdown of human endometrium can be mimicked in vitro and is selectively and reversibly blocked by inhibitors of matrix metalloproteinases. Proc. Natl. Acad. Sci. U. S. A. 93 (17), 9120–9125.
Markee, J.E., 1978. Menstruation in intraocular endometrial transplants in the Rhesus monkey. Am. J. Obstet. Gynecol. 131 (5), 558–559.

Marsh, M.M., et al., 1997. Endothelin and neutral endopeptidase in the endometrium of women with menorrhagia. Hum. Reprod. 12 (9), 2036–2040.
Martinez, F.O., Gordon, S., 2014. The M1 and M2 paradigm of macrophage activation: time for reassessment. F1000prime Rep. 6, 13.
Martinez, F.O., et al., 2008. Macrophage activation and polarization. Front. Biosci. 13, 453–461.
Martínez-Aguilar, R., et al., 2020. Menstrual blood-derived stromal cells modulate functional properties of mouse and human macrophages. Sci. Rep. 10 (1), 21389.
Maybin, J.A., Critchley, H.O.D., 2016. Medical management of heavy menstrual bleeding. Women's Health (Lond. Engl.) 12 (1), 27–34.
Maybin, J.A., Critchley, H.O., Jabbour, H.N., 2011. Inflammatory pathways in endometrial disorders. Mol. Cell. Endocrinol. 335 (1), 42–51.
Maybin, J.A., et al., 2011a. The regulation of vascular endothelial growth factor by hypoxia and prostaglandin $F_2\alpha$ during human endometrial repair. J. Clin. Endocrinol. Metab. 96 (8), 2475–2483.
Maybin, J.A., et al., 2011b. Novel roles for hypoxia and prostaglandin E2 in the regulation of IL-8 during endometrial repair. Am. J. Pathol. 178 (3), 1245–1256.
Maybin, J.A., et al., 2012. The presence and regulation of connective tissue growth factor in the human endometrium. Hum. Reprod. 27 (4), 1112–1121.
Maybin, J.A., et al., 2017a. Steroids regulate CXCL4 in the human endometrium during menstruation to enable efficient endometrial repair. J. Clin. Endocrinol. Metab. 102 (6), 1851–1860.
Maybin, J.A., et al., 2017b. Reduced Transforming Growth Factor-β Activity in the Endometrium of Women With Heavy Menstrual Bleeding. J. Clin. Endocrinol. Metab. 102 (4), 1299–1308.
Maybin, J.A., et al., 2018. Hypoxia and hypoxia inducible factor-1α are required for normal endometrial repair during menstruation. Nat. Commun. 9 (1), 295.
McLaren, J., et al., 1996. Vascular endothelial growth factor is produced by peritoneal fluid macrophages in endometriosis and is regulated by ovarian steroids. J. Clin. Invest. 98 (2), 482–489.
Meduri, G., Bausero, P., Perrot-Applanat, M., 2000. Expression of vascular endothelial growth factor receptors in the human endometrium: modulation during the menstrual cycle. Biol. Reprod. 62 (2), 439–447.
Mints, M., et al., 2007. Wall discontinuities and increased expression of vascular endothelial growth factor-A and vascular endothelial growth factor receptors 1 and 2 in endometrial blood vessels of women with menorrhagia. Fertil. Steril. 88 (3), 691–697.
Mowat, A.M., Scott, C.L., Bain, C.C., 2017. Barrier-tissue macrophages: functional adaptation to environmental challenges. Nat. Med. 23 (11), 1258–1270.
Munro, M.G., Critchley, H.O., Fraser, I.S., 2011. The FIGO classification of causes of abnormal uterine bleeding in the reproductive years. Fertil. Steril. 95 (7), 2204–2208. e1-3.
Munro, M.G., Critchley, H.O.D., Fraser, I.S., 2018. The two FIGO systems for normal and abnormal uterine bleeding symptoms and classification of causes of abnormal uterine bleeding in the reproductive years: 2018 revisions. Int. J. Gynaecol. Obstet. 143 (3), 393–408.
Murray, P.J., et al., 2014. Macrophage activation and polarization: nomenclature and experimental guidelines. Immunity 41 (1), 14–20.
Nahrendorf, M., Swirski, F.K., 2016. Abandoning M1/M2 for a network model of macrophage function. Circ. Res. 119 (3), 414–417.
Ning, F., Liu, H., Lash, G.E., 2016. The role of decidual macrophages during normal and pathological pregnancy. Am. J. Reprod. Immunol. 75 (3), 298–309.

Orecchioni, M., et al., 2019. Macrophage polarization: different gene signatures in M1(LPS+) vs. classically and M2(LPS–) vs. alternatively activated macrophages. Front. Immunol. 10, 1084.

Parameswaran, N., Patial, S., 2010. Tumor necrosis factor-α signaling in macrophages. Crit. Rev. Eukaryot. Gene Expr. 20 (2), 87–103.

Payne, K.J., et al., 2012. Residency and activation of myeloid cells during remodeling of the prepartum murine cervix. Biol. Reprod. 87 (5), 106.

Pepe, G., et al., 2017. Self-renewal and phenotypic conversion are the main physiological responses of macrophages to the endogenous estrogen surge. Sci. Rep. 7, 44270.

Prentice, A., 1999. Health care implications of dysfunctional uterine bleeding. Baillieres Best Pract. Res. Clin. Obstet. Gynaecol. 13 (2), 181–188.

Quillay, H., et al., 2015. Distinct characteristics of endometrial and decidual macrophages and regulation of their permissivity to HIV-1 infection by SAMHD1. J. Virol. 89 (2), 1329–1339.

Reavey, J.J., et al., 2021. Markers of human endometrial hypoxia can be detected in vivo and ex vivo during physiological menstruation. Hum. Reprod. 36 (4), 941–950.

Richter, O.N., et al., 2005. Tumor necrosis factor alpha secretion by peritoneal macrophages in patients with endometriosis. Arch. Gynecol. Obstet. 271 (2), 143–147.

Salamonsen, L.A., 2003. Tissue injury and repair in the female human reproductive tract. Reproduction 125 (3), 301–311.

Salamonsen, L.A., Lathbury, L.J., 2000. Endometrial leukocytes and menstruation. Hum. Reprod. Update 6 (1), 16–27.

Salamonsen, L.A., Woolley, D.E., 1999. Menstruation: induction by matrix metalloproteinases and inflammatory cells. J. Reprod. Immunol. 44 (1-2), 1–27.

Salamonsen, L.A., et al., 1997. Production of endometrial matrix metalloproteinases, but not their tissue inhibitors, is modulated by progesterone withdrawal in an in vitro model for menstruation. J. Clin. Endocrinol. Metab. 82 (5), 1409–1415.

Salamonsen, L.A., et al., 2000. Regulation of matrix metalloproteinases in human endometrium. Hum. Reprod. 15 (suppl_3), 112–119.

Saunders, P.T.K., Horne, A.W., 2021. Endometriosis: etiology, pathobiology, and therapeutic prospects. Cell 184 (11), 2807–2824.

Serhan, C.N., Savill, J., 2005. Resolution of inflammation: the beginning programs the end. Nat. Immunol. 6 (12), 1191–1197.

Shalaby, M.R., et al., 1989. Tumor necrosis factor-alpha-associated uterine endothelial injury in vivo. Influence of dietary fat. Lab. Invest. 61 (5), 564–570.

Shapley, M., Jordan, K., Croft, P.R., 2004. An epidemiological survey of symptoms of menstrual loss in the community. Br. J. Gen. Pract. 54 (502), 359–363.

Sharpe-Timms, K.L., et al., 2002. Endometriotic haptoglobin binds to peritoneal macrophages and alters their function in women with endometriosis. Fertil. Steril. 78 (4), 810–819.

Shipley, J.M., et al., 1996. Metalloelastase is required for macrophage-mediated proteolysis and matrix invasion in mice. Proc. Natl. Acad. Sci. U. S. A. 93 (9), 3942–3946.

Simitsidellis, I., Saunders, P.T.K., Gibson, D.A., 2018. Androgens and endometrium: new insights and new targets. Mol. Cell. Endocrinol. 465, 48–60.

Smith, O.P., Jabbour, H.N., Critchley, H.O., 2007. Cyclooxygenase enzyme expression and E series prostaglandin receptor signalling are enhanced in heavy menstruation. Hum. Reprod. 22 (5), 1450–1456.

Soares, M.P., Hamza, I., 2016. Macrophages and iron metabolism. Immunity 44 (3), 492–504.

Song, J.Y., Fraser, I.S., 1995. Effects of progestogens on human endometrium. Obstet. Gynecol. Surv. 50 (5), 385–394.

Starkey, P.M., Clover, L.M., Rees, M.C., 1991. Variation during the menstrual cycle of immune cell populations in human endometrium. Eur. J. Obstet. Gynecol. Reprod. Biol. 39 (3), 203–207.
Stewart, J.A., Bulmer, J.N., Murdoch, A.P., 1998. Endometrial leucocytes: expression of steroid hormone receptors. J. Clin. Pathol. 51 (2), 121–126.
Stowell, S.B., et al., 1997. Cytologic diagnosis of peritoneal fluids. Applicability to the laparoscopic diagnosis of endometriosis. Acta Cytol. 41 (3), 817–822.
Strunz, B., et al., 2021. Continuous human uterine NK cell differentiation in response to endometrial regeneration and pregnancy. Sci. Immunol. 6 (56).
Tabibzadeh, S., 1990. Proliferative activity of lymphoid cells in human endometrium throughout the menstrual cycle*. J. Clin. Endocrinol. Metab. 70 (2), 437–443.
Tabibzadeh, S., 1996. The signals and molecular pathways involved in human menstruation, a unique process of tissue destruction and remodelling. Mol. Hum. Reprod. 2 (2), 77–92.
Tabibzadeh, S., et al., 1995. Site and menstrual cycle-dependent expression of proteins of the tumour necrosis factor (TNF) receptor family, and BCL-2 oncoprotein and phase-specific production of TNF alpha in human endometrium. Hum. Reprod. 10 (2), 277–286.
Takahashi, K., Yamamura, F., Naito, M., 1989. Differentiation, maturation, and proliferation of macrophages in the mouse yolk sac: a light-microscopic, enzyme-cytochemical, immunohistochemical, and ultrastructural study. J. Leukoc. Biol. 45 (2), 87–96.
Takebayashi, A., et al., 2015. Subpopulations of macrophages within eutopic endometrium of endometriosis patients. Am. J. Reprod. Immunol. 73 (3), 221–231.
Thiruchelvam, U., et al., 2013. The importance of the macrophage within the human endometrium. J. Leukoc. Biol. 93 (2), 217–225.
Thiruchelvam, U., et al., 2016. Cortisol regulates the paracrine action of macrophages by inducing vasoactive gene expression in endometrial cells. J. Leukoc. Biol. 99 (6), 1165–1171.
Travaglione, S., et al., 2002. Epithelial cells and expression of the phagocytic marker CD68: scavenging of apoptotic bodies following Rho activation. Toxicol. in Vitro 16 (4), 405–411.
Tremellen, K.P., Russell, P., 2012. The distribution of immune cells and macrophages in the endometrium of women with recurrent reproductive failure. II: adenomyosis and macrophages. J. Reprod. Immunol. 93 (1), 58–63.
Vallvé-Juanico, J., et al., 2019. Macrophages display proinflammatory phenotypes in the eutopic endometrium of women with endometriosis with relevance to an infectious etiology of the disease. Fertil. Steril. 112 (6), 1118–1128.
Van Langendonckt, A., Casanas-Roux, F., Donnez, J., 2002. Iron overload in the peritoneal cavity of women with pelvic endometriosis. Fertil. Steril. 78 (4), 712–718.
Van Roon, J.A.G., et al., 2003. Selective elimination of synovial inflammatory macrophages in rheumatoid arthritis by an Fcγ receptor I–directed immunotoxin. Arthritis Rheum. 48 (5), 1229–1238.
Vassilev, V., et al., 2005. Response of matrix metalloproteinases and tissue inhibitors of metalloproteinases messenger ribonucleic acids to ovarian steroids in human endometrial explants mimics their gene- and phase-specific differential control in vivo. J. Clin. Endocrinol. Metab. 90 (10), 5848–5857.
Verreck, F.A., et al., 2006. Phenotypic and functional profiling of human proinflammatory type-1 and anti-inflammatory type-2 macrophages in response to microbial antigens and IFN-gamma- and CD40L-mediated costimulation. J. Leukoc. Biol. 79 (2), 285–293.
Yovich, J.L., et al., 2020. Pathogenesis of endometriosis: Look no further than John Sampson. Reprod. BioMed. Online 40 (1), 7–11.
Zetterberg, E., et al., 2007. Pericyte coverage of abnormal blood vessels in myelofibrotic bone marrows. Haematologica 92 (5), 597–604.

FSC
www.fsc.org
MIX
Papier | Fördert gute Waldnutzung
FSC® C083411

Kolibri 360 GmbH
Ferdinand-Jühlke-Straße 7
99095 Erfurt, Deutschland
produktsicherheit@kolibri360.de